Dynamics of Controlled Mechanical Systems

International Union of Theoretical
and Applied Mechanics

International Federation of Automatic Control

G. Schweitzer, M. Mansour (Eds.)

Dynamics of Controlled Mechanical Systems

IUTAM/IFAC Symposium, Zurich,
Switzerland, May 30 - June 3, 1988

Springer-Verlag
Berlin Heidelberg New York
London Paris Tokyo

Prof. G. Schweitzer

Institute of Mechanics
ETH-Zentrum
CH-8092 Zurich
Switzerland

Prof. M. Mansour

Institute of Automatic Control
ETH-Zentrum
CH-8092 Zurich
Switzerland

ISBN 3-540-50201-7 Springer-Verlag Berlin Heidelberg NewYork
ISBN 0-387-50201-7 Springer-Verlag NewYork Berlin Heidelberg

Library of Congress Cataloging-in-Publication Data
IUTAM/IFAC Symposium (1988: Zurich, Switzerland)
Dynamics of controlled mechanical systems / IUTAM/IFAC Symposium,
Zurich, Switzerland, May 30 - June 3, 1988; G. Schweitzer, M. Mansour, eds.
(IUTAM symposia)
At head of title: International Union of Theoretical and Applied Mechanics.
 ISBN 0-387-50201-7 (U.S.)
1. Automatic control--Congresses.
2. Machinery, Dynamics of--Congresses.
I. Schweitzer, G. (Gerhard)
II. Mansour, M.
III. International Union of Theoretical and Applied Mechanics.
IV. International Federation of Automatic Control.
V. Title.
VI. Series: IUTAM-Symposien.
TJ212.2.187   1988
629.8--dc 19   88-31205

This work is subject to copyright. All rights are reserved, whether the whole or part of the material is concerned, specifically the rights of translation, reprinting, re-use of illustrations, recitation, broadcasting, reproduction on microfilms or in other ways, and storage in data banks. Duplication of this publication or parts thereof is only permitted under the provisions of the German Copyright Law of September 9, 1965, in its version of June 24, 1985, and a copyright fee must always be paid. Violations fall under the prosecution act of the German Copyright Law.

© Springer-Verlag, Berlin Heidelberg 1989
Printed in Germany

The use of registered names, trademarks, etc. in this publication does not imply, even in the absence of a specific statement, that such names are exempt from the relevant protective laws and regulations and therefore free for general use.

Offsetprinting: Mercedes-Druck, Berlin; Bookbinding: B. Helm, Berlin
2161/3020 5 4 3 2 1 0

Preface

Many mechanical systems are actively controlled in order to improve their dynamic performance. Examples are elastic satellites, active vehicle suspension systems, robots, magnetic bearings, automatic machine tools.

Problems that are typical for mechanical systems arise in the following areas:

- Modeling the mechanical system in such a way that the model is suitable for control design
- Designing multivariable controls to be robust with respect to parameter variations and uncertainties in system order of elastic structures
- Fast real-time signal processing
- Generating high dynamic control forces and providing the necessary control power
- Reliability and safety concepts, taking into account the growing role of software within the system

The objective of the Symposium has been to present methods that contribute to the solutions of such problems. Typical examples are demonstrating the state of the art. It intends to evaluate the limits of performance that can be achieved by controlling the dynamics, and it should point to gaps in present research and areas for future research. Mainly, it has brought together leading experts from quite different areas presenting their points of view.

The International Union of Theoretical and Applied Mechanics (IUTAM) has initiated and sponsored, in cooperation with the International Federation of Automatic Control (IFAC), this Symposium on Dynamics of Controlled Mechanical Systems, held at the Swiss Federal Institute of Technology (ETH) in Zurich, Switzerland, May 30-June 3, 1988. It is the first time that these two scientific institutions have been jointly sponsoring such an event. And there are reasons to assume that common interests will lead the IFAC and the IUTAM to cosponsor another symposium on this interdisciplinary topic within the next years.

A Scientific Committee has been appointed consisting of

J. Ackermann, Germany;   P. Coiffet, France;   T.R. Kane, USA;
D.M. Klimov, USSR; M. Mansour (Co-Chairman), Switzerland;
W. Schiehlen, Germany; G. Schweitzer (Co-Chairman), Switzerland;
K. Yoshimoto, Japan

The Committtee suggested the participants to be invited and the papers to be presented at the Symposium. As a result of this process, 65 active scientific participants from 11 countries followed the invitation and 29 papers were presented. The lectures were devoted to the following main topics:

>Modeling, Typical Examples for the Dynamics of Controlled Mechanical Systems, Design Tools, Graphical Tools, Sensors and Actuators, Aerospace, Vehicles, and Robotics.

Some of the papers are related to more than one of these main topics, but in order to assist the reader we have structured this volume according to the main topics, thus maintaining the structure of the Symposium.

The lectures, giving a survey on the state of the art and presenting recent research results, show the high level of performance and sophistication already obtained when dealing with the control of mechanical systems. The lectures were extensively discussed, and it is expected that the Symposium will have a stimulating effect on further research in this important and interdisciplinary field of mechanics and control. Discussions and statements of the members of the Scientific Committee indicate that there are necessary and promising directions where future efforts will have to go:

- Improvements of the man-machine interface, including high level application oriented programming languages, graphics, and safety aspects.
- Extension of the role of software both at the design stage and as part of the controlled system itself making it more intelligent, capable of learning, safer and adaptable to the needs of the human user.
- Modeling of complex mechanical systems, especially for control purposes.

The organizers gratefully acknowledge the financial support and effective help of the following institutions and industrial companies in the preparation of the Symposium:

>International Union of Theoretical and Applied Mechanics (IUTAM)
>International Federation of Automatic Control (IFAC)
>Eidgenössische Technische Hochschule Zürich (ETH Zürich)
>European Research Office of the US Army
>Sulzer Brothers Ltd.

A main contribution to the success of the Symposium is due to the help and excellent work of the staff of the Institute of Mechanics of the ETH, and the Local Organizing Committee. We thank especially Mrs G. Junker.

The editorial work of the Proceedings was supported by the Institute of Mechanics of the ETHZ. In essence the original manuscripts submitted by the authors are reproduced. Thanks to the Springer-Verlag are due for an agreeable and efficient cooperation.

Zurich, July 1988   G. Schweitzer   M. Mansour

Participants

(Authors are identified by a * chairman are identified by a ©)

*© Ackermann, J., Institut für Dynamik der Flugsysteme, DFVLR Oberpfaffenhofen, D-8031 Oberpfaffenhofen, Germany

* Asaka, K., Showa Electric Wire and Cable Co. Ltd., 4-1-1 Minamihashimoto, Sagamihara-shi, Kanagawa, Japan

Badreddin, E., Institut für Automatik und Industrielle Elektronik, Eidgenössische Technische Hochschule, ETH-Zentrum/ETL I 28, CH-8092 Zürich, Switzerland

Betti, F., COPE SP, Cidade Universitaria USP, Av. Prof. Lineu Prestes 2242, São Paulo SP, Brasil

© Brauchli, H., Institut für Mechanik, Eidgenössische Technische Hochschule, ETH-Zentrum/HG F41, CH-8092 Zürich, Switzerland

Buck, U., Abteilung MEMB, Dornier System GmbH, Postfach 1360, D-7990 Friedrichshafen, Germany

* Chatila, R., Laboratoire d'Automatique et d'Analyse des Systemes du CNSR, 7 av. du Colonel Roche, F-31077 Toulouse Cedex, France

* Chiba, M., Dept. of Mechanical Engineering, Iwate University, 4-3-5 Ueda, Morioka, Japan

Cuny, B., Institut National des Sciences et Techniques Nucléaire, F-91191 Gif-sur-Yvette Cedex, France

© De Carli, A., Dipartimento di Informatica e Sistematica, Via Eudossiana 18, I-Roma, Italy

* Diez, D., Institut für Mechanik, Eidgenössische Technische Hochschule, ETH-Zentrum/LEO B1.2, CH-8092 Zürich, Switzerland

* Fässler, H. P., Institut für Mechanik, Eidgenössische Technische Hochschule, ETH-Zentrum/LEO C13, CH-8092 Zürich, Switzerland

Fayé, I. C., Institut für Dynamik der Flugsysteme, DFVLR, D-8031 Weßling, Germany

*© Friedmann, P. P., Mechanical, Aerospace and Nuclear Eng. Dept., University of California, 5732 Boelter Hall, Los Angeles, CA 90024, USA

*© Glattfelder, A. H., Corporate R & D, Sulzer Bros. Ltd., Postfach, CH-8401 Winterthur, Switzerland

* Guzzella, L., Corporate R & D, Sulzer Bros. Ltd., Postfach, CH-8401 Winterthur, Switzerland

*© Hiller, M., Fachgebiet Mechanik, Universität -GH- Duisburg, FB 7, Lotharstraße 41, D-4100 Duisburg 1, Germany

* Horiuchi, E., Mechanical Engineering Lab., Agency of Ind. Sci. and Techn., Namiki 1-2, Sakura-mura, Niihari-gun, Ibaraki-ken 305, Japan

*© Hugel, J., Elektrot. Entwicklungen und Konstruktionen, Eidgenössische Technische Hochschule, ETH-Zentrum/ETZ F96, CH-8092 Zürich,Switzerland

*© Kane, T. R., Dept. of Applied Mechanics, Stanford University, Stanford, CA 94305, USA

*© Khatib, O., Robotics Lab., Computer Sciences Dept., Stanford University, Stanford, CA 94305, USA

*© Khosla, P. K., Dept. of El. and Comp. Eng. (Robotics Inst.), Carnegie Mellon University, Pittsburgh, PA 15213, USA

© Klimov, D. M., Institute of Mechanics, The USSR Academy of Sciences, Pr. Vernadskogo 101, UdSSR-117526 Moscow, USSR

* Kokkinis, Th., Center for Robotic Systems in Microelectronics, University of California, Santa Barbara, CA 93106, USA

* Komatsu, T., Mechanical Engineering Lab., R&D Center, Toshiba Corporation, 4-1 Ukishima-cho, Kawasaki-ku, Kawasaki 210, Japan

*© Kortüm, W., Institut für Dynamik der Flugsysteme, DFVLR Oberpfaffenhofen, D-8031 Weßling, Germany

* Kruise, L., Control Systems & Computer Eng. Lab., Twente University of Technology, P. O. Box 217, NL-7500 AE Enschede, The Netherlands

* Maier, G. E., Corporate Research CRBC, ASEA Brown Boveri, CH-5405 Baden, Switzerland

© Mansour, M., Institut für Automatik und Industrielle Elektronik, Eidgenössische Technische Hochschule, ETH-Zentrum/ETL I 24, CH-8092 Zürich, Switzerland

Marcuard, J.-D., Institut d'Automatique, Eidgenössische Technische Hochschule, EPFL-DME-Ecublens, CH-1015 Lausanne, Switzerland

* Meirovitch, L., Dept. of Engineering Science & Mechanics, Virginia Polyt. Institute and State University, Blacksburg, VA 24061, USA

*© Miura, H., Dept. of Mechanical Engineering, The University of Tokyo, Hongo 7-3-1, Bunkyo-Ku, Tokyo 113, Japan

* Mizuno, T., Faculty of Engineering, Saitama University, 225 Shimo-okubo, Urawa-city, Saitama 338, Japan

*© Müller, P.C., Sicherheitstechnische Regelungs- und Messtechnik, Bergische Universität - GH Wuppertal, Postfach 100 127, D-5600 Wuppertal 1, Germany

Partoni, M., Institut d'Automatique, Eidgenössische Technische Hochschule, EPFL-DME-Ecublens, CH-1015 Lausanne, Switzerland

Pierri, P. S., COPE SP, Cidade Universitaria USP, Av. Prof. Lineu Prestes 2242, 05508 São Paulo SP, Brasil

*   Putz, P., Dornier System GmbH, Postfach 1360, D-7990 Friedrichshafen, Germany

*   Qian, Z. Y., Dept. Automatic Control, Shanghai Jiao-Tong University,
    Shanghai, 200030, P. R. China

    Ruf, W. D., Robert Bosch GmbH, Postfach 30 02 60, D-7000 Stuttgart 30, Germany

*© Sarychev, V. A., Keldysh Institute of Applied Mathematics, USSR Academy of
    Sciences, Miusskaja Sq. 4, UdSSR-125047 Moscow, USSR

    Schäfer, P., Institut B für Mechanik, Universität Stuttgart, Pfaffenwaldring 9,
    D-7000 Stuttgart 80, Germany

*© Schiehlen, W., Institut B für Mechanik, Universität Stuttgart, Pfaffenwaldring 9,
    D-7000 Stuttgart 80, Germany

    Schrama, R. J. P., Laboratory for Measurement and Control, Delft University of
    Technology, Mekelweg 2, NL-2628 CD Delft, The Netherlands

*© Schweitzer, G., Institut für Mechanik, Eidgenössische Technische Hochschule,
    ETH-Zentrum/HG F41, CH-8092 Zürich, Switzerland

©   Shubin, A. N., Institute of Control Sciences, Profsojuznaja ul. 65,
    UdSSR-117806 Moscow, USSR

*© Skelton, R. E., School of Aeronautics and Astronautics, Purdue University,
    West Lafayette, IN 47907, USA

©   Thomson, B., EW–402, BMW AG, Petuelring 130, D-8000 München 40, Germany

*   Tsuchiya, K., Central Research Laboratory, Mitsubishi Electr. Corp.,
    1-1 Tsukaguchi-Honmachi 8 Chome, Amagasaki, Hyogo 661, Japan

    Tuncelli, A. C., Institut d'Automatique, Eidgenössische Technische Hochschule,
    EPFL-DME-Ecublens, CH-1015 Lausanne, Switzerland

    Ulbrich, H., Technische Universität München, Postfach 20 24 20, D-8000 München 2,
    Germany

    von Hagen, A., Engineering Editor, Springer Verlag, Tiergartenstraße 17,
    D-6900 Heidelberg, Germany

©   Weber, H. I., Lab. de Projeto Mecanico, UNICAMP-FEC-DEM, Caixa Postal 6122,
    13081 Campinas–São Paulo, Brazil

©   Wehrli, Ch., Institut für Mechanik, Eidgenössische Technische Hochschule,
    ETH-Zentrum/HG F41, CH-8092 Zürich, Switzerland

*   Yamakita, M., Dept. of Control Engineering, Tokyo Institute of Technology,
    2-12-1 Oh-Okayama, Meguro-Ku, Tokyo 152, Japan

    Yano, S., Institut für Mechanik (c/o Prof. Hagedorn), Technische Hochschule
    Darmstadt, Postfach, D-6100 Darmstadt, Germany

©   Yoshimoto, K., Dept. of Mechanical Engineering, University of Tokyo,
    7-3-1 Hongo, Bunkyo-ku, Tokyo 113, Japan

Zhao, H., Dept. of Engineering Physics, Tsinghua University, Beijing, P. R. China

Local Participants

Ledwozyw, M., Institut f. Biomedizinische Technik und Medizinische Informatik, Eidgenössiche Technische Hochschule, Moussonstrasse 18, CH-8044 Zürich, Switzerland

Junker, G., Institut für Mechanik, Eidgenössische Technische Hochschule, ETH-Zentrum/HG F42.1, CH-8092 Zürich, Switzerland

Scherrer, J.K., Institut für Mechanik, Eidgenössische Technische Hochschule, ETH-Zentrum/HG F37.6, CH-8092 Zürich, Switzerland

Herzog, R., Institut für Mechanik, Eidgenössiche Technische Hochschule, ETH-Zentrum/HG F38.3, CH-8092 Zürich, Switzerland

Zumbach, M., Institut für Mechanik, Eidgenössische Technische Hochschule, ETH-Zentrum/LEO B1.2, CH-8092 Zürich, Switzerland

Bleuler, H., Institut für Mechanik, Eidgenössische Technische Hochschule, ETH-Zentrum/HG F38.2, CH-8092 Zürich, Switzerland

Blech, J., Faculty of Mechanical Engineering, Technion - Israel Institute of Technology, 32000 Haifa, Israel

Knobloch, H.W., Universität Würzburg, Würzburg, Germany

Contents

Modeling

Ackermann, J. and P. Wirth
    Model Verification by Experiments with Finite Effect Sequences.(FES) 3

Hiller, M.
    Modeling the Dynamics of a Complete Vehicle with Nonlinear Wheel
    Suspension Kinematics and Elastic Hinges 15

Kane, T.R.
    Computer Aided Formulation of Equations of Motion 29

Meirovitch, L.
    State Equations of Motion for Flexible Bodies in Terms of Quasi-Coordinates 37

Design Tools

Diez, D., and G. Schweitzer
    Simulation, Test and Diagnostics Integrated for a Safety Design of Magnetic
    Bearing Prototypes 51

Schiehlen, W.O.
    Hardware - Software Interfaces for Dynamical Simulations 63

Graphical Tools

Maier, G.E.
    Towards Graphical Programming in Control of Mechanical Systems 77

Putz, P.
    Graphical Verification of Complex Multibody Motion in Space Applications 91

Examples for the Dynamics of Controlled Mechanical Systems

Chiba, M., Tani, J., Liu, G., Takahashi, F., Kodama, S., and H. Doki
    Active Vibration Control of a Cantilever Beam by a Piezoelectric Ceramic
    Actuator 107

Furuta, K., Yamakita, M. Sugiyama, N., and K. Asaka
    Fiber Connected Tug of War 119

Mizuno, T., and T. Higuchi
    Structure of Magnetic Bearing Control System for Compensating Unbalance
    Force 135

Sensors and Actuators

Norris G.A., and R.E. Skelton
    Placing Dynamic Sensors and Actuators on Flexible Space Structures 149

Aerospace

Friedmann, P.P., and M.D. Takahashi
    A Simple Active Controller to Supress Helicopter Air Resonance in Hover
    and Forward Flight 163

Komatsu, T., Uenohara, M., Iikura, S.,
Miura, H., and I. Shimoyama
    Active Vibration Control for Flexible Space Environment Use Manipulators 181

Sarychev, V.A., Belyaev, M.Yu., Sazonov, V.V., and T.N. Tyan
    Orientation of Large Orbital Stations 193

Tsuchiya, K., Yamada K., and B.N. Agrawal
    Attitude Stability of a Flexible Asymmetric Dual Spin Spacecraft 207

Robotics

Fässler, H.P.
    Robot Control in Cartesian Space with Adaptive Nonlinear Dynamics
    Compensation 221

Gürgöze, M., and P.C. Müller
    Modeling and Control of Elastic Robot Arm with Prismatic Joint 235

Guzzella, L., and A.H. Glattfelder
    A Decentralized and Robust Controller for Robots 247

Khatib, O., and S. Agrawal
    Isotropic and Uniform Inertial and Acceleration Characteristics: Issues in the
    Design of Redundant Manipulators 259

Khosla, P.K.
    Effect of Sampling Rates on the Performance of Model-Based Control
    Schemes 271

Kruise, L., van Amerongen, J., Löhnberg, P., and M.J.L. Tiernego
    Modeling and Control of a Flexible Robot Link 285

Lu, D., Qian, Z.-Y., and Z.-J. Zhang
    Decomposed Parameter Identification Approach of Robot Dynamics 297

Wehrli, E., and T. Kokkinis
    Dynamic Behavior of a Flexible Robotic Manipulator 309

Vehicles

Horiuchi, E., Usui, S., Tani, K., and N. Shirai
    Control of an Active Suspension System for a Wheeled Vehicle 323

Kortüm, W., Schwartz, W., and I. Fayé
    Dynamic Modeling of High Speed Ground Transportation Vehicles for Control
    Design and Performance Evaluation 335

Laumond, J.-P., Simeon, T., Chatila, R., and G. Giralt
    Trajectory Planning and Motion Control of Mobile Robots 351

Miura, H.
    Researches of the Biped Robot in Japan 367

Modeling

Model Verification by Experiments with Finite Effect Sequences (FES)

J. ACKERMANN, P. WIRTH

DFVLR-Institut für Dynamik der Flugsysteme
Oberpfaffenhofen, D 8031 Wessling

Summary

A finite effect sequence (FES) is a good input signal to verify agreement between a linear plant and its model. The FES theory is reviewed, the influence of nonlinearities in the plant is studied and their influence on the test is reduced by a modification of the FES. Practical problems arising in the application to a robot arm are discussed and recommendations for further investigations are given.

Introduction

Assume a linear model of a system is known, e.g. a local linearization of a nonlinear simulation model. Also assume that the system is available for undisturbed input-output measurements. What is a good input signal to verify agreement between model and system? The answer is: A finite effect sequence (FES). The FES theory [1] is reviewed with emphasis on the alternatives that the system is only the plant or a control system containing the plant. Modifications of FESs are discussed, which reduce the effect of nonlinear distortions in the simulation model and the plant.

A robot arm is studied as an example. Some practical problems are discussed and recommendations for further investigations are given.

Finite Effect Sequences

Consider an n-th order linear discrete-time siso system

$$\mathbf{x}(k+1) = \mathbf{A}\mathbf{x}(k) + \mathbf{b}u(k)$$
$$y(k) = \mathbf{c'}\mathbf{x}(k) \tag{1}$$

with the characteristic polynomial

$$\bar{A}(z) = \det(z\mathbf{I}-\mathbf{A}) = a_0 + a_1 z + \ldots + a_{n-1} z^{n-1} + z^n \tag{2}$$

Apply the following input sequence to the system

$$\begin{aligned} u(0) &= 1 \\ u(1) &= a_{n-1} \\ &\vdots \\ u(n) &= a_0 \\ u(k) &= 0 \quad \text{for } k > n \end{aligned} \tag{3}$$

the response is

$$\begin{aligned} y(0) &= \mathbf{c'x}(0) \\ y(1) &= \mathbf{c'Ax}(0) + \mathbf{c'b} \\ y(2) &= \mathbf{c'A^2 x}(0) + \mathbf{c'Ab} + a_{n-1}\mathbf{c'b} \\ &\vdots \\ y(n+1) &= \mathbf{c'A^{n+1}x}(0) + \mathbf{c'A^n b} + a_{n-1}\mathbf{c'A^{n-1}b} + \ldots + a_0\mathbf{c'b} \end{aligned} \tag{4}$$

By the Cayley-Hamilton theorem we have
$$\mathbf{A}^n + a_{n-1}\mathbf{A}^{n-1} + \ldots + a_0\mathbf{A}^0 = \mathbf{0} \text{ and thus}$$

$$y(n+1) = \mathbf{c'A^{n+1}x}(0) \tag{5}$$

For $k > n$ the input is zero and the system follows its homogeneous solution

$$y(k) = \mathbf{c'A^k x}(0) \qquad k > n \tag{6}$$

This is the same homogeneous solution as we obtain it with a zero input. The sequence (3) has an effect on y(k) only over a finite time, therefore (3) is called a "Finite Effect Sequence" (FES). Some useful properties of FESs [1] are summarized here.

1) For a controllable and observable system, (3) is the FES of minimal duration (in short "minimal FES"), otherwise the coefficients of the observable and controllable subsystem constitute a minimal FES. For simplicity we assume here (1) to be controllable and observable.

2) Other FESs can be generated by the three operations i) multiplication by a scalar factor, ii) time shift, iii) superposition of FESs. By these operations FESs of arbitrary length > n+1 can be generated.

3) The z-transform of (3) yields
$$u_z(z) = 1 + a_{n-1}z^{-1} + \ldots + a_0 z^{-n} = A(z^{-1}) = z^{-n}\bar{A}(z) \tag{7}$$

The three operations under 2) correspond to a multiplication of $A(z^{-1})$ by an arbitrary polynomial $R(z^{-1})$. If $A(z^{-1})$ is a FES then also $A(z^{-1})R(z^{-1})$ is a FES.

4) The z-transfer function of (1) is
$$h_z(z) = c'(zI-A)^{-1}b = \bar{B}(z)/\bar{A}(z) \tag{8}$$
$$\bar{A}(z) = a_0 + a_1 z + \ldots + a_{n-1}z^{n-1} + z^n$$
$$\bar{B}(z) = b_0 + b_1 z + \ldots + b_{n-1}z^{n-1}$$

Polynomials in z^{-1} like in (7) are obtained by multiplication of numerator and denominator of $h_z(z)$ by z^{-1}. Let
$$A(z^{-1}) = z^{-n}\bar{A}(z) = 1 + a_{n-1}z^{-1} + \ldots + a_0 z^{-n}$$
$$B(z^{-1}) = z^{-n}\bar{B}(z) = \phantom{1 + {}} b_{n-1}z^{-1} + \ldots + b_0 z^{-n} \tag{9}$$

Now $h_z(z) = B(z^{-1})/A(z^{-1})$, $y_z(z) = h_z(z)u_z(z)$ and for a minimal FES input $u_z(z) = A(z^{-1})$

$$y_z(z) = B(z^{-1}) \qquad (10)$$

The FES response consists of the numerator coefficients of the z-transfer function. By comparison with (4)

$$b_i = c'A^{n-i-1}b + a_{n-1}c'A^{n-i-2}b + \ldots + a_{i+1}c'b \qquad (11)$$

(This relation may also be obtained by applying Leverrier's algorithm to (8).)

5. If the system is a closed loop with compensator $D(z^{-1})/C(z^{-1})$ and plant $B(z^{-1})/A(z^{-1})$, then the sequences indicated in fig. 1 occur.

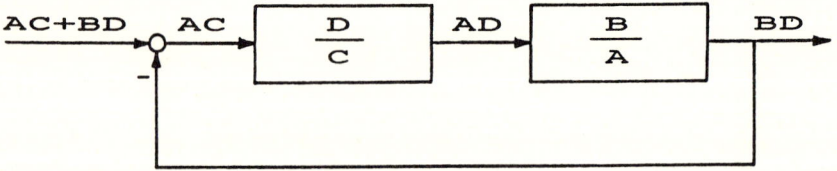

Fig. 1 FES responses in a closed loop

Input is the closed-loop characteristic polynomial. AC + BD, the response at the feedback error is the open-loop characteristic polynomial and the response at the output is the open-loop numerator polynomial.

6. In the multivariable case the FES input matrix $P(z^{-1})$ and response matrix $Q(z^{-1})$ form a prime factorization

$$C(zI-A)^{-1}B = Q(z^{-1})P^{-1}(z^{-1}) \qquad (12)$$

Thus the FES and its response constitute a complete minimal system description. The structure of $P(z^{-1})$ is feedback invariant, but its coefficients can be arbitrarily changed by state feedback. The closed-loop properties may be specified by the closed-loop FESs.

Model verification by a FES

We now come back to the initial question of model verification. In principle we can use step responses, frequency responses (for stable systems only) or other input-output measurements for the comparison of model and system. There are, however, some advantages of FESs for this purpose.

1) We only have to know the eigenvalues or poles of the transfer function. If they are correct, then the response is finite and the correct numerator can be read off from the experiment. If the response is not finite, then only the model poles must be adjusted, not the zeros. This separates numerator and denominator determination.

2) The signal energy at input and output is concentrated to a short time interval. In a plant with changing operating conditions and changing local linearizations many such short-time experiments may be performed along a trajectory.

3. The short-time test is feasible for unstable plants, if we can make the initial state $x(0)$ very small. For $k > n$ the response is very sensitive to a mismatch of unstable eigenvalues. An alternative is the closed-loop test at a stabilized plant. For known $C(z^{-1})$ and $D(z^{-1})$ in fig. 1 the relationship between the responses and $A(z^{-1})$ and $B(z^{-1})$ is almost as simple as in the open loop.

If the compensator shifts the eigenvalues close to the origin of the z-plane, then the output signal for $k > n$ becomes small and insensitive to a mismatch of $A(z^{-1})$. A tradeoff is a compensator that barely stabilizes the plant. The signal $y(k)$, $k > n$ is then still sensitive to a mismatch of the most critical eigenvalues near the unit circle and the experiment can be performed with a stable system.

Influence of nonlinearities

Frequently the linear model describes a local linearization of a nonlinear system and all previous results are only approximations. In this section some modifications of the FES experiment are derived with the aim to reduce the influence of nonlinearities.

In the simplest case u is generated by an actuator with a nonlinear characteristic. If this characteristic is monotonically increasing, then it has a unique inverse and the FES at the actuator input can be modified such that the desired FES occurs at the actuator output. However real actuator nonlinearities like backlash and saturation do not have an inverse.

Small signals are distorted by backlash and friction. In order to avoid this effect, the FES should be multiplied by a large scalar factor. This factor is, however, limited by saturation effects; also the state variables should not leave the region, where a local linearization is valid. There are two ways to reduce the maximum input amplitude: Longer sampling intervals and nonminimal FESs.

To some extent the input energy can be injected into the system by smaller amplitude and longer duration of the impulses. This shifts the excitation energy towards lower frequencies. Practically the amplitude is kept constant over N sampling intervals and the continuous plant is discretized with a sampling interval NT. N is limited by the fact that the controllability and observability of high frequency complex eigenvalues is reduced.

An alternative approach is the use of nonminimal FESs which are determined such that the maximum amplitude is reduced. This is illustrated by the following example.

A loading bridge [1] has the following parameters. Crab mass = 1 t (= 1000 kg), load mass = 3 t, rope length = 10 m, sampling interval $T = \pi/8$ seconds. The z-transfer function from u = "force accelerating the crab" to y = "crab position" is

$$h_z(z) = \frac{y_z(z)}{u_z(z)} = \frac{0.0742z^3 - 0.0629z^2 - 0.0629z + 0.0742}{z^4 - 3.414z^3 + 4.828z^2 - 3.414z + 1} \quad (13)$$

The maximum absolute value of the denominator coefficients can be reduced by multiplication of numerator and denominator by $z + 1$ or even more by $z^2 + 1.757z + 1$. The resulting expanded z-transfer function in the latter case is

$$h_z(z) = \frac{0.0742z^5 + 0.0675z^4 - 0.0992z^3 - 0.0992z^2 + 0.0675z + 0.0742}{z^6 - 1.657z^5 - 0.172z^4 + 1.657z^3 - 0.172z^2 - 1.657z + 1} \quad (14)$$

Normalizing both FESs corresponding to (13) and (14) to $|u| \leq 1$ the resulting input sequences are

k	u(k), eq. (13)	u(k), eq. (14)
0	0.207	0.603
1	-0.707	-1
2	1	-0.104
3	-0.707	1
4	0.207	-0.104
5	0	-1
6	0	0.603
7	0	0
$\Sigma\, u^2(k)$	2.085	3.749
$\Sigma\, y^2(k)$	0.0189	0.0398

The input energy has been increased by a factor 1.8 without violating the amplitude constraint; the output energy was increased by a factor 2.1. The resulting output y(k) according to the numerator of (14) must be divided by $(z^2 - 1.757z + 1)$ in order to obtain the numerator of (13).

Conclusions from a study on a robot application

A preliminary study on the application of a FES test to a robot arm was made [2]. It does not give a neat illustration, but it shows where the practical problems are. A detailed nonlinear simulation model for a Manutec r3 robot was derived in [3]. Fig. 2 visualizes a simplified model assuming that the arms 3, 4, 5 and 6 are one rigid unit called arm h that rotates around axis 3. For this study also the joints 1 and 2 were fixed.

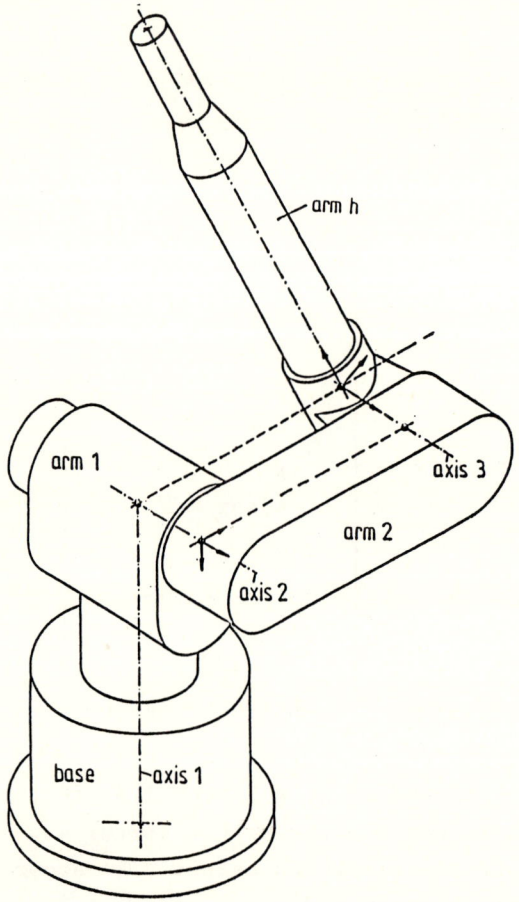

Fig. 2 Robot Manutec r3

The model has two states for arm position and velocity and two states for rotor position and velocity. The arm and the rotor are connected by a force law describing elasticity, damping and backlash in the gear. A second nonlinear force law describes the friction acting on the rotor. A saturation arises from the maximum motor torque of 9 Nm.

At the time of this writing the robot was not yet fully instrumented. Therefore only linearized model and nonlinear simulation model could be compared. This is recommended also as a first part of a continuing study, because then the influence of each nonlinearity can be studied separately.

The construction of the robot does not allow a stable equilibrium position of arm h. Therefore the reference position, for which the model is linearized, must be held by a robot controller. The standard controller has the transfer function

$$\frac{v(1+T_0 s)}{s(1+T_1 s)} \left\{ [r(s) - \text{rotorposition}(s)]k_v + k_d sr(s) - \frac{1+T_a s}{1+T_b s} \text{rotorvelocity}(s) \right\}$$

(15)

Thus the controller is of third order and the total system is of order seven. The eigenvalues of the linearized closed loop are:

$\lambda_{1,2} = -11.5 \pm 7.23j$

$\lambda_{3,4} = -30.2 \pm 83.5j$

$\lambda_{5,6} = -114 \pm 291j$

$\lambda_7 = -1110$

The response of the rotor position to a step reference input $r(s) = 1/s$ is shown in fig. 3.

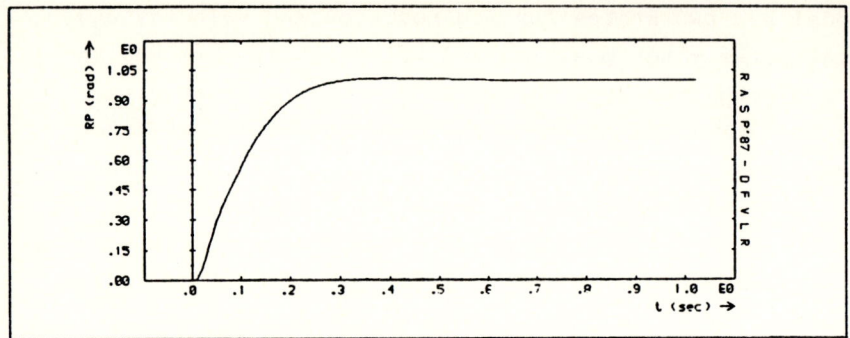

Fig. 3 Step response of the controlled robot arm

Here we see one of the difficulties with the original FES test or with other tests like step responses. The absolute values of the eigenvalues differ by a factor of about 100. The dominant behavior with a mild overshoot is due to eigenvalues $\lambda_{1,2}$. The faster modes $\lambda_{3,4}$ appear as small wiggles during the rise time of the response and the remaining eigenvalues $\lambda_{5,6}$ and λ_7 have an effect only for very small t, they are invisible in the resolution of fig. 3. It is difficult to excite and measure all modes with a single FES input. For the slow modes a sufficient excitation requires a sampling interval in the order of magnitude of 50 ... 100 ms. The FES for T = 50 ms is shown in fig. 4.

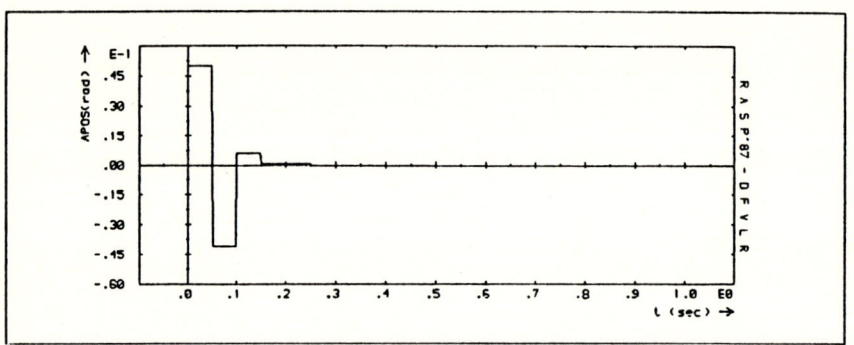

Fig. 4 Finite effect sequence with a sampling interval T = 50 ms

The form of the FES suggests, that a reduced model of third or fifth order could be used as well.
For the test two FESs for T = 50 ms were superimposed such that a constant input was applied during the first 100 ms, see fig. 5.

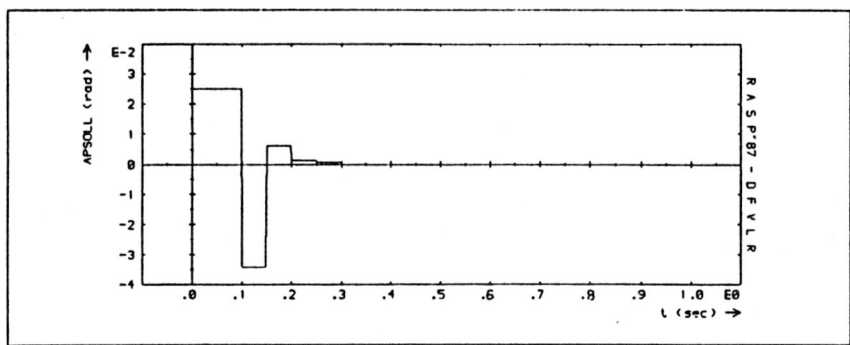

Fig. 5 Two superimposed FESs for T = 50 ms

Fig. 6a shows the FES response of the linear model and fig. 6b the FES response of the nonlinear simulation model.

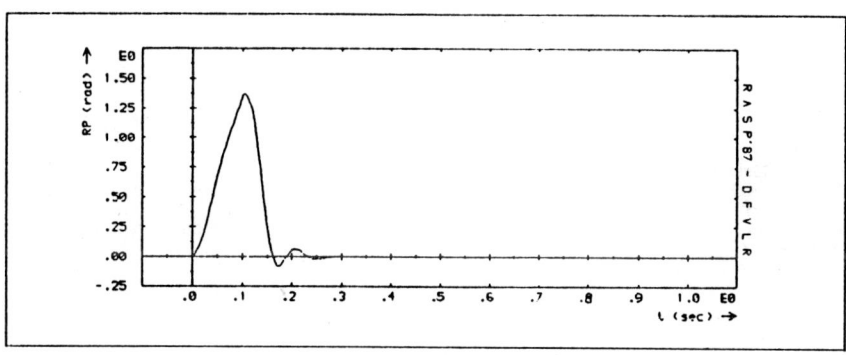

a) linear model

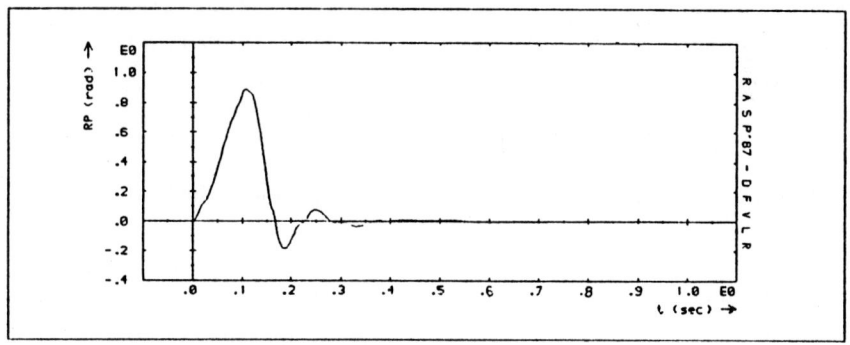

b) nonlinear model

Fig. 6 Response of the rotor position to the FES of fig. 5

The higher frequency modes are sufficiently damped, such that their response after 50 ms is negligible. In fact this experiment was used for a gripper mass calculation from the response of the nonlinear simulation model [2]. The gripper mass primarily enters into the slowest mode. It was possible to determine an uncertain mass m in the interval [0 , 15kg] with an accuracy of 200 g. The remaining uncertainty is due to uncertainty of friction parameters.

For a verification of the high frequency modes a much higher frequency of excitation is required, e.g. a sampling interval of 1 ms. Obviously their effect is obscured by the fact that we use a good controller that provides sufficient damping. A sensitive model verification or parameter estimation would require a bad controller, that results in an undamped oscillation or even gives a mild instability.

A sufficient response amplitude is required because otherwise the signal is distorted by the small signal nonlinearities backlash and friction. A tradeoff must be made between this lower signal level constraint and the upper constraint by saturation of the motor current.

The resulting recommendation for a continuation of this study is: Use a 50 ms FES for the verification of low frequency behavior. For each high frequency mode find a controller that makes it mildly unstable (e.g. by changing the original controller attached to the nonlinear simulation model). Do FES tests with sufficiently small sampling intervals.

References

1. Ackermann, J.: Sampled-data Control Systems, Berlin: Springer 1985.

2. Wirth, P.: Modelltest mit Folgen endlicher Systemantwort. Diplomarbeit TU München 1988.

3. Türk, S.: Dynamische Robotermodelle am Beispiel des Manutec R3, DFVLR-Mitteilung 1988.

Modeling the Dynamics of a Complete Vehicle with Nonlinear Wheel Suspension Kinematics and Elastic Hinges

MANFRED HILLER

Universität Duisburg, Fachgebiet Mechanik
Lotharstr. 1
D-4100 Duisburg

Summary

By a geometrical approach, the complex equations of motion of a passenger car which represents a complex spatial multiloop multibody system can be stated analytically in minimum coordinates. In particular, the nonlinear constraint equations arising from the closed loops can be stated explicitly in recursive form. In addition, significant elasticities of the vehicle are considered.

The corresponding simulation program requires a minimum number of operations. The program is applied for extended simulation runs. It has to serve as a basis for the control design of anti-block-systems (ABS), drive-slide control systems (ASR) and active suspension systems.

1 Introduction

In the design process of modern passenger cars simulation models for the representation of the complete vehicle are a desirable instrument which will be applied to shorten the developmental period and to reduce the costs. This is also valid for the design of particular car components like anti-block-systems (ABS = Antiblockiersystem), drive-slide-control systems (ASR = Antriebs-Schlupfregelung) and active suspension systems. Simulation techniques enable the variation of parameters in a manifold which can never be provided by experiments with the real vehicle; and this to a substantial reduced expenditure once the simulation programs are available. The validity of the simulation results depends mainly on the quality of the mechanical model and on the reliability of the vehicle data.

The driving performance of a modern passenger car is influenced by different parameters. Of main importance is the guided displacement of the wheel carriers due to the suspension system. Thus the stability of the vehicle when changing lanes or driving through curves as well as the passenger comfort can be influenced in a desired manner. The wheel suspension systems of modern cars are realized as spatial multibody systems with closed multibody loops. Furthermore, the kinematical behaviour of the wheel suspensions can be influenced by desired elasticities in the hinges. These provide a certain flexibility of the wheel carrier in the longitudinal direction which increases the passenger comfort and decreases the material stress.

Difficulties arise in the modeling of mechanical subsystems like the nonlinear kinematics of the wheel suspension systems or the consideration of the elasticities mentioned above, as for example the elasticities in certain hinges of the wheel suspensions or the elasticities of the tires. In the presented paper it will be shown that an effective analytical model in minimum coordinates with a recursive structure of the constraint equations can be derived using the following three concepts:

- "The characteristic pair of joints" to state constraint equations of the individual multibody loop in the always most recursive form [7];

- "the kinematical transformer" to represent the kinematical transmission behaviour of the individual loops which are connected linearly to a kinematical net and represented by a block-diagram [6];

- "kinematical differentials" to provide the partial derivates of the joint coordinates with respect to the independent coordinates by purely kinematical expressions without using analytical differentiations [5].

2 Mechanical setup of the vehicle

The vehicle under consideration consists of a car body with McPherson strut front suspension and a so-called trailing arm torsion-beam rear suspension (Verbundlenkerachse [2]). This mechanical setup is realized in many modern middle-class passenger cars (Fig. 1).

The corresponding mechanical system is built up of rigid bodies interconnected by ideal hinges and arbitrary lines of forces. The topological structure of the arising multibody system is characterized by closed multibody loops which appear in the wheel suspension systems. In addition, desired elasticities in the front and rear suspension can be represented by particular rigid-body subsystems.

The contact between tire and road is described by means of a particular kinematical model. Thus a general calculation of position and velocity at the contact domain is possible which is independent of the particular tire forces. The characteristics of the complex multibody system under consideration can be summarized as follows:

- The car body is a rigid body represented by a general tensor of inertia;

- the McPherson strut front suspension consists of wheel carriers which are guided by a strut (spring-damper unit), a lateral control arm, and a steering rod. Of great influence is the lateral elasticity of the rear hinge of the lateral control arm, which can be modeled by a local – kinematically compatible – rigid-body subsystem (Fig. 1);

- the steering mechanism is realized by a tooth rack which is interconnected to the steering rods on each side by spherical joints;

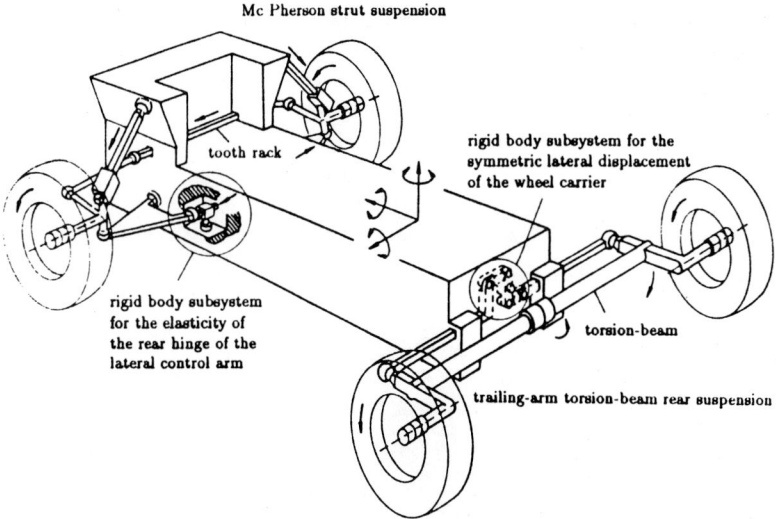

Figure 1: Mechanical setup of a passenger car with McPherson strut suspensions and trailing arm torsion beam rear suspension

- the wheel carriers of the rear axis are connected to an elastic torsion beam which is mounted to the car body by rubber elements. This so-called trailing arm torsion-beam rear suspension can be described by the rigid-body subsystem shown in Fig. 1, which guarantees the symmetric as well as the antisymmetric vertical suspension modes of the rear wheel carriers.

3 Kinematical Analysis

The topological structure of the complete vehicle is shown in Fig. 2. All rigid bodies of the multibody system are drawn in black; the connecting joints by little circles. Here, joints with more than one degree of freedom are replaced by a corresponding number of joints with one degree of freedom. The six degrees of freedom of the car-body with respect to the inertial frame can be represented by a fictitious mechanism consisting of three prismatic joints (translation) and three revolute joints (rotation). The overall system consists of $n_B = 25$ rigid bodies with $k = 133$ constraints. It is a mixed structure containing independent multibody loops (L_1 to L_8) and tree-type parts. The number of degrees of freedom of the system is $f = 17$.

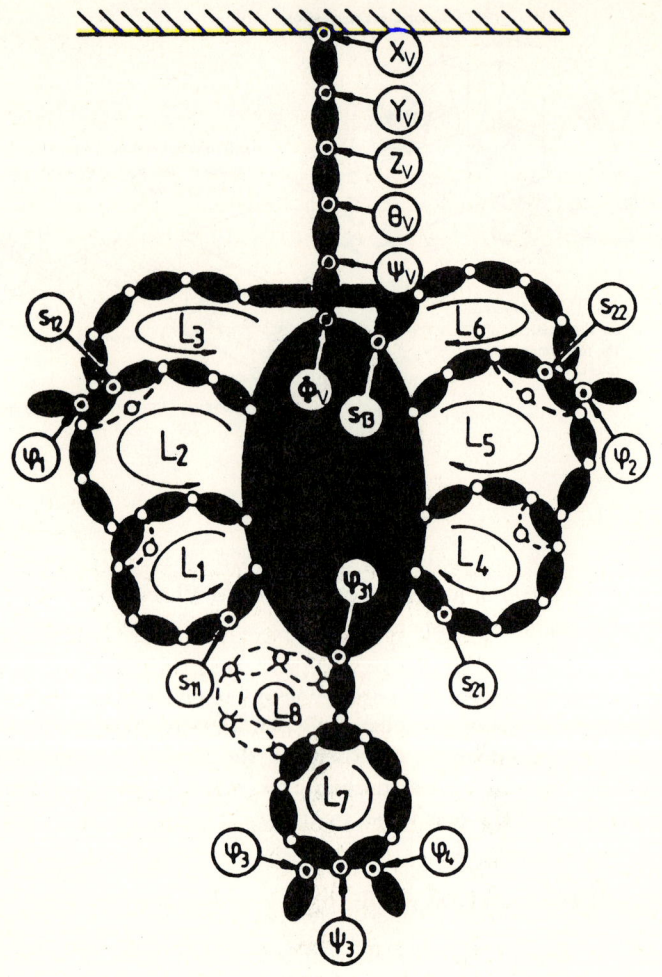

Figure 2: Topological structure of the multibody system

One problem arises now from the question how to choose the independent coordinates to get the equations of constraints in a suitable form for the later on statement of the equations of motion. Here, the main difficulties arise from the analysis of the kinematical loops due to the strongly nonlinear interdependency of the joint coordinates. Every individual kinematical loop shows a particular input-output behaviour which depends on the number of joints and joint coordinates, the geometry of the loop, but not the location of the loop. This input-output behaviour can be described by a so-called "kinematical transformer" which represents the nonlinear dependency of the six output coordinates with respect to the locally independent input coordinates, i.e. degrees of freedom of the loop [6]. The interconnection of the individual loops to a kinematical net can now be illustrated by a block-diagram where the degrees of freedom of the complete system depends on the way the loops are arranged. As the connecting nodal equations are linear, the constraint equations of the overall system can be split up into two groups:

- The nonlinear equations of the locally independent loop, i.e. the "kinematical transformer",
- the linear equations at the connecting nodes.

The number of degrees of freedom as well as the choice of the independent coordinates in the kinematical net can be determined by means of methods of graph theory. By this, an optimal solution flow in the sense that the kinematical loops can be solved recursively or as recursively as possible is guaranteed [1]. Furthermore, the individual kinematical loop can be analyzed using the concept of the "characteristic pair of joints" which enables the most recursive structure of the constraint equations of the individual loop. Depending on the type and the degrees of freedom of the joints in the loop the constraint equations in many technical examples are completely recursive [7].

By the two concepts "characteristic pair of joints" and "kinematical transformer" which are discussed in detail in the references it is possible to state the equations of constraints of even very complex multiloop multibody systems in – to a great extent – recursive form. This holds also for the vehicle model regarded in this paper where the multibody loops occur mainly in the wheel suspension systems.

The modeling of the McPherson strut suspension which includes the lateral elasticity in the rear hinge of the lateral control arm is more detailed than the investigations given in Ref. [10] and [3]. The mechanical setup is illustrated by Fig. 3a. Due to the hinge elasticity mentioned above, the corresponding multibody system consists of three independent loops on the left and right hand side (L_1 to L_3 and L_4 to L_6) respectively. In Fig. 3b only the left hand side is represented. The loops L_1 to L_3 have the following properties:

- Loop L_1: $\beta_{L_1} = 7$ joint coordinates; $f_{L_1} = 1$ d.o.f.,
- loop L_2: $\beta_{L_2} = 9$ joint coordinates; $f_{L_2} = 3$ d.o.f.,
- loop L_3: $\beta_{L_3} = 10$ joint coordinates; $f_{L_3} = 4$ d.o.f..

Due to the particular coupling of the loops L_1 to L_3 a local kinematical net is built up with three degrees of freedom. By the quantities s_{11}, s_{12} and s_{13} (see Fig. 3 and also Fig. 2) as independent coordinates the constraint equations of this subsystem

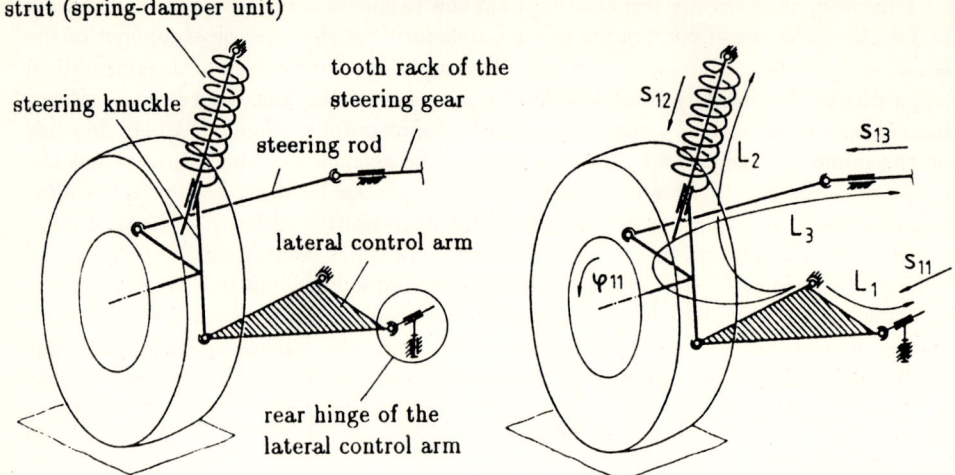

Figure 3: Mechanical model of the McPherson strut suspension with lateral elasticity in the rear hinge of the lateral control arm

can be stated in recursive form. The same holds for the symmetric McPherson strut suspension on the right hand side, described by the loops L_4 to L_6. The connection of the subsystems is given by the tooth rack of the steering mechanism which is actuated by the common input coordinate s_{13}. The kinematical structure of the complete front suspension system with altogether five degrees of freedom can now be represented by six kinematical transformers L_1 to L_6 arranged to the block-diagram shown in Fig. 4.

In a simpler way, the trailing arm torsion-beam rear suspension can be modeled. The symmetric and the antisymmetric vertical suspension mode is enabled by the the detailed arrangement of Fig. 5. The corresponding multibody system consists of the single loop L_7 with $f_{L_7} = 1$ d.o.f.. In addition, a massless multibody loop L_8 is realized

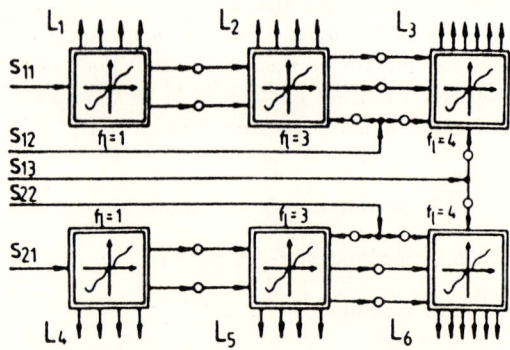

Figure 4: Block-diagram of the McPherson suspension system

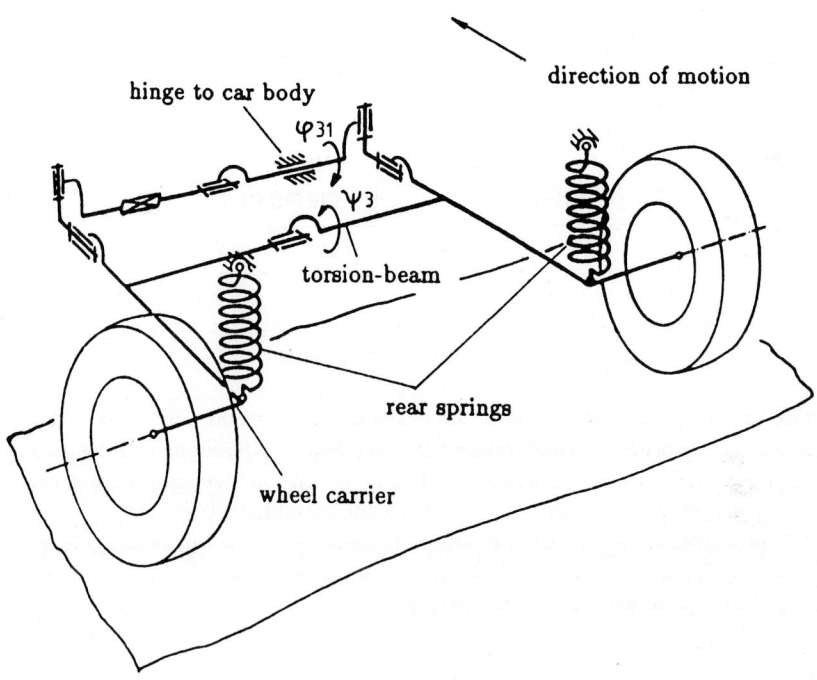

Figure 5: Mechanical model of the trailing arm torsion-beam rear suspension

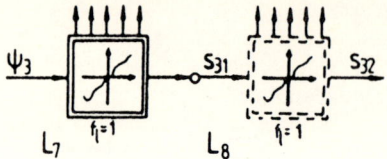

Figure 6: Block-diagram of the trailing arm torsion-beam rear suspension

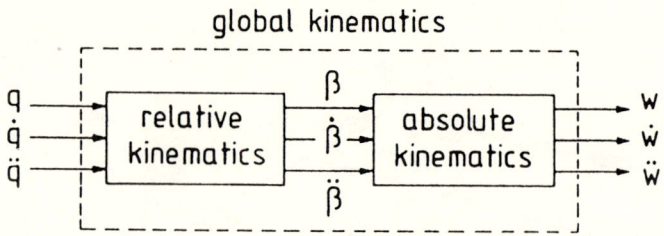

Figure 7: Global kinematics

to provide the symmetric lateral deflection of the connecting rear axis (see also Fig. 1 and 2). The rear axis represents a subsystem with one degree of freedom and the generalized coordinate ψ_3; the block diagram is given by Fig. 6. The subsystem is connected to the car body by a revolute joint with the independent coordinate ψ_{31}.

The topological structure of the complete multibody system is already given by Fig. 2, where the independent coordinates corresponding to the $f = 17$ degrees of freedom of the system are marked by circles.

4 Kinematical differentials

The kinematic analysis of the previous section provides the relationship of all relative joint coordinates β and its time derivatives with respect to the independent coordinates q and its time derivatives. For the equations of motion the absolute first and second time derivatives of the coordinates of all bodies of the multibody system are required, i.e. the relative coordinates β and its time derivatives have to be transmitted into the absolute body coordinates w and its derivatives. The absolute kinematics can be calculated explicitly in recursive form. Thus the kinematics of the multibody system can be separated into two parts: the "relative kinematics" and the "absolute kinematics" put together in the "global kinematics" (Fig. 7).

For the relationship between generalized coordinates q and the absolute coordinates of body i we have:

$$w_i = w_i(q) \ . \tag{1}$$

One obtains for the first and second time derivatives:

$$\dot{w}_i = J_{w_i} \cdot \dot{q} \ , \tag{2}$$

$$\dot{w}_i = J_{w_i} \cdot \dot{q} + a_{w_i} \quad . \tag{3}$$

The $(6 \times f)$ Jacobians and the (6×1) vectors of generalized gyroscopic forces might be calculated by analytical differentiations:

$$J_{w_i} = \frac{\partial w_i}{\partial q} \quad , \tag{4}$$

$$a_{w_i} = \sum_j \sum_k \frac{\partial^2 w_i}{\partial q_j \partial q_k} \dot{q}_j \dot{q}_k \quad . \tag{5}$$

Due to the highly implicit character of the functions $w_i(q)$ the analytical formulation of the partial derivatives for a complex system like the vehicle model is a tiresome or even impossible undertaking.

To overcome this problem, the kinematical analysis proposed above can be used and the analytical expression required in Eqs. (2) to (5) can be replaced by purely kinematical expressions: The time derivatives $\dot{w}_i$ can be stated from global kinematics for any set of generalized velocities $\dot{q}$. In particular, one can evaluate *pseudo-velocities* $\tilde{\dot{w}}_i^{(j)}$ defined by particular velocity inputs

$$\tilde{\dot{q}}^{(j)} = e^{(j)} \quad , \quad e^{(j)} = [0,\ldots,0,\overset{\downarrow j}{1},0,\ldots,0] \quad . \tag{6}$$

Here, the $e^{(j)}$ are $(f \times 1)$ unit-vectors having vanishing components except in the i-th row, which is 1. As the actual time-derivatives $\dot{w}_i$ are linear combinations of the independent generalized velocities $\dot{q}_j$, it holds:

$$\dot{w}_i = \sum_j \tilde{\dot{w}}_i^{(j)} \dot{q}_j \quad . \tag{7}$$

By comparison of Eq. (7) with Eq. (2) one obtains the simple rule:

$$j\text{-th column } \{J_{w_i}\} = \tilde{\dot{w}}_i^{(j)} \quad . \tag{8}$$

For given position- and velocity-state of the system, one can state the acceleration $\ddot{w}_i$ for any set of generalized accelerations $\ddot{q}$ again by purely kinematical expressions. Particularly, one can evaluate a *pseudo-acceleration* $\tilde{\ddot{w}}_i$ which is given for vanishing generalized accelerations, i.e. for $\ddot{q} = 0$. By Eq. (3) one then immediately obtains:

$$a_{w_i} = \tilde{\ddot{w}}_i \quad . \tag{9}$$

Eqs. (8) and (9) now state the complete partial derivatives by virtue of the already defined global kinematics. As they are based on elementary kinematical expressions – i.e. basically the laws of relative kinematics already applied in the previous section – they shall be designated here as "kinematical differentials" [5].

The time derivatives of the absolute coordinates w can now be separated into the translational parts $\dot{\underline{s}}_i, \ddot{\underline{s}}_i$ and the rotational parts $\underline{\omega}_i, \dot{\underline{\omega}}_i$ which are physical vectors. The corresponding equations are:

$$\dot{\underline{s}}_i = \sum_j \tilde{\dot{\underline{s}}}_i^{(j)} \dot{q}_j \quad ; \quad \ddot{\underline{s}}_i = \sum_j \tilde{\dot{\underline{s}}}_i^{(j)} \ddot{q}_j + \tilde{\ddot{\underline{s}}}_i \quad , \tag{10}$$

$$\underline{\omega}_i = \sum_j \tilde{\underline{\omega}}_i^{(j)} \dot{q}_j \quad ; \quad \dot{\underline{\omega}}_i = \sum_j \tilde{\underline{\omega}}_i^{(j)} \ddot{q}_j + \tilde{\dot{\underline{\omega}}}_i \quad . \tag{11}$$

From Eqs. (10) and (11) a further advantage of this representation becomes obvious: Due to the kinematical representation the arising expressions are of purely physical character and thus not dependent on particular coordinate systems. These are only needed in the very last step of the calculation.

5 Dynamics

The equations of motion are based on d'Alembert's principle. For n_B rigid bodies holds:

$$\sum_{i=1}^{n_B}[(m_i\ddot{\underline{s}}_i - \underline{F}_i) \cdot \delta\underline{s}_i + (\underline{\underline{\theta}}_{s_i}\dot{\underline{\omega}}_i + \underline{\omega}_i \times \underline{\underline{\theta}}_{s_i}\underline{\omega}_i - \underline{T}_i) \cdot \delta\underline{\phi}_i] = 0 , \qquad (12)$$

body "i":
$m_i, \underline{\underline{\theta}}_{s_i}$ — mass and tensor of inertia,
$\ddot{\underline{s}}_i$ — acceleration of mass center,
$\underline{\omega}_i, \dot{\underline{\omega}}_i$ — angular velocity and acceleration,
$\underline{F}_i, \underline{T}_i$ — resulting applied forces and torques,
$\delta\underline{s}_i, \delta\underline{\phi}_i$ — virtual displacements.

In Eq. (12) the dependent virtual displacements $\delta\underline{s}_i$, $\delta\underline{\phi}_i$ as well as the accelerations $\ddot{\underline{s}}_i$, $\dot{\underline{\omega}}_i$ have to be related to the independent virtual displacements and accelerations of the generalized coordinates. Noticing that the virtual displacements transform in the same way as the velocities, it follows for the corresponding translational and rotational parts from the previous section:

$$\delta\underline{s}_i = \sum_j \tilde{\underline{s}}_i^{(j)} \delta q_j \quad ; \quad \ddot{\underline{s}}_i = \sum_j \tilde{\underline{s}}_i^{(j)} \ddot{q}_j + \tilde{\ddot{\underline{s}}}_i , \qquad (13)$$

$$\delta\underline{\phi}_i = \sum_j \tilde{\underline{\omega}}_i^{(j)} \delta q_j \quad ; \quad \dot{\underline{\omega}}_i = \sum_j \tilde{\underline{\omega}}_i^{(j)} \ddot{q}_j + \tilde{\dot{\underline{\omega}}}_i . \qquad (14)$$

Inserting Eqs. (13) and (14) into Eq. (12) and considering the independence of the virtual displacements δq, one obtains the equations of motion in the reduced form:

$$M\ddot{q} + b = Q . \qquad (15)$$

The coefficients for the $(f \times f)$ generalized mass-matrix M, the $(f \times 1)$ vector of generalized gyroscopic forces b and the $(f \times 1)$ vector of generalized applied forces Q are:

$$M_{j,k} = \sum_i \{m_i \tilde{\underline{s}}_i^{(j)} \cdot \tilde{\underline{s}}_i^{(k)} + \tilde{\underline{\omega}}_i^{(j)} \cdot (\underline{\underline{\theta}}_{s_i} \tilde{\underline{\omega}}_i^{(k)})\} ,$$

$$b_j = \sum_i \{m_i \tilde{\underline{s}}_i^{(j)} \cdot \tilde{\ddot{\underline{s}}}_i + \tilde{\underline{\omega}}_i^{(j)} \cdot (\underline{\underline{\theta}}_{s_i} \tilde{\dot{\underline{\omega}}}_i + \underline{\omega}_i \times \underline{\underline{\theta}}_{s_i}\underline{\omega}_i)\} , \qquad (16)$$

$$Q_j = \sum_i \{\tilde{\underline{s}}_i^{(j)} \cdot \underline{F}_i + \tilde{\underline{\omega}}_i^{(j)} \cdot \underline{T}_i\} .$$

As all terms in Eqs. (16) are known from previous sections, the equations of motion are now stated in closed from. Here, a further advantage of using "kinematical differentials" becomes obvious: All coefficients can be calculated from scalar products of "physical" vectors, making the formulation independent of the reference frames in which they are evaluated. Thus the described approach is a simple tool for the derivation of the equations of motion. It can be applied for the automatic generation and solution of the dynamics of complex multiloop mechanisms.

6 Kinematical model of the tire-road contact

The modeling of the contact between the rolling wheel and the road is one of the most complex problems in vehicle dynamics. Mainly the model of a tire in connection with a stationary or non-stationary driving performance is still an unsolved problem. Today, two major possibilities are taken into account:

- Tire models based on an approximation which represents the physical properties as closely as possible [8];
- Tire models based on experimental characteristics which are approximated by mathematical curves [9].

In both cases the geometry of the tire and its contact surface are required. Therefore, a kinematical model of the tire-road contact can be stated which can be easily integrated into the modeling techniques of the vehicle mentioned above. The model contains the following ideas [11]:

- The contact geometry "tire-road" is described by a simple rigid-body subsystem, i.e. a mechanism with elementary joints which reproduces the displacements of the contact surface with respect to the road;
- tire models of different complexity – based on the velocity of the contact surface – determine the longitudinal and the lateral forces with respect to slip and slip angle;
- for more complex tire models with non-stationary driving performance the contact surface can be discretized with the kinematics available for every point.

In Fig. 8 the kinematical model of the tire-road contact is shown. By this model the kinematical quantities slip and slip angle can be calculated. Together with characteristic curves obtained from experiments the required forces can be determined.

7 Program system and simulation results

The analytical model of the complete vehicle stated in the previous sections represents a highly nonlinear system of coupled second-order differential equations, but due to its compact formulation it requires a number of operations. The numerical integration routine – based on the method of Shampine and Gordon [12] – is the core of a simulation program written in FORTRAN-77 and implemented on the mainframe computer IBM 3081, the mini-computer VAX 11/785, the workstation APOLLO DN 3000, and parts of it on a personal computer ATARI ST 1024 [4]. The program has a modular structure, it consists of about 20000 statements and requires a number of about 12000 operations.

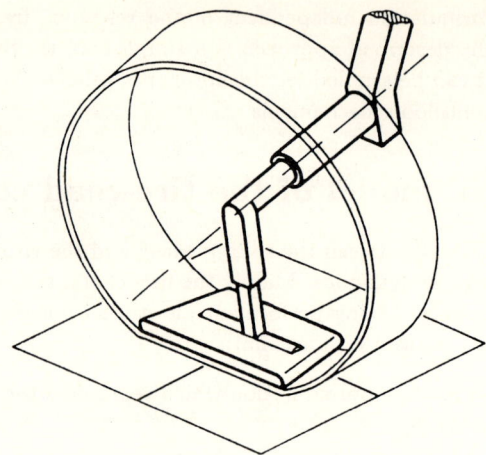

Figure 8: Kinematical model of the tire-road contact

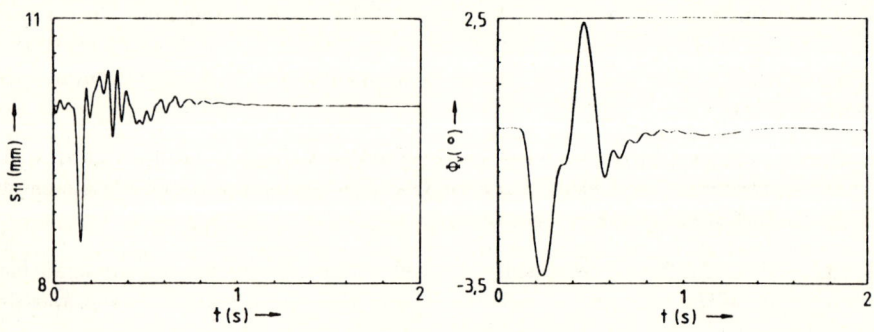

Figure 9: Time response on a run over a sinusoidal bump

A typical example for the driving performance of the vehicle is a straight run over a sinusoidal bump (length: $2\,m$, height: $10\,cm$) with a velocity of $36 km/h$. Fig. 9 shows the time response of the lateral elastic deflection of the rear hinge of the lateral control arm and of the pitch angle of the vehicle respectively. The simulation time is $2\,sec$. Remarkable is the frequency of about $20\,Hz$ due to the lateral elasticity in the lateral control arm hinge. A second example is given by Fig. 10, which shows the behaviour of the vehicle in a swerving manoeuvre to the left. A typical ratio of CPU-time (IBM 3081) to simulation time is 150 : 1.

8 Conclusion

By the three concepts "characteristic pair of joints", "kinematical transformer" and "kinematical differentials" the equations of motion of a complete passenger car which

Figure 10: Vehicle in a swerving manoeuvre to the left

represents a complex spatial multiloop multibody system can be stated analytically in a very compact way. The system with $f = 17$ degrees of freedom is described in minimum coordinates and the constraint equations of the inherent $n_L = 8$ multibody loops can be solved recursively in explicit form. The efficiency of the method is illustrated by numerical results of a simulation program based on the proposed method.

The program is applied for extended simulations of a real passenger car in industry. The results are used for comparison with experimental data. The model has to serve as an exact reference for comparison with subsequent simplified models. By model reduction simplified models will be derived using techniques like partial linearization, neglecting mass properties of small masses or omitting coordinates with small displacements. The integration process has to be accelerated by particular techniques. The simplified model will be applied for on-line simulations and they are the basis for the design of control devices like anti-block systems (ABS), drive-slide control systems (ASR) or active suspension systems.

Acknowledgement

This investigation is under contract with the Robert Bosch GmbH, Stuttgart FRG. The passenger car under consideration is the Volkswagen Golf II.

References

[1] Anantharaman, M.P.; Hiller, M.: Systematische Strukturierung der Bindungsgleichungen mehrschleifiger Mechanismen. *ZAMM* 69, 1989, to appear.

[2] Banholzer, D.: The design of the running gear of light passenger cars for comfort and safety. *Int. J. of Vehicle Design*, 129–146, 1986. Special issue on Vehicle Safety.

[3] Cronin, D.L.: McPherson strut kinematics. *Mechanism and Machine Theory*, 16:631–644, 1981.

[4] Frik, S.; Hiller, M.: Kinematik und Dynamik einer McPherson-Vorderradaufhängung mit elastischem hinterem Querlenkerlager. *ZAMM* 69, 1989, to appear.

[5] Hiller, M.; Kecskeméthy, A.: A computer-oriented approach for the automatic generation and solution of the equations of motion of complex mechanisms. In *Proc. of the 7th World Congress, The Theory of Machines and Mechanisms*, pages 425–430, 1987.

[6] Hiller, M.; Kecskeméthy, A.; Woernle, C.: A loop-based kinematical analysis of spatial mechanisms. 1986. ASME Paper 86-DET-184.

[7] Hiller, M.; Woernle, C.: A systematic approach for solving the inverse kinematic problem of robot manipulators. In *Proc. of the 7th World Congress, The Theory of Machines and Mechanisms*, pages 1135–1139, 1987.

[8] Pacejka, H.B.: *Modelling of the Pneumatic Tire and its Impact on Vehicle Dynamic Behaviour*. Lecture V 2.03, Lecture Series V, Carl Cranz Gesellschaft (CCG), Oberpfaffenhofen, 1985.

[9] Schieschke, R.; Gnadler, R.: *Modellbildung und Simulation von Reifeneigenschaften*. VDI-Bericht Nr.650, 1987.

[10] Schmidt, A.; Wolz, U. Nichtlineare räumliche Kinematik von Radaufhängungen - kinematische und dynamische Untersuchungen mit dem Programmsystem MESA VERDE. *Automobilindustrie*, 6:639–644, 1987.

[11] Schnelle, K.-P.: Die Kinematik des Rad-Straße-Kontakts. *ZAMM* 69, 1989, to appear.

[12] Shampine, L.F.; Gordon, M.K.: *Computer-Lösung gewöhnlicher Differentialgleichungen*. Vieweg Verlag, Braunschweig, 1984.

Computer Aided Formulation of Equations of Motion

T. R. KANE

Stanford University
Stanford, California

Summary

As part of the process of designing a control system for a mechanical device, one frequently must formulate the equations of motion of the device, which is a task that can be very laborious, especially if the device under consideration has a relatively large number of moving parts. This paper deals with a computer program intended to enable an analyst to formulate equations of motion with minimal labor. The name of the program is AUTOLEV.

The principal concept underlying the program is that one can create symbol manipulation functions that carry out many of the operations one normally performs by hand when formulating equations of motion. In practice, the dynamicist makes use of such functions by typing instructions on a computer terminal; the computer responds with lines of text representing equations needed to continue the analysis. Ultimately, the equations of motion appear on the screen, and one additional command then leads to a FORTRAN simulation program.

Illustrative Example

The most direct way to illustrate the use of the program is to discuss a specific example in some detail. Hence, consider the system depicted in Fig. 1, where N designates a Newtonian reference frame, B is a rigid body, and P is a particle fastened to C. Body B represents a man-made Earth satellite equipped with a pendulum-like device formed by C and P. A motor at O, connecting B to C, can cause θ, the angle between C and a line fixed in B, to vary, and the attitude of B in N is affected by such variations, which means that it may be possible to vary θ in such a way as to control the attitude of B in N to some extent. Specifically, suppose that B is axisymmetric and that point O lies on one of the central principal axes of inertia of B. Then, if, throughout some time interval, P, O, and B^*, the mass center of B, form a straight line, the system formed by B and C is an axisymmetric rigid body throughout this time interval, and must, therefore, move in N in such a way that ϕ, the angle between line $O - B^*$ and H, the inertial, central angular momentum of the system, remains constant; and, by varying θ suitably, one may be able to reduce ϕ to zero, that is, to impart to B a motion of simple spin. To explore this idea, simulations of the motion of B in N are to be performed, with θ specified as a function of t.

The numbered lines on the next page represent text typed by the user of the program. The first two lines are simply the name of the file that is being created and a brief description of its purpose. Line (3) informs the program that the system under consideration has six

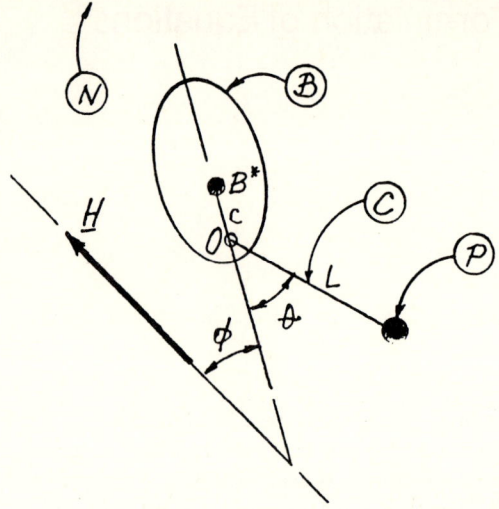

Fig. 1 Satellite with Control Boom

degrees of freedom. The names of the two rigid bodies that form the system are entered in line (4), and lines (5) - (7) let the program know that C is massless, that B has a mass to be called MB, and that the central principal moments of inertia of B are $I1$, $I2$, and $I3$. Next, in lines (8) - (10), the program is told that O and P are points of interest, but that O is massless, whereas P has a mass MP. The description of the system is continued in lines (11) and (12), which assign the letters C and L to the distance from B^* to O and the length of the pendulum, respectively, and record the fact that theta is to be a specified function of time, rather than a dependent variable. Line (13) tells the program what this function is to be; and the fact that A and $PERIOD$ in this line stand for constants is communicated to the program via line (14).

(1) ! IUTAM

(2) ! ILLUSTRATIVE EXAMPLE: A RIGID BODY + A PARTICLE PENDULUM

(3) DOF(6)

(4) FRAMES(B,C)

(5) NOMASS(C)

(6) MASS(B,MB)

(7) INERTIA(B,I1,I2,I3,0,0,0)

(8) POINTS(O,P)

(9) NOMASS(O)

(10) MASS(P,MP)

(11) CONST(C,L)

(12) SPECIFIED(THETA)

(13) THETA=A*(1-COS(2*PI*T/PERIOD))^3

(14) CONST(A,PERIOD)

(15) SIMPROT(B,C,3,THETA)

-> (16) DIRCOS(B,C,COS(THETA),-SIN(THETA),0,SIN(THETA),COS(THETA),0,0,0,1)

(17) WBN=U1*B1+U2*B2+U3*B3

(18) VBSTARN=U4*B1+U5*B2+U6*B3

(19) ALFBN=U1'*B1+U2'*B2+U3'*B3

(20) ABSTARN=DERIV(VBSTARN,T,N)

-> (21) Z1=U2*U6-U3*U5

-> (22) Z2=-U1*U6+U3*U4

-> (23) Z3=U1*U5-U2*U4

-> (24) ABSTARN=(U4'+Z1)*B1+(U5'+Z2)*B2+(U6'+Z3)*B3

(25) WCB=THETADOT*B3

(26) WCN=ADD(WBN,WCB)

-> (27) WCN=U1*B1+U2*B2
+THETADOT+U3)*B3

(28) PBSTARO=C*B1

(29) V2PTS(N,B,BSTAR,O)

-> (30) VON=U4*B1+(C*U3+U5)*B2+(-C*U2+U6)*B3

In Fig. 2, $B1$, $B2$, $B3$ and $C1$, $C2$, $C3$ designate sets of mutually perpendicular unit vectors fixed in B and C, respectively. Line (15) notifies the program that, after aligning $B1$ with $C1$, $B2$ with $C2$, and $B3$ with $C3$, one can perform a simple rotation of amount θ of C relative to B about an axis parallel to $B3$ to bring C into a general orientation relative to B. Once this line has been entered by the user of the program, line (16) appears on the screen. Note the symbol to the left of this line. This indicates that the line is supplied by the computer rather than by the user. Thus, line (16) is a *result*; specifically, it reports the elements of the direction cosine matrix relating the sets of unit vectors fixed in B and C.

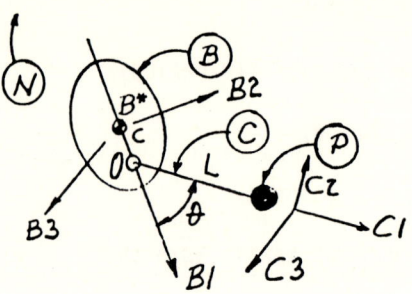

Fig. 2 Unit vectors fixed in B and C

Kinematical considerations play a major role in the formulation of equations of motion. For the system at hand, the kinematical analysis begins with line (17), in which the angular velocity of B in N, called WBN, is expressed in terms of the unit vectors $B1$, $B2$, $B3$ and *generalized speeds* $U1$, $U2$, $U3$. Similarly, in line (18), the velocity of point B^* in N is expressed in terms of generalized speeds $U4$, $U5$, $U6$; and the angular acceleration of B in N is recorded in line (19), where $U1'$ stands for the first time-derivative of $U1$, etc. The first two of these lines really *define* the symbols $U1$, ... $U6$, and the third line then represents a well known consequence of the definition of angular acceleration. Line (20), on the other hand, begins to show the power of the program. This line deals with the acceleration of B^* in N; but, instead of simply entering an expression for this acceleration, the user tells the program to find the acceleration by forming the derivative of the velocity of B^* in N with respect to time T in N. The program responds with lines (21) - (24) [note the symbol to the left of each of lines (21) - (24)], the first three of which constitute definitions of symbols $Z1$ - $Z3$, in terms of which the program then reports the desired acceleration in line (24).

The capability to perform vector additions is demonstrated by lines (25) - (27). In the first of these, the user inputs the angular velocity of C in B; in the second, he instructs the computer to add this angular velocity to the angular velocity of B in N, available in line (17); and in the third, line (27), he finds the result produced by the computer.

Lines (28) - (30) show how the program can help one to find the velocity of a point of a rigid body when one already knows the velocity of another point of this body. Specifically, the velocity of B^* in N is given in line (18). To find the velocity of O in N, one begins by introducing the position vector from B^* to O as in line (28), where this vector is called $PBSTARO$. Next, one issues the command set forth in line (29), whereupon one obtains the velocity of O in N in line (30).

Proceeding in this manner, one can construct an expression for the velocity of P in N, then use the $DERIV$ function [see line (20)] to obtain the acceleration of P in N. Once this has been done, everything required for the formation of expressions for the six generalized inertia forces for the system is in hand, so one issues the command shown in line (52), which causes the program to construct $Z17 - Z22$, the inertia torque for B in N, and lines (60) - (65), which contain the desired expressions.

```
-> (60) F1STAR=(-I1-MP*Z5*Z5)*U1'+MP*Z5*Z6*U2'-MP*Z5*U6'-MP*Z16*
        Z5-Z20

-> (61) F2STAR=MP*Z5*Z6*U1'+(-I2-MP*Z6*Z6)*U2'+MP*Z6*U6'+MP*Z16*
        Z6-Z21

-> (62) F3STAR=((-Z5*Z5-Z6*Z6)*MP-I3)*U3'+MP*Z5*U4'-MP*Z6*U5'-(-
        Z13*Z5+Z15*Z6)*MP-Z22

-> (63) F4STAR=MP*Z5*U3'+(-MB-MP)*U4'-MB*Z1-MP*Z13

-> (64) F5STAR=-MP*Z6*U3'+(-MB-MP)*U5'-MB*Z2-MP*Z15

-> (65) F6STAR=-MP*Z5*U1'+MP*Z6*U2'+(-MB-MP)*U6'-MB*Z3-MP*Z16
```

Since the generalized active forces for the present system vanish identically, all that remains to be done to write the equations of motion is to set the generalized inertia forces equal to zero. Before doing this, however, it is helpful to add a few steps that will prove useful in the sequel. For instance, one can issue the command shown in line (66), which causes the program to find the center of mass of the system and to construct the position vector from B^* to the center of mass, expressing it in the $B1$, $B2$, $B3$ basis, as indicated in lines (67) - (70); and the central angular momentum of the system, also expressed in terms of the unit vectors $B1$, $B2$, $B3$, is found by typing line (71), which leads to lines (72) - (74). Finally, the simple instruction of line (75) causes the program to find the kinetic energy of the system, reported in lines (76) - (78).

```
   (66) CM(BSTAR,B)

-> (67) TOTALMASS=MB+MP

-> (68) PBSTARCM1=(C+COS(THETA)*L)*MP/TOTALMASS

-> (69) PBSTARCM2=L*MP*SIN(THETA)/TOTALMASS

-> (70) PBSTARCM=PBSTARCM1*B1+PBSTARCM2*B2

   (71) ANGMOM(B)

-> (72) ZH1=C+COS(THETA)*L-PBSTARCM1

-> (73) ZH2=L*SIN(THETA)-PBSTARCM2
```

-> (74) ANGMOM=(-MB*PBSTARCM2*U6+MP*Z11*ZH2+Z17)*B1+(MB*PBSTARCM
    1*U6-MP*Z11*ZH1+Z18)*B2+((-PBSTARCM1*U5+PBSTARCM2*U4)*MB+(ZH1
    *Z10-ZH2*Z9)*MP+Z19)*B3

   (75) KE

-> (76) ZKE1=(U4*U4+U5*U5+U6*U6)*MB+U1*Z17+U2*Z18+U3*Z19

-> (77) ZKE2=(Z10*Z10+Z11*Z11+Z9*Z9)*MP

-> (78) KE=.5*(ZKE1+ZKE2)

Given all of the expressions that have been generated so far, one can write a computer program for the evaluation of these expressions, and thus for the numerical solution of the differential equations of motion. However, it is unnecessary to do this: by issuing just one more command, the one shown in line (86), one causes the computer to create a FORTRAN program called IUTAM.FOR, a program that can be used to integrate the equations of motion and evaluate both the system's central angular momentum and kinetic energy for any instant of time. And results generated by this program then can be used directly to generate graphs such as the one shown in Fig. 3, where ϕ is plotted as a function of time.

   (86) CODE(IUTAM,ANGMOM,ENERGY)

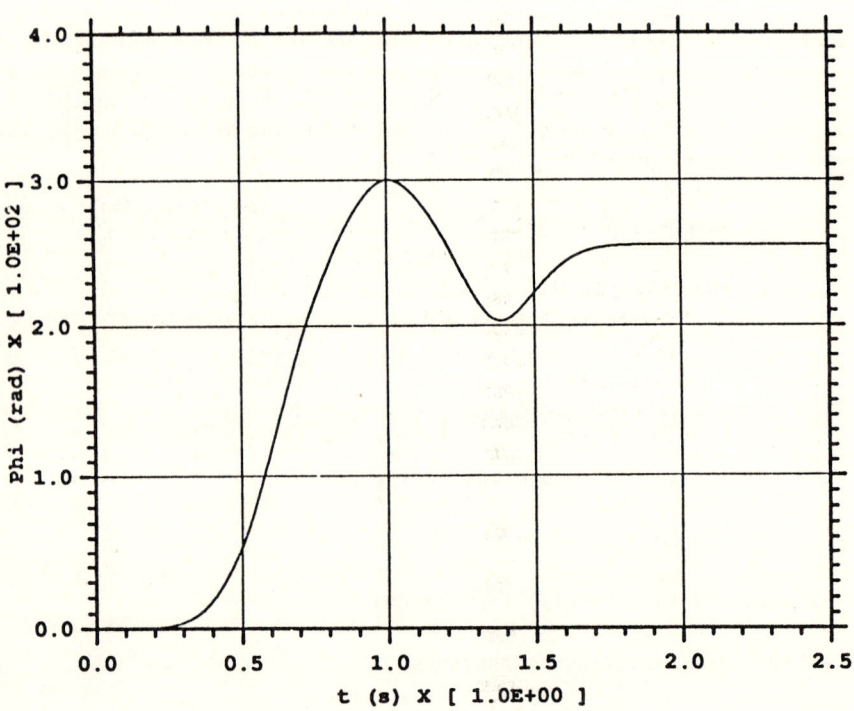

Fig. 3 Angle ϕ as a function of time t

Discussion

As has been shown, one takes the following steps when using AUTOLEV to produce simulations of motions of a mechanical system:

(1) Draw a sketch of the system to be analyzed [see Fig. 2], showing on it the names assigned to rigid bodies (e.g., B and C in Fig. 2) and points or particles (e.g., O and P), as well as geometric quantities, such as lengths (e.g., C and L) and angles (e.g., θ). The names used for this purpose can be chosen at will; that is, they need not be single letters. For example, B and P could be called *SATELLITE* and *PARTICLE*, respectively. Line (4) of the AUTOLEV program then would read (4) FRAMES(SATELLITE,C), and line (6) could become, say, (6) MASS(SATELLITE,MS). In other words, AUTOLEV gives the user considerable latitude in the choice of names.

(2) Use AUTOLEV commands to create an AUTOLEV program, such as the one that follows, which shows the user inputs for the problem considered in the illustrative example.

```
! IUTAM
! ILLUSTRATIVE EXAMPLE:  A RIGID BODY + A PARTICLE PENDULUM
DOF(6)
FRAMES(B,C)
NOMASS(C)
MASS(B,MB)
INERTIA(B,I1,I2,I3,0,0,0)
POINTS(O,P)
NOMASS(O)
MASS(P,MP)
CONST(C,L)
SPECIFIED(THETA)
THETA=A*(1-COS(2*PI*T/PERIOD))^3
CONST(A,PERIOD)
SIMPROT(B,C,3,THETA)
WBN=U1*B1+U2*B2+U3*B3
VBSTARN=U4*B1+U5*B2+U6*B3
ALFBN=U1'*B1+U2'*B2+U3'*B3
ABSTARN=DERIV(VBSTARN,T,N)
WCB=THETADOT*B3
WCN=ADD(WBN,WCB)
PBSTARO=C*B1
V2PTS(N,B,BSTAR,O)
POP=L*C1
EXPRESS(POP,B)
V2PTS(N,C,O,P)
APN=DERIV(VPN,T,N)
FRSTAR
CM(BSTAR,B)
ANGMOM(B)
KE
KANE
CODE(IUTAM,ANGMOM,ENERGY)
```

This file can be prepared with the use of a text editor, rather than in order to produce the FORTRAN program. While one is creating an AUTOLEV program interactively, one can see what commands are available by typing the word WHAT and then pressing the ENTER key, which causes the following to appear on the screen:

```
The commands that AUTOLEV recognizes are:
A1PT        A2PTS       ADD         ANGMOM      AUTOZ
AXI .       CLEAR       CM          CODE        CONST
CONSTRAIN   CONTROLS    CROSS       DERIV       DIRCOS
DOF         DOT         EULERP      EXEC        EXIT
EXPRESS     FIND        FR          FRAMES      FRSTAR
HELP        INERTIA     KANE        KE          LINE
LIST        LOAD        MASS        NOMASS      PANGVEL
POINTS      PRINCIPAL   PRINT       PVEL        RECORD
SAVE        SIMPROT     SPECIFIED   SUSPEND     VAR
V2PTS       WHAT
```

An explanation of a particular command is obtained on the screen by typing the word HELP followed by the name of the command. Thus, it is unnecessary to memorize AUTOLEV commands.

(3) Prepare an input file for the FORTRAN program created by AUTOLEV in response to the CODE command, and execute the program.

Conclusion

By freeing him from the burden of performing tedious algebraic operations, AUTOLEV enables a dynamicist to formulate equations of motion and to produce numerical simulations of motions of mechanical systems in a highly effective way.

Note

The originator of AUTOLEV, as well as the author of the underlying computer code, ist David B. Schaechter. Many of the algorithms implemented in the program were furnished by David A. Levinson. The theoretical basis for this work is set forth in the book *DYNAMICS: Theory and Applications* by T. R. Kane and David A. Levinson, McGraw-Hill Book Company, 1985.

State Equations of Motion for Flexible Bodies in Terms of Quasi-Coordinates

LEONARD MEIROVITCH**

Department of Engineering Science and Mechanics
Virginia Polytechnic Institute and State University
Blacksburg, VA 24061

Summary

This paper is concerned with the general motion of a flexible body in space. Using the extended Hamilton's principle for distributed systems, standard Lagrange's equations for hybrid systems are first derived. Then, the equations for the rigid-body motions are transformed into a symbolic vector form of Lagrange's equations in terms of general quasi-coordinates. The hybrid Lagrange's equations of motion in terms of general quasi-coordinates are subsequently expressed in terms of quasi-coordinates representing rigid-body motions. Finally, the second-order Lagrange's equations for hybrid systems are transformed into a set of state equations suitable for control. An illustrative example is presented.

Introduction

The derivation of the equations of motion has preoccupied dynamicists for many years, as can be concluded from the texts by Whittaker [1], Pars [2] and Meirovitch [3]. References 1-3 consider the motion of systems of particles and rigid bodies, and the equations of motion are presented in a large variety of forms. In this paper, we concentrate on a certain formulation, namely, Lagrange's equations. For an n-degree-of-freedom system, Lagrange's equations consist of n second-order ordinary differential equations for the system displacements.

In the control of dynamical systems, it is often convenient to work with first-order rather than second-order differential equa-

---

* Sponsored in part by the AFOSR Research Grant F49620-88-C-0044 monitored by Dr. A. K. Amos, whose support is greatly appreciated.

** University Distinguished Professor.

tions. Introducing the velocities as auxiliary variables, it is possible to transform the n second-order equations into 2n first-order state equations. The state equations are widely used in modern control theory [4].

With the advent of man-made satellites, there has been a renewed interest in the derivation of the equations of motion. The motion of rigid spacecraft can be defined in terms of translations and rotations of a reference set of axes embedded in the body and known as body axes. The equations of motion for such systems can be obtained with ease by means of Lagrange's equations. It is common practice to define the orientation of the body relative to an inertial space in terms of a set of rotation about nonorthogonal axes [3]. However, the kinetic energy has a simpler form when expressed in terms of angular velocity components about the orthogonal body axes than in terms of angular velocities about nonorthogonal axes. Moreover, for feedback control, it is more convenient to work with angular velocity components about the body axes, as sensors measure angular motions and actuators apply torques in terms of components about the body axes. In such cases, it is often advantageous to work not with standard Lagrange's equations but with Lagrange's equations in terms of quasi-coordinates [1,3]. If the body contains discrete parts, such as lumped masses connected to a main rigid body by massless springs, it is convenient to work with a set of axes embedded in the undeformed body. The equations of motion consist entirely of ordinary differential equations and can be obtained by a variety of approaches, including the standard Lagrange's equations and Lagrange's equations in terms of quasi-coordinates [5]*.

In the more general case, the body can be regarded as being either entirely flexible with distributed mass and stiffness properties or as consisting of a main rigid body with distributed elastic appendages. Unlike the previous case, the equations of motion are hybrid, in the sense that the equations for the rigid

---

* Note that Ref. 5 refers to Lagrange's equations in terms of quasi-coordinates as Boltzmann-Hamel equations.

body motions are ordinary differential equations and those for
the elastic motions are partial differential equations. Hybrid
equations were obtained for the first time in Ref. 6. Moreover,
the formulation of Ref. 6 was obtained by using Lagrange's equations in terms of quasi-coordinates, but some generality was lost
in that the body considered was assumed to be symmetric and to
undergo antisymmetric elastic motion. As a result, the rigid-body translations were zero.

This paper is concerned with the general motion of a flexible
body in space. Using the extended Hamilton's principle for distributed systems [7], standard Lagrange's equations for hybrid
systems are first derived. Then, using the approach of Ref. 3,
the equations for the rigid-body motions are transformed into a
symbolic vector form of Lagrange's equations in terms of general
quasi-coordinates. The hybrid Lagrange's equations of motion in
terms of general quasi-coordinates are subsequently expressed in
terms of quasi-coordinates representing rigid-body motions. This
is a very important step, as the latter form permits the derivation of the hybrid equations of motion with relative ease, thus
eliminating a great deal of tedious work. These hybrid equations
represent an extension to flexible bodies of Lagrange's differential equations in terms of quasi-coordinates derived in Ref. 3
for rigid bodies. The second-order equations are then used to
derive the hybrid state equations.

As an illustration, the hybrid equations of motion of a spacecraft consisting of a rigid hub with a flexible appendage simulating an antenna are derived.

Standard Lagrange's Equations for Hybrid Systems

Let us consider a flexible body and assume that the Lagrangian $L = T - V$, in which T is the kinetic energy and V is the potential
energy, can be written in the general form $L = L(q_i, \dot{q}_i, u_j, \dot{u}_j, u'_j, u''_j, \ldots, u_j^{(p)})$, where $q_i = q_i(t)$ ($i = 1, 2, \ldots, m$) are generalized coordinates
describing rigid-body motions of the body and $u_j(P,t)$ ($j = 1, 2, \ldots, n$) are generalized coordinates describing elastic motions
relative to the rigid-body motions of a typical point in the body

identified by the spatial position P. Dots designate derivatives with respect to time and primes derivatives with respect to the spatial position. For convenience, we express the Lagrangian in terms of the Lagrangian density $\hat{L}$ in the form $L = \int_D \hat{L} \, dD$, where D is the domain of extension of the body.

We propose to derive Lagrange's equations by means of the extended Hamilton's principle [7], which can be stated as

$$\int_{t_1}^{t_2} \int_D (\delta\hat{L} + \delta\hat{W}) dDdt = 0, \quad \delta q_i = \delta u_j = 0 \text{ at } t = t_1, t_2 \quad (1)$$

where $\delta\hat{W}$ is the nonconservative virtual work density, which is related to the virtual work by $\delta W = \int_D \delta\hat{W} \, dD$. The virtual work can be written in the form

$$\delta W = \sum_{i=1}^{m} Q_i \delta q_i + \sum_{j=1}^{n} \int_D \hat{U}_j \delta u_j \, dD \quad (2)$$

where Q_i are nonconservative generalized forces associated with the rigid body motions and $\hat{U}_j$ are nonconservative generalized force densities associated with the elastic motions; δq_i and δu_j are associated virtual displacements. Following the usual steps [7], we obtain Lagrange's equations of motion, which can be expressed in the symbolic vector form

$$\frac{d}{dt}\left(\frac{\partial L}{\partial \dot{q}}\right) - \frac{\partial L}{\partial q} = Q, \quad \frac{\partial}{\partial t}\left(\frac{\partial \hat{L}}{\partial \dot{u}}\right) - \frac{\partial \hat{L}}{\partial u} + \mathcal{L} u = \hat{U} \quad (3a,b)$$

where q and Q are m-vectors, u and $\hat{U}$ are n-vectors and $\mathcal{L}$ is an $n \times n$ operator matrix. Because of the mixed nature of the differential equations, we refer to the set (3) as hybrid. The elastic displacements are subject to given boundary conditions.

Equations in Terms of Quasi-Coordinates for the Rigid-Body Motions

Quite often it is convenient to express the Lagrangian not in terms of the velocities $\dot{q}_i$ but in terms of linear combinations w_ℓ ($\ell=1,2,\ldots,m$) of $\dot{q}_i$. The difference between $\dot{q}_i$ and w_ℓ is that the former represent time derivatives dq_i/dt, which can be integrated with respect to time to obtain the displacements q_i, whereas w_ℓ cannot be integrated to obtain displacements. It is customar to refer to w_ℓ as derivatives of quasi-coordinates [3]. The relation between w_ℓ and $\dot{q}_i$ can be expressed in the compact matrix form $\underset{\sim}{w} = A^T \underset{\sim}{\dot{q}}$, where the notation is obvious. Similarly, we express the velocities $\dot{q}_i$ in terms of the variables w_ℓ as $\underset{\sim}{\dot{q}} = B\underset{\sim}{w}$, from which it follows that the m × m matrices A and B are related by $A^T B = B^T A = I$, where I is the identity matrix of order m.

Our object is to derive Lagrange's equations in terms of w_ℓ instead of $\dot{q}_i$. Using the relations indicated above, it can be shown [3] that Eqs. (3a) can be replaced by

$$\frac{d}{dt}\left(\frac{\partial L}{\partial \underset{\sim}{w}}\right) + B^T E \frac{\partial L}{\partial \underset{\sim}{w}} - B^T \frac{\partial L}{\partial \underset{\sim}{q}} = \underset{\sim}{N} \tag{4}$$

where

$$E = [\underset{\sim}{w}^T B^T \frac{\partial a_{k\ell}}{\partial \underset{\sim}{q}}] - [\underset{\sim}{w}^T B^T \frac{\partial A}{\partial q_k}], \quad \underset{\sim}{N} = B^T \underset{\sim}{Q} \tag{5a,b}$$

and we note that the first matrix in E is obtained by first carrying out a triple matrix product for every one of the m^2 entries in A and then arranging the resulting scalars in a square matrix. On the other hand, the second matrix in E is obtained by first generating a row matrix for every generalized coordinate q_k (k = 1,2,...,m) and then arranging the row matrices in a square matrix. Equation (4) represents a symbolic vector form of the Langrange equations for quasi-coordinates. The complete formulation is obtained by adjoining to Eq. (4), the equations for the elastic motion, Eq. (3b), as well as the associated boundary conditions.

General Equations in Terms of Quasi-Coordinates for a Translating and Rotating Flexible Body.

Let us consider the body depicted in Fig. 1. The motion of the body can be described by attaching a set of body axes xyz to the body in undeformed state. The origin of the body axes coincides with an arbitrary point 0. Then, the motion can be defined in terms of the translation of point 0, and the rotation of the body axes xyz relative to the inertial axes XYZ. The position of 0 relative to XYZ is given by the radius vector $\underset{\sim}{R} = \underset{\sim}{R}(R_X, R_Y, R_Z)$. The rotation can be defined in terms of a set of angles θ_1, θ_2 and θ_3 (Fig. 2). Hence, the generalized coordinates are $q_1 = R_X$, $q_2 = R_Y$, $q_3 = R_Z$, $q_4 = \theta_1$, $q_5 = \theta_2$, $q_6 = \theta_3$. In addition, there are the elastic displacement components $u_x(P,t)$, $u_y(P,t)$, $u_z(P,t)$. The displacements R_X, R_Y, R_Z are measured relative to the inertial axes XYZ. On the other hand, the displacements u_x, u_y, u_z are measured relative to the body axes xyz. Moreover, the components $\dot{R}_X$, $\dot{R}_Y$, $\dot{R}_Z$ of the velocity vector $\dot{R}$ are also measured relative to XYZ. On the other hand, the angular velocity vector ω has components ω_x, ω_y, ω_z, measured relative to the body axes xyz. It will prove convenient to express all motions in terms of components along the body axes. To this end, if we denote the velocity of point 0 in terms of components along the body axes by V, then it can be shown that $V = C\dot{R}$, where $C = C(\theta_1, \theta_2, \theta_2)$ is a rotation matrix. Moreover, the angular velocity vector ω can be expressed in terms of the angular velocities $\dot{\theta}_1$, $\dot{\theta}_2$ and $\dot{\theta}_3$ in the form $\omega = D\dot{\theta}$, where $D = D(\theta_1, \theta_3)$ is a transformation matrix. We note that the angular velocity components ω_x, ω_y and ω_z cannot be integrated with respect to time to yield angular displacements α_x, α_y and α_z about axes x, y and z, respectively. Hence, ω_x, ω_y, ω_z can be regarded as time derivatives of quasi-coordinates and treated by the procedure presented in the preceding section. Although it is not very common to regard the velocity components V_x, V_y and V_z as time derivatives of quasi-coordinates, they can still be treated as such. In view of this, if we introduce the generalized velocity vector $\dot{q} = [\dot{R}_X \; \dot{R}_Y \; \dot{R}_Z \; \dot{\theta}_1 \; \dot{\theta}_2 \; \dot{\theta}_3]^T$, as well as the "quasi-velocity" vector $w = [V_x \; V_y \; V_z \; \omega_x \; \omega_y \; \omega_z]^T$, we conclude that the coefficient matrices are defined by

$$A^T = \begin{bmatrix} C & 0 \\ \hline 0 & D \end{bmatrix}, \quad B^T = A^{-1} = \begin{bmatrix} C & 0 \\ \hline 0 & (D^T)^{-1} \end{bmatrix} \quad (6a,b)$$

where we recognized that $C^{-1} = C^T$, because rotation matrices are orthonormal. It can be shown, after lengthy algebraic manipulations, that

$$B^T E = \begin{bmatrix} \tilde{\omega} & 0 \\ \hline \tilde{V} & \tilde{\omega} \end{bmatrix} \quad (7)$$

where $\tilde{\omega}$ and $\tilde{V}$ are skew-symmetric matrices corresponding to ω and V [3], respectively.

Using Eqs. (3b) and (4) in conjunction with the above relations, we obtain the hybrid Lagrange's equations in terms of quasi-coor dinates

$$\frac{d}{dt}\left(\frac{\partial L}{\partial \underline{V}}\right) + \tilde{\underline{\omega}} \frac{\partial L}{\partial \underline{V}} - C \frac{\partial L}{\partial \underline{R}} = \underline{F} \tag{8a}$$

$$\frac{d}{dt}\left(\frac{\partial L}{\partial \underline{\omega}}\right) + \tilde{\underline{V}} \frac{\partial L}{\partial \underline{V}} + \tilde{\underline{\omega}} \frac{\partial L}{\partial \underline{\omega}} - (D^T)^{-1} \frac{\partial L}{\partial \underline{\theta}} = \underline{M} \tag{8b}$$

$$\frac{\partial}{\partial t}\left(\frac{\partial \hat{L}}{\partial \underline{v}}\right) - \frac{\partial \hat{T}}{\partial \underline{u}} + \mathcal{L}\underline{u} = \hat{\underline{U}} \tag{8c}$$

where $\underline{F}$ and $\underline{M}$ are external nonconservative force and torque, respectively, in terms of components about the body axes, $\partial L/\partial \underline{\theta} = [\partial L/\partial \theta_1 \ \partial L/\partial \theta_2 \ \partial L/\partial \theta_3]^T$ and $\underline{v} = \underline{\dot{u}}$. Note that $\underline{\theta}$ does not really represent a vector and must be interpreted as a mere symbolic notation. We recall that the components of $\underline{u}$ are still subject to given boundary conditions.

It should be pointed out that, in deriving Eqs. (8), no explicit use was made of the angles θ_1, θ_2 and θ_3, so that Eqs. (8) are valid for any set of angles describing the rotation of the body axes, such as Euler's angles, and they are not restricted to the angles used here. Moreover, point 0 is an arbitrary point, not necessarily the mass center of the undeformed body, and axes xyz are not necessarily principal axes of the undeformed body. Clearly, if xyz are chosen as the principal axes with the origin at the mass center, then the equations of motion can be simplified.

State Equations in Terms of Quasi-Coordinates

Equations (8), and in particular Eqs. (8a) and (8b), can be expressed in more detailed form. To this end, we write the velocity vector of a typical point P in the body in terms of components along the body axes as follows:

$$\underline{v}_P = \underline{V} + \underline{\omega} \times (\underline{r} + \underline{u}) + \underline{v} = \underline{V} + (\tilde{\underline{r}} + \tilde{\underline{u}})^T \underline{\omega} + \underline{v} \tag{9}$$

where $\underline{r}$ is the nominal position of P relative to 0. Moreover, $\tilde{\underline{r}}$ and $\tilde{\underline{u}}$ represent skew-symmetric matrices associated with the vectors $\underline{r}$ and $\underline{u}$, respectively. Then, denoting by ρ the mass density, the kinetic energy can be shown to have the expression

$$T = \frac{1}{2} \int_D \rho \underline{v}_P^T \underline{v}_P \, dD = \frac{1}{2} m \underline{V}^T \underline{V} + \underline{V}^T \tilde{S}^T \underline{\omega} + \underline{V}^T \int_D \rho \underline{v} \, dD$$

$$+ \underline{\omega}^T \int_D \rho(\tilde{r} + \tilde{u})\underline{v} \, dD + \frac{1}{2} \underline{\omega}^T J \underline{\omega} + \frac{1}{2} \int_D \rho \underline{v}^T \underline{v} \, dD \qquad (10)$$

where $\tilde{S} = \int_D \rho(\tilde{r} + \tilde{u}) dD$, $J = \int_D \rho(\tilde{r} + \tilde{u})(\tilde{r} + \tilde{u})^T dD$, in which $\tilde{S}$ is recognized as a skew-symmetric matrix of first moments and J as a symmetric matrix of mass moments of inertia, both corresponding to the deformed body. Moreover, we assume that the potential energy has the functional form $V = V(\underline{R}, \underline{\theta}, \underline{u}, \underline{u}', \ldots, \underline{u}^{(p)})$.

Inserting Eq. (10) into Eqs. (8) and rearranging, we obtain the explicit Lagrange's equations in terms of hybrid coordinates

$$m\underline{\dot{V}} + \tilde{S}^T \underline{\dot{\omega}} + \int_D \rho \underline{\dot{v}} \, dD = (2\tilde{S}_V + m\tilde{V} + \tilde{\omega}\tilde{S})\underline{\omega} - C \frac{\partial V}{\partial \underline{R}} + \underline{F} \qquad (11a)$$

$$\tilde{S}\underline{\dot{V}} + J\underline{\dot{\omega}} + \int_D \rho(\tilde{r} + \tilde{u})\underline{\dot{v}} \, dD = [2 \int_D \rho(\tilde{r} + \tilde{u})\tilde{v} \, dD + \tilde{S}\tilde{V} - \tilde{\omega}J]\underline{\omega}$$
$$- (D^T)^{-1} \frac{\partial V}{\partial \underline{\theta}} + \underline{M} \qquad (11b)$$

$$\rho \underline{\dot{V}} + \rho(\tilde{r} + \tilde{u})^T \underline{\dot{\omega}} + \rho \underline{\dot{v}} = - \rho \tilde{V}^T \underline{\omega} - \rho \tilde{\omega}^2 (\underline{r} + \underline{u}) - 2\rho \tilde{v}^T \underline{\omega} - \mathcal{L}\underline{u} + \underline{\hat{U}} \qquad (11c)$$

where $\tilde{S}_V = \int_D \rho \tilde{v} \, dD$. The state equations are completed by adjoining the kinematical relations

$$\underline{\dot{R}} = C^T \underline{V}, \quad \underline{\dot{\theta}} = D^{-1}\underline{\omega}, \quad \underline{\dot{u}} = \underline{v} \qquad (11d,e,f)$$

Illustrative Example

As an illustration, we consider a spacecraft consisting of a rigid hub and a flexible appendage, as shown in Fig. 3. From the figure, we can write

$$\underline{r} = x\underline{i}, \quad \underline{u} = u_y \underline{j} + u_z \underline{k}, \quad \underline{v} = v_y \underline{j} + v_z \underline{k} \qquad (12)$$

so that

$$\tilde{S} = \begin{bmatrix} 0 & -\int \rho u_z dx & \int \rho u_y dx \\ \int \rho u_z dx & \int \rho u_z dx & -m\bar{x} \\ -\int \rho u_y dx & m\bar{x} & 0 \end{bmatrix} \qquad (13a)$$

where ρ is the mass density of the appendage, m is the total mass and $\bar{x}$ is the position of the mass center of the appendage. Moreover,

$$J = \begin{bmatrix} J_{xx}+\int\rho(u_y^2+u_z^2)dx & -\int\rho xu_y dx & -\int\rho xu_z dx \\ -\int\rho xu_y dx & J_{yy}+\int\rho u_z^2 dx & -\int\rho u_y u_z dx \\ -\int\rho xu_z dx & -\int\rho u_y u_z dx & J_{zz}+\int\rho u_y^2 dx \end{bmatrix} \quad (13b)$$

where J_{xx}, J_{yy} and J_{zz} are the mass moments of inertia of the spacecraft regarded as rigid.

Using Eqs. (12) and (13), the state equations, Eqs. (11), can be written in the explicit forms

$$\dot{R}_X = (c\theta_2 c\theta_3 + s\theta_1 s\theta_2 s\theta_3)V_x - (c\theta_2 c\theta_3 - s\theta_1 s\theta_2 c\theta_3)V_y + c\theta_1 s\theta_2 V_z \quad (14a)$$

$$\dot{R}_Y = c\theta_1 s\theta_3 V_x + c\theta_1 c\theta_3 V_y - s\theta_1 V_z \quad (14b)$$

$$\dot{R}_Z = -(s\theta_2 c\theta_3 - s\theta_1 c\theta_2 s\theta_3)V_x + (s\theta_2 s\theta_3 + s\theta_1 c\theta_2 c\theta_3)V_y + c\theta_1 s\theta_2 V_z \quad (14c)$$

$$\dot{\theta}_1 = c\theta_3 \omega_x - s\theta_3 \omega_y, \quad \dot{\theta}_2 = \frac{s\theta_3}{c\theta_1}\omega_x + \frac{c\theta_3}{c\theta_1}\omega_y \quad (14d,e)$$

$$\dot{\theta}_3 = \frac{s\theta_1 s\theta_3}{c\theta_1}\omega_x + \frac{s\theta_1 c\theta_3}{c\theta_1}\omega_y + \omega_z, \quad \dot{u}_y = v_y, \quad \dot{u}_z = v_z \quad (14f,g,h)$$

$$m\dot{V}_x + \dot{\omega}_y\int\rho u_z\,dx - \dot{\omega}_z\int\rho u_y\,dx = mV_y\omega_z - mV_z\omega_y + m_1\bar{x}(\omega_y^2 + \omega_z^2) - \omega_x\omega_y\int\rho u_y\,dx$$
$$- \omega_x\omega_z\int\rho u_z\,dx + 2\omega_z\int\rho v_y\,dx - 2\omega_x\int\rho v_z\,dx - (c\theta_2 c\theta_3 + s\theta_1 s\theta_2 s\theta_3)\frac{\partial V}{\partial R_X}$$
$$- c\theta_2 s\theta_3\frac{\partial V}{\partial R_Y} + (s\theta_2 c\theta_3 - s\theta_1 c\theta_2 s\theta_3)\frac{\partial V}{\partial R_Z} + F_x \quad (14i)$$

$$m\dot{V}_y - \dot{\omega}_x\int\rho u_z\,dx + m_1\bar{x}\dot{\omega}_z = mV_z\omega_x - mV_x\omega_z - m_1\bar{x}\omega_x\omega_y + (\omega_x^2 + \omega_z^2)\int\rho u_y\,dx$$
$$- \omega_y\omega_z\int\rho u_z\,dx + 2\omega_x\int\rho v_z\,dx + (c\theta_2 s\theta_3 - s\theta_1 s\theta_2 c\theta_3)\frac{\partial V}{\partial R_X} - c\theta_2 c\theta_3\frac{\partial V}{\partial R_Y}$$
$$- (s\theta_2 s\theta_3 + s\theta_1 c\theta_2 c\theta_3)\frac{\partial V}{\partial R_Z} + F_y \quad (14j)$$

$$m\dot{V}_z + \dot{\omega}_x\int\rho u_y\,dx - m_1\bar{x}\dot{\omega}_y = mV_x\omega_y - mV_y\omega_x - m_1\bar{x}\omega_x\omega_z + (\omega_x^2 + \omega_y^2)\int\rho u_z\,dx$$
$$- \omega_y\omega_z\int\rho u_y\,dx + 2\omega_x\int\rho v_y\,dx - c\theta_1 s\theta_2\frac{\partial V}{\partial R_X} + s\theta_1\frac{\partial V}{\partial R_Y} - c\theta_1 c\theta_2\frac{\partial V}{\partial R_Z} + F_z$$

$$(14k)$$

$$-(\int \rho u_z \, dx)\dot{V}_y + (\int \rho u_y \, dx)\dot{V}_z + [J_{xx} + \int \rho(u_y^2 + u_z^2)dx]\dot{\omega}_x - (\int \rho x u_y \, dx)\dot{\omega}_y$$

$$- (\int \rho x u_z \, dx)\dot{\omega}_z + \int \rho(u_y \dot{v}_z - u_z \dot{v}_y)dx = - (V_y \int \rho u_y \, dx + V_z \int \rho u_z \, dx)$$

$$+ V_x \omega_y \int \rho u_y \, dx + V_x \omega_z \int \rho u_z \, dx + \omega_x \omega_y \int \rho x u_z \, dx - \omega_x \omega_z \int \rho x u_y \, dx$$

$$+ (\omega_y^2 - \omega_z^2)\int \rho u_y u_z \, dx + \omega_y \omega_z [J_{yy} - J_{zz} - \int \rho(u_y^2 - u_z^2)dx]$$

$$- 2\omega_x \int \rho(u_y v_y + u_z v_z)dx - c\theta_3 \frac{\partial V}{\partial \theta_1} - \frac{s\theta_3}{c\theta_1}\frac{\partial V}{\partial \theta_2} - \frac{s\theta_1 s\theta_3}{c\theta_1}\frac{\partial V}{\partial \theta_3} + M_x \quad (14\ell)$$

$$(\int \rho u_z dx)\dot{V}_x - m_1 \bar{x} \dot{V}_z - (\int \rho x u_z \, dx)\dot{\omega}_x + (J_{yy} + \int \rho u_z^2 \, dx)\dot{\omega}_y - (\int \rho u_y u_z \, dx)\dot{\omega}_z$$

$$+ \int \rho(u_z \dot{v}_x - x\dot{v}_z)dx = m_1 \bar{x} V_y \omega_x - (V_z \int \rho u_z \, dx + m_1 \bar{x} V_x)\omega_y + V_y \omega_z \int \rho u_z \, dx$$

$$- \omega_z \omega_x (J_{xx} - J_{zz} + \int \rho u_z^2 \, dx) - (\omega_x^2 - \omega_z^2)\int \rho x u_z \, dx + \omega_y \omega_z \int \rho x u_y \, dx$$

$$- \omega_x \omega_y \int \rho u_y u_z \, dx + 2\omega_x \int \rho x v_y \, dx - 2\omega_y \int \rho u_z v_z \, dx$$

$$+ 2\omega_z \int \rho u_z v_y \, dx + s\theta_3 \frac{\partial V}{\partial \theta_1} - \frac{c\theta_3}{c\theta_1}\frac{\partial V}{\partial \theta_2} - \frac{s\theta_1 c\theta_3}{c\theta_1}\frac{\partial V}{\partial \theta_3} + M_y \quad (14m)$$

$$-(\int \rho u_y \, dx)\dot{V}_x + m_1 \bar{x} \dot{V}_y - (\int \rho x u_z \, dx)\dot{\omega}_x - (\int \rho u_y u_z \, dx)\dot{\omega}_y + (J_{zz} + \int \rho u_y \, dx)\dot{\omega}_x$$

$$+ \int \rho(x\dot{v}_y - u_y \dot{v}_x)dx = m_1 \bar{x} V_z \omega_x + V_z \omega_y \int \rho u_y \, dx - (V_y \int \rho u_y \, dx - V_x m_1 \bar{x})\omega_z$$

$$+ \omega_x \omega_y (J_{xx} - J_{yy} + \int \rho u_y^2 \, dx) + (\omega_x^2 - \omega_y^2)\int \rho x u_y \, dx - \omega_y \omega_z \int \rho x u_z \, dx$$

$$+ \omega_x \omega_z \int \rho u_y u_z \, dx + 2\omega_x \int \rho x v_z \, dx + 2\omega_y \int \rho u_y v_z \, dx - 2\omega_z \int \rho u_y v_y \, dx - \frac{\partial V}{\partial \theta_3} + M_z$$

$$(14n)$$

$$\rho \dot{V}_y - \rho u_z \dot{\omega}_x + \rho x \dot{\omega}_z + \rho \dot{v}_y = \rho V_z \omega_x - \rho V_x \omega_z - \rho x \omega_x \omega_y + \rho(\omega_x^2 + \omega_z^2)u_y$$

$$- \rho \omega_y \omega_z u_z + 2\rho v_z \omega_x - \mathcal{L}_y u_y + \hat{U}_y \quad (14o)$$

$$\rho \dot{V}_z + \rho u_y \dot{\omega}_x - \rho x \dot{\omega}_y + \rho \dot{v}_z = - \rho V_y \omega_x + \rho V_x \omega_y - \rho x \omega_x \omega_z - \rho \omega_y \omega_z u_y$$

$$+ \rho(\omega_x^2 + \omega_y^2)u_z - 2\rho v_y \omega_x - \mathcal{L}_z u_z + \hat{U}_z \quad (14p)$$

where m_1 is the mass of the appendage, $s\theta_i = \sin \theta_i$, $c\theta_i = \cos \theta_i$ ($i = 1,2,3$) and

$$\mathcal{L}_y = \frac{\partial^2}{\partial x^2}(EI_y \frac{\partial^2}{\partial x^2}) - \frac{\partial}{\partial x}[(\int_x^L \rho\omega_y^2 \varsigma \, d\varsigma)\frac{\partial}{\partial x}] \qquad (15a)$$

$$\mathcal{L}_z = \frac{\partial^2}{\partial x^2}(EI_z \frac{\partial^2}{\partial x^2}) - \frac{\partial}{\partial x}[(\int_x^L \rho\omega_z^2 \varsigma \, d\varsigma) \frac{\partial}{\partial x}] \qquad (15b)$$

in which E is the modulus of elasticity and I_y and I_z are area moments of inertia. The operators $\mathcal{L}_y$ and $\mathcal{L}_z$ include the effects of bending and of the axial force on the appendage [7].

Summary and Conclusions

In deriving the equations of motion for flexible bodies by the Lagrangian approach, it is common practice to express the rotational motion in terms of angular velocities about nonorthogonal axes, which tends to complicate the equations. Moreover, this creates difficulties in feedback control, in which the torque actuators apply moments about body axes and the output of sensors measuring angular motion is also expressed in terms of components about the body axes. The same can be said about force actuators and translational motion sensors. It turns out that the equations of motion are appreciably simpler when the rigid-body translations and rotations are expressed in terms of components about the body axes. Such equations can be obtained by introducing the concept of quasi-coordinates. The concept of quasi-coordinates was used earlier by this author to derive equations of motion of rotating bodies with flexible appendages, but never in the general context considered here. Indeed, in this paper, Lagrange's equations in terms of quasi-coordinates are derived for a distributed flexible body undergoing arbitrary rigid-body translations and rotations, in addition to elastic deformations. The second-order differential equations in time for the hybrid system are then transformed into a set of hybrid state equations suitable for control design. The approach is demonstrated by deriving the hybrid state equations of motion for a spacecraft consisting of a rigid body with a flexible appendage in the form of a beam.

References

1. Whittaker, E. T., A Treatise on the Analytical Dynamics of Particles and Rigid Bodies, 4th Edition, Cambridge University Press, London, 1937.

2. Pars, L. A., *A Treatise on Analytical Dynamics*, William Heinemann, Ltd., London, 1965.
3. Meirovitch, L., *Methods of Analytical Dynamics*, McGraw-Hill Book Co., New York, 1970.
4. Brogan, W. L., *Modern Control Theory*, Quantum Publishers, Inc., New York, 1974.
5. Kane, T. R. and Levinson, D. A., "Formulation of Equations of Motion for Complex Spacecraft," *Journal of Guidance and Control*, Vol. 3, No. 2, 1980, pp. 99-112.
6. Meirovitch, L. and Nelson, H. D., "On the High-Spin Motion of a Satellite Containing Elastic Parts," *Journal of Spacecraft and Rockets*, Vol. 3, No. 11, 1966, pp. 1597-1602.
7. Meirovitch, L., *Computational Methods in Structural Dynamics*, Sijthoff & Noordhoff, The Netherlands, 1980.

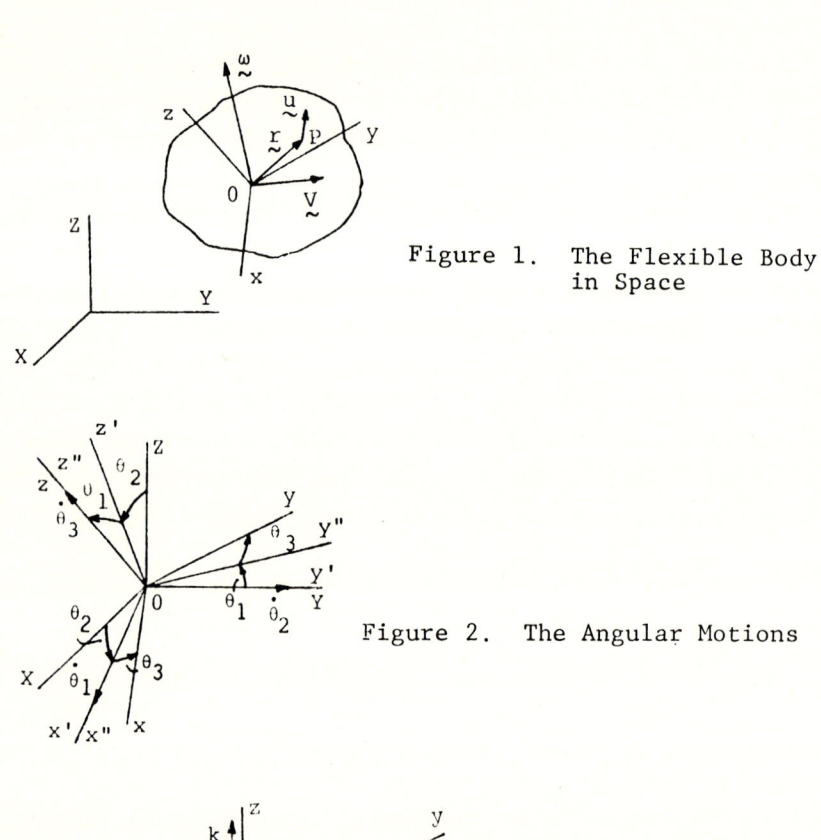

Figure 1. The Flexible Body in Space

Figure 2. The Angular Motions

Figure 3. A Rigid Spacecraft with a Flexible Appendage

Design Tools

Simulation, Test and Diagnostics Integrated for a Safety Design of Magnetic Bearing Prototypes

D. Diez and G. Schweitzer
Institute of Mechanics
ETH Zurich

Abstract

The objective of this work is to provide already in the design phase the basic procedures for a systematic verification of reliability and safety of a complex mechatronic product, consisting of hardware and of software. These basic procedures form a selfcontained software package - the "safety development system" SDS - closely linked to the actual product. As an example we will apply this development system to the design of magnetic bearings.

Introduction

In this paper the magnetic bearing system stands for a typical mechatronic system, consisting of mechanical elements, electronics and built-in software, where safety requirements are essential. Magnetic bearings are used for the contact free suspension of rotors. They operate on the basis of a closed loop control system, and their typical features allow to tackle some of the problems of classical rotor dynamics in a new way. Quite a number of detailed and specific measures are known to enhance reliability and safety (redundancy of the electronic hardware, robustness of the control software, etc.), but the actual efficiency of each such measure cannot be assessed easily. There are no general rules for the overall safety design of a mechatronic product. Therefore a strategy for diagnostics and failure control is necessary for these hardware/software products with safety requirements. Simulation and test methods are required for validation of theoretical concepts. Monitoring of data is necessary, at least for the prototype to improve the modeling on-line and off-line. For the magnetic bearing system, which works as a feedback control system, it is necessary to detect and to distinguish where controller, sensor and actuator failures occur while preserving system stability. In order to do that at the design stage already, we present our concept of a safety development system (SDS) which creates a true working environment for the controller design and for flexible programming of diagnostic strategies for a mechatronic system.

An important feature of the safety development system is that it is closely linked to the actual product. From the design stage it carries over safety properties and even hardware elements into the actual product, in our case the magnetic bearing. It allows to systematically assess and check safety properties of that product, even during operation, and it will have specific features and interfaces that enable us to introduce diagnostics and modifications. The system consists of interacting blocks and is designed as a functional object oriented system with interactively defined procedure calls. The interface to the user is implemented on a personal computer and gives interactive access to the other procedures, for example the interactive configurator or the diagnosis block. As a high level programming language Modula 2 is used.

The safety development system is being implemented now for a magnetic bearing at the ETH.

Magnetic Bearings: Function and Application

Let us first introduce the magnetic bearing which we want to refer to as an example and use it as background for the technical application. Fig. 1 shows the principle of the electromagnetic suspension: any deviation of the rotor from a reference position is measured by a suitable sensor, the sensor signal is processed in a controller; the control signal is amplified and fed to the coils of the electromagnet, thus generating electromagnetic forces which keep the rotor in a stable hovering position.

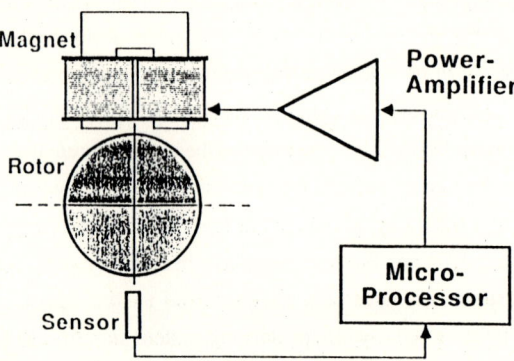

Fig. 1: Principle of the magnetic suspension

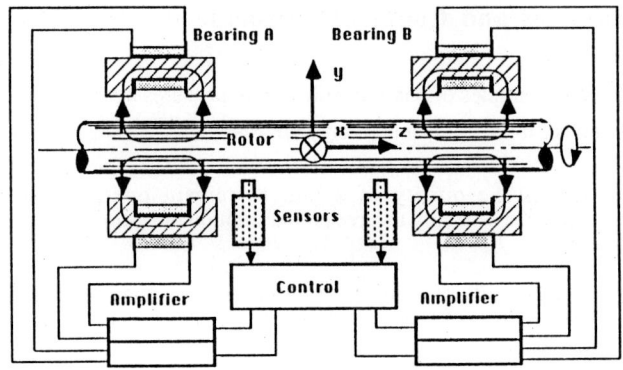

Fig. 2: Schematic of the radial suspension of a rotor

Of course for practical applications the set-up has to be more sophisticated, and it usually includes a multivariable digital control by a microprocessor system as indicated in Fig. 2. The axial bearing is not shown there. The control laws can be quite demanding as to robustness and realtime requirements.

The application areas for magnetic bearings make use of their inherent features:

- vacuum techniques, clean or sterile rooms, space applications (no lubriation, no mechanical wear)
- turbomachinery, machine tool spindles, centrifuges (high speed, controllable dynamics, high loads, low energy consumption, low maintenance)

A recent survey on theory and application is given in the Proceedings of the First International Symposium on Magnetic Bearings /SCH 88/.

Obviously some of the applications require high reliability and safety standards. Magnetic bearings have qualified for space applications already, demonstrating their potential for excellent reliability. However, strategies for designing and operating an inherently safe bearing system in a systematic and econonmic way, are not yet available. As in most mechatronic systems the contents of built-in software is already high, and it appears to be a profitable way to make even better use of this already availabe "intelligence" by letting it contribute to improve safety properties of the product.

Failure Examples and Counter Measures

Before giving some examples of possible failure and measures against them it is useful to recall the definitions for reliability and safety /BIR 85/:

> Reliability is the quality of a unit to remain operational. It characterizes the probality to have no interruption of operation during a certain time.
>
> Safety is the quality of a unit to represent no danger to humans or environment when the unit fails (technical safety). It is investigated with reliability theory.

The two terms are related to one another, but there are essential distinctions. A completly safe system may be the one that does not work at all and is totally unreliable, and a magnetic bearing that unreliably fails to operate may still coast down safely. Both areas, however, require extensive investigation of the potential failure sources, their consequences and the eventual counter-measures.

Typical failure examples for the built-in software are a system breakdown through incorrect operation, run-time exceptions (division by zero, address error, bus timeout,..), incompatible program version, or as no complete program test is possible there may be cases like an endless loop or a wrong branch. Hardware failures within the sensors are most consequential as the sensors give the primary information. They may be due to external disturbances, to incorrect adjustment or a defect in the sensor electronics. Other hardware failures include the breakdown of mechanical parts, defects in the microcomputer or disturbances in the power supply. All these failures are especially important when we are dealing with controlled mechanical systems. They usually are built to transmit forces and motions, that is power, and therefore they are inherently hazardous.

Measures for increasing the safety and reliabilty are emergency actions and stop strategies, failure detection, robust control, redundancy, fall back actions with recovery, major risk area and weak area evaluation, and diagnostics during operation as well as post mortem.

All these measures certainly do contribute to reduce danger and risks and to support functioning and operation. But how much do they contribute, and are they really necessary or only desirable? Implementing all these measures could make the product too expensive. Therefore these measures have to be checked in a systematic way. We suggest to do it in the socalled safety development system (SDS), a strongly software oriented tool, that assists in doing experiments, to try out ideas, at all levels of the design process, and that in the end can partially become part of the product itself.

Concept of the Safety Design System

Emphasizing the role of software in the design process and in the mechatronic product allows us to make the system more "intelligent", and to address the following most desirable objectives: detection of complicated failures, optimization of control strategies, diagnosis and recovery actions, design flexibility and economy. Certainly the increasing role of built-in software in any product raises new questions connected with the assessment of software quality. Compared to hardware failures new problems for example are: there will be no sign of imminent failure (because there is no wear either), minor repairs may change the whole system, no full tests are possible (because we do not foresee all possibilities of future use). This means that that the software quality has to be very high, and this is achieved by using a high level language (in our case Modula-2 with cross software tools and high level debugger), by using specific libraries with qualified

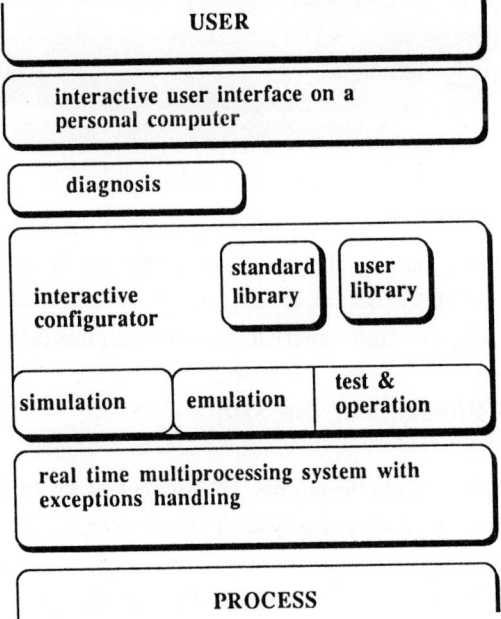

Fig. 3: Concept of the Safety Design System SDS

standard components, by using the same basic programs for simulation, emulation and testing, and by incorporating validation tests.

The structure of the SDS is given in Fig. 3, demonstrating the block configuration of the SDS and its location between the user and the process. And Fig. 4 gives an overview on the corresponding experimental setup with a magnetically suspended highly elastic rotor in the front. The next section specifies some blocks of the SDS, and their tasks will be presented in some more detail.

Fig. 4: Experimental setup for the magnetic suspension of a highly elastic rotor with the rotor-bearing system (a), the sensor unit (b), the process computer consisting of two MOTOROLA 68000 (c), the amplifiers (d), and an IBM compatible personal computer with cross-software /HOL 87/ as the user interface (e)

Structure and Elements of the SDS

The SDS connects the user and the process. The user has access to the system through a PC or a workstation, the process usually is addressable through the process computer, often being a multiprocessor system. The tasks are shared so that the PC is the design computer with the interactive user interface, the management and programming of libraries, the interfacing to some host with powerful design and simulation programs (Promatlab, ACSL), the cross software tools, the interactive configuration, the simulation of tasks, the off-line diagnosis, the mass memory and the target interface. The process computer contains the real time operating system with exceptions handling and synchronization mechanism, the high speed data monitoring and the peripheral interface.

As an example the block for the diagnosis system with its modules is shown in Fig. 5. It gets its information from the process or from simulation where the relevant data have to be retained in a predescribed manner in a ring buffer. In that way any on-line or post-mortem diagnosis can be performed. After the design phase some modules of the diagnosis can be permanently assigned to the process computer for further on-line diagnosis.

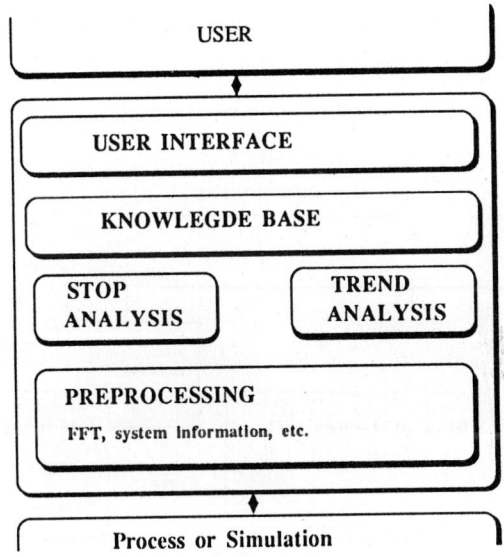

Fig. 5: Diagnosis system overview

Other examples for the blocks of the SDS are the library system and the interactive configurator. They have specific and very useful features, adopted from /MAI 88/, which facilitates their use by an still inexperienced user:
- your work only with blocks, characterized graphically and by name
- the "copy and paste" method is implemented
- the variables are taken from buffers, which can be defined as needed
- a block reads a variable and by doing so connects to another block, thus supporting a systematic and self-controlling configuration of the blocks
- a block has parameters, and the program asks for them and you only have to enter the values

- the import from the libraries into the configurator is made through the reference files. Through this way the libraries are automatically taken over from the source code without intermediate steps.

The application of these concepts and some details are shown by the following example connected to the design of a control element for a current-controlled magnetic suspension. Fig. 6 shows the block configuration for testing the controller, which could be used for the suspension of the mass in fig. 1. It demonstrates the simple procedures for switching between real time operation and simulation. In the upper part of Fig.6 the controller connects to the AD and the DC converter being part of the real time hardware, in the lower part the simulation blocks with their interconnections and the relevant notations are shown. These blocks are laid down in the library and can be looked up there.

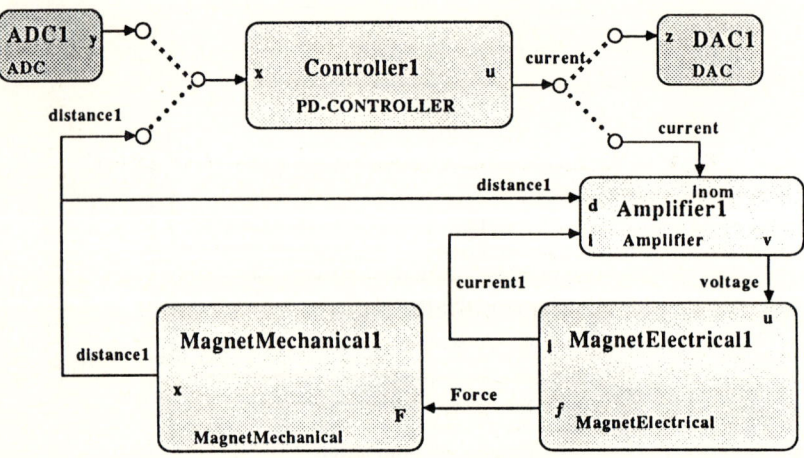

Fig. 6: Block configuration for the switching between simulation and real time operation of the controller

These blocks can be called by the interactive mouse-technique, for example the "Controller" and the "Magnetic Bearing" with their modules as shown in Fig. 7. A menue line indicates the kind of operations that can be performed on these blocks and modules. For building up the realtime test of the controller as suggested in fig. 6, only the modules "ADC", "PD-CONTROLLER" and "DAC" have to be called. They are displayed, automatically together with the variables they write and which they represent. You only have to assign suitable names to these modules and variables (for example "Controller1") just as you want to use it in your layout. Fig. 8 finally shows the "Controller1" with its input and output variables and its parameters where the values again have to be entered by the user. For the output variable the number of samples to be written into the ring buffer, too, has to be specified.

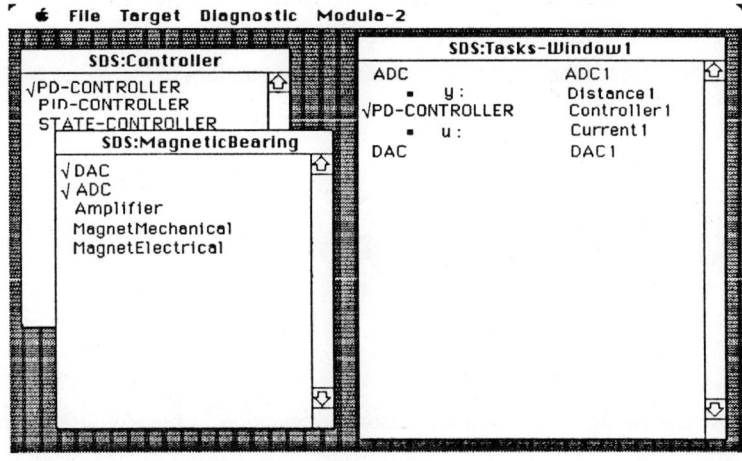

Fig. 7: Window for the blocks needed in the task of fig. 6

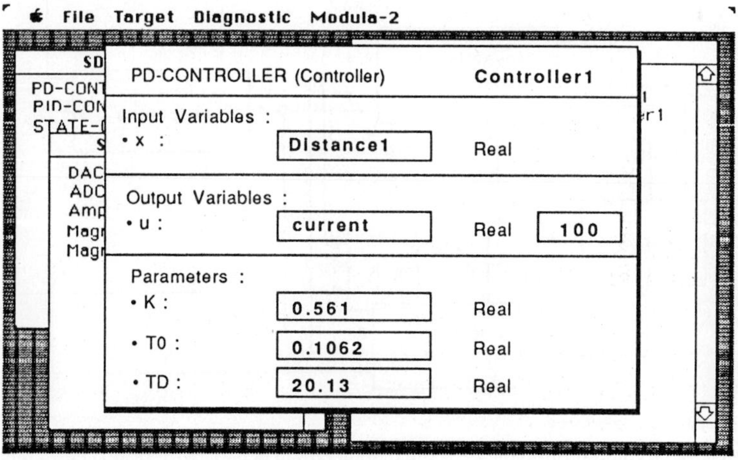

Fig. 8: Window with specifications of the PD-CONTROLLER

Design of Analytical Redundancy with Observer for a Magnetic Bearing

As an example for the application of the SDS, a suggestion for improving sensor redundancy is investigated. The sensor in the magnetic suspension in fig.1 for measuring displacements of the rotor from the reference position should give redundant information. This can be achieved for example by one of the configurations of Fig. 9.

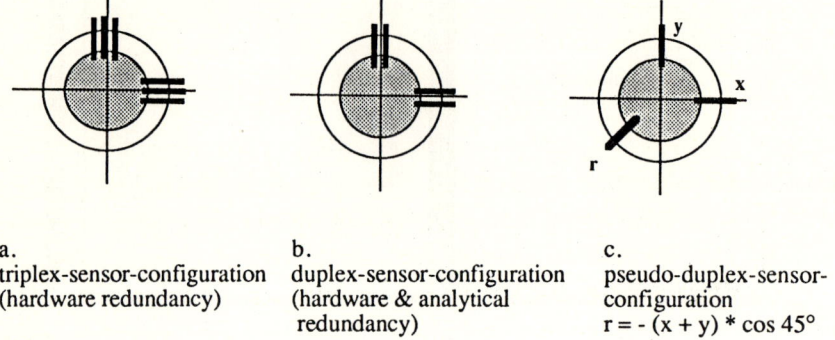

a.
triplex-sensor-configuration
(hardware redundancy)

b.
duplex-sensor-configuration
(hardware & analytical redundancy)

c.
pseudo-duplex-sensor-configuration
$r = -(x + y) * \cos 45°$

Fig. 9: Redundancy configurations for the dispacement sensor of the magnetically suspended rotor

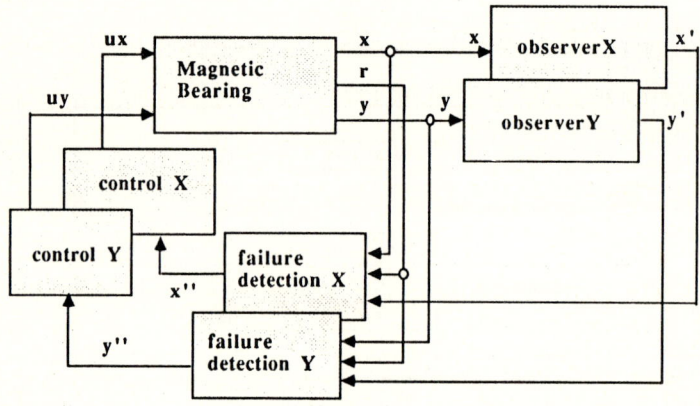

Fig. 10: Analytical redundancy for the displacement sensor with observer

Considering hardware costs the solution c is more desirable, and it has to be investigated whether it will work. Following a suggestion of /STU 85/ its function can possibly be improved by adding an observer as shown in Fig. 10. Of course a major problem lies in defining a suitable strategy for the failure detection, taking into account sensor inaccuracies and noise. But at least the simulation of ideas and variations and their consequences can be investigated easily using the SDS with its configuration support. The current results indicate that the failure detection is possible, and now the redundancy system will be implemented. However, a careful calibration of the sensors is necessary, which requires an additional effort in software. These results have been verified with an experimental magnetic suspension setup, available at the Institute of Mechanics at the ETH.

Hardware for the SDS and Portability

The programs for the SDS are written in MODULA-2. Only a few modules are hardware dependent: for the process computer, usually a multiprocessorsystem, a few hardware chips, for the PC on the user side the user interfaces. The MODULA-2 software tools are necessary for additional local programming and for cross programming.

The SDS is being implemented now on the user side with Macintosh II and MacMETH, a MODULA-2 software package for Macintosh. The process computer consists of a VME-System for the 32-bit Motorola Processor family. The additional cross-software with MacMETH adaption has some very useful time and effort saving features like incremental linker and cross-debugging tools.

Conclusions

The concept for a Safety Design System (SDS) has been presented which facilitates the systematic design of a safe mechatronic product. Its application has been shown for the example of an electromagnetic rotor-bearing system.

The following steps characterize the systematic application of the SDS:

- derive the mathematical models for the elements of the mechatronic product and chose control strategies as usual
- implement it to the SDS with the Interactive Configurator
- simulate
- improve your system based on simulation results and on diagnostic results
- build your hardware and connect it to the SDS

- test your hardware and use your on-line or your off-line diagnostics to check and to improve safety features

These steps can be easily followed as the SDS creates a true working environment fot the controller optimization and diagnostic strategies. It allows to eliminate safety relevant failure sources already in the design phase and to carry over some safety relevant features like diagnostics into the actual mechatronic product istself. Thereby time and effort for developing a safe product will be reduced essentially.

References

/BIR 85/ Birolini, A.: Qualitaet und Zuverlaessigkeit technischer Systeme. Springer-Verlag, 1985.

/HOL 87/ Holliger, R. and Meister, W. : MODES, MODULA-2 cross-software-tools for IBM compatible personal computer. Institute of Mechanics ETH Zurich, 1987.

/MAI 88/ Maier, G.: Towards Graphical Programming in Control of Mechanical Systems. In: Schweitzer, G. and M. Mansour, eds: Dynamics of Controlled Mechanical Systems. IUTAM/IFAC Symposium, ETH Zurich, June 1988, Springer-Verlag

/PAU 81/ Pau, L.F.: Failure Diagnosis and Performance Monitoring. Marcel Dekker, Inc, 1981.

/SCH 88/ Schweitzer, G., ed.: Magnetic Bearings. Proc. First International Symposium, ETH Zurich, June 1988, Springer Verlag. To appear.

/STU 85/ Stuckenberg, N.: Ein Beitrag zur Erkennung und Isolation von Sensorfehlern in Flugregelsystemen unter Verwendung von Beobachtern. DFVLR Forschungbericht, 1981.

/VDI 87/ IMEKO Symposium 1987, Paderborn : Technical Diagnostics. VDI Bericht Nr. 644.

Hardware – Software Interfaces for Dynamical Simulations

W. O. SCHIEHLEN

Institute B of Mechanics,
University of Stuttgart,
Stuttgart, FRG

Summary

The modeling of mechanical parts in controlled systems is well developed. Numerical and symbolical formalisms are available for the generation of equations of motion like ADAMS or NEWEUL, respectively. However, the dynamical behavior of the actively controlled servomechanisms and the corresponding electronic control devices cannot be modeled with adequate accuracy. Therefore, a combined dynamical simulation using a software model of the mechanical parts and a hardware design of the active elements is an economic strategy. However, the problem of the interfaces between hardware and software has to be solved.

Introduction

In the dynamics of controlled mechanical systems the approach of multibody systems is most appropriate. The mechanical parts are modeled as rigid bodies interconnected by bearings, springs, dampers and actively controlled servomechanisms. Typical examples for such active mechanical systems are found in robotics, walking machines, advanced vehicles and magnetically supported high speed rotors. For the controller design often the state space approach is used and the devices are realized by electronic components. From this point of view active mechanical systems represent an interdisciplinary science also known as mechatronics.

The method of multibody systems has been developed during the last two decades and the state-of-the-art is presented in the proceedings of IUTAM Symposia edited by Magnus [1] and Bianchi and Schiehlen [2]. The state space approach is widely used in control theory for a long time and, therefore, only the recent book of Mansour [3] will be mentioned. The fundamentals of mechatronics are presented in a survey by Schweitzer [4].

The paper presents the approach of module design of multibody systems. The design offers especially the possibility to partition a total system in mechanical parts and controlled elements. Then, the complicated software of the mechanical parts and the actively controlled servomechanisms interact in a natural way. The essential variables for the interfaces are shown and the problem of realtime simulation of the motion of the mechanical parts is adressed. From another point of view, the approach can be also interpreted as a more intelligent test rig for hardware elements.

Mechanical Part Modelling

The mechanical parts of a multibody system are given by rigid bodies with inertia as well as bearings, springs and dampers without inertia, Fig. 1. According to the free body principle, each rigid body of the mechanical system is treated separately and all elements without inertia are replaced by forces. The system's position is given relative to an inertial frame by the 3x1-translation vector $r_i(t)$ of the center of mass C_i and the 3x3-rotation tensor $S_i(t)$ written down for each body of the system, $i = 1(1)p$.

A free system of p bodies without any mechanical constraint holds 6p degrees of freedom. Thus, the position of the system can be uniquely described by 6px1-position vector

$$x(t) = [x_1 \ x_2 \ \ldots \ x_{6p}]^T. \tag{1}$$

Typical generalized coordinates of a free system are translational coordinates, Euler angles or relative distances. Then, the system's position can also be represented by

$$r_i = r_i(x) \quad , \quad S_i = S_i(x) \quad , \quad i = 1(1)p . \tag{2}$$

Further, the translational und rotational velocity and acceleration, respectively, of the system are found by differentation with respect to the inertial frame as 3x1-vectors:

$$v_i = H_{Ti}(x)\, \dot{x}(t) \quad , \quad \omega_i = H_{Ri}(x)\, \dot{x}(t) \quad , \tag{3}$$

$$a_i = H_{Ti}(x)\, \ddot{x}(t) + \frac{\partial v_i}{\partial x^T}\, \dot{x}(t) \quad ,$$

$$\alpha_i = H_{Ri}(x)\, \ddot{x}(t) + \frac{\partial \omega_i}{\partial x^T}\, \dot{x}(t) \quad . \tag{4}$$

The 3x6p-matrices are Jacobians introduced for abbreviation. For more details see Ref. [5].

A holonomic system of p bodies and q holonomic, rheonomic constraints due to rigid bearings and/or active kinematical elements results in f = 6p-q positional degrees of freedom. The constraint equation and its derivative

$$x = x(y,t) \quad , \quad \dot{x} = I(y)\, \dot{y} + \frac{\partial x}{\partial t} \tag{5}$$

represent an explicit relation between the 6px1-position vector x(t) and the reduced fx1-position vector

$$y(t) = [\, y_1\ y_2\ \cdots\ y_f\,]^T \tag{6}$$

summarizing the generalized coordinates of the holonomic system. From (2) and (5) it follows for the system's position

$$r_i = r_i(y,t) \quad , \quad s_i = s_i(y,t) \quad , \quad i = 1(1)p \quad , \tag{7}$$

and the accelerations read as

$$a_i = J_{Ti}(y,t)\, \ddot{y}(t) + \frac{\partial v_i}{\partial y^T}\, \dot{y}(t) + \frac{\partial v_i}{\partial t} \quad ,$$

$$\alpha_i = J_{Ri}(y,t)\, \ddot{y}(t) + \frac{\partial \omega_i}{\partial y^T}\, \dot{y}(t) + \frac{\partial \omega_i}{\partial t} \quad . \tag{8}$$

The 3xf-Jacobian matrices can be also obtained from

$$J_{Ti} = H_{Ti} I \quad , \quad J_{Ri} = H_{Ri} I \qquad (9)$$

reducing the computational work in some cases.

Additionally to the holonomic constraints, in nonholonomic systems there exist r nonholonomic constraints. The resulting number of motional degrees of freedom is g = f-r . The nonholonomic constraint equation and its derivative

$$\dot{y} = \dot{y}(y,z,t), \quad \ddot{y} = K(y,z)\dot{z} + \frac{\partial \dot{y}}{\partial y^T}\dot{y} + \frac{\partial \dot{y}}{\partial t} \qquad (10)$$

show the relation between the fx1-velocity vector $\dot{y}(t)$ and the reduced gx1-velocity vector

$$z(t) = [z_1 \; z_2 \; \cdots \; z_g]^T \qquad (11)$$

characterizing the generalized velocities of the nonholonomic system. From (3), (5) and (10) it follows for the system's velocity

$$v_i = v_i(y,z,t) \quad , \quad \omega_i = \omega_i(y,z,t) \qquad (12)$$

and for the system's acceleration it remains

$$a_i = L_{Ti}(y,z,t)\dot{z}(t) + \frac{\partial v_i}{\partial y^T}\dot{y} + \frac{\partial v_i}{\partial t} ,$$

$$\alpha_i = L_{Ri}(y,z,t)\dot{z}(t) + \frac{\partial \omega_i}{\partial y^T}\dot{y} + \frac{\partial \omega_i}{\partial t} . \qquad (13)$$

The nonholonomic constraints are rarely found in engineering mechanics. Nevertheless, the approach of the generalized velocities can be applied to holonomic systems with great advantage. Generalized velocities are widely used in gyrodynamics.

For the application of Newton's and Euler's equations to multibody systems, the free body principle has to be used again. For each body of the system these equations read as

$$m_i a_i = f_i^{am} + f_i^r + f_i^{ac} , \quad i = 1(1)p \qquad (14)$$

$$I_i \alpha_i + \tilde{\omega}_i I_i \omega_i = l_i^{am} + l_i^r + l_i^{ac} . \qquad (15)$$

The inertia is represented by the scalar mass m_i and the 3x3-inertia tensor I_i with respect of the center of mass C_i of each body. The forces and torques are 3x1-vectors, all torques have to be related to the center of mass C_i of each body. The applied mechanical forces f_i^{am} and torques l_i^{am}, respectively, depend on the motion by physical laws. Further, the applied control forces and torques are added. The reaction forces f_i^r and the reaction torques l_i^r, respectively, are due to the constraints given by (5) and/or (10).

The proportional forces are characterized by the system's position and time functions:

$$f_i^a = f_i^a(x,t) . \qquad (16)$$

Conservative forces due to gravity and springs as well as purely time-varying forces are proportional forces. The proportional-differential forces depend on position and velocity:

$$f_i^a = f_i^a(x,\dot{x},t) \qquad (17)$$

A parallel spring-damper configuration is a typical example for this class of forces.

The reaction forces and torques originate from bearings and supports. They can be reduced to the generalized constraint forces summarized in a $(q+r) \times 1$-vector as

$$g(t) = [g_1\ g_2\ \cdots\ g_{q+r}]^T . \tag{18}$$

Then, it yields

$$f_i^r = F_i(y,z)\ g(t), \quad l_i^r = L_i(y,z)\ g(t) \tag{19}$$

where F_i, L_i are $3 \times (q+r)$-distribution matrices. The generalized constraint forces are characteristic design parameters of bearings and supports. The distribution matrices can be found by geometrical considerations, too.

Controlled Element Modeling

The controlled elements in multibody systems may be kinematical or dynamical elements, respectively. A kinematical controlled element is nothing else than a rheonomic constraint as introduced by (5). The time history of such a rheonomic constraint is due to a time dependent control function. A possible delay between the control function and the kinematical position of the active element can be modeled by appropriate differential equations.

A dynamical controlled element results in applied forces depending not only on position and velocity but also the control function. Usually dynamical elements show some delay between the control function and the forces generated. Therefore, additional differential equations are necessary.

The controlled element E_k, $k=1(1)p$, is acting between body K_i and K_j, $i,j=1(1)p$, see Fig. 1. The nodes P_{ik} and P_{jk} are characterized by the body-fixes quantities u_{ik}, V_{ik} and u_{jk}, V_{jk} representing translational vectors and rotational tensors, respectively. Then, the corresponding kinematical equations of the element read as

$$b_k(y) = r_i + s_i u_{ik} - r_j - s_j u_{jk} ,$$

$$c_k(y) = v_{jk}^T s_j^T s_i v_{ik} , \qquad (20)$$

relating node P_{ik} to node P_{jk}. In addition, the forces and torques acting to node P_{jk} are introduced as f_{jk}^c, l_{jk}^c. Then, according to the reaction principle, the forces and torques at node P_{ik} are given as

$$f_{ik}^c = - f_{jk}^c , \quad l_{ik}^c = - l_{jk}^c + \tilde{b}_k f_{jk} . \qquad (21)$$

The forces and torques, respectively, generated by element E_k depend on the kinematics of the multibody system and the control function $u_k(t)$ as

$$f_{jk}^c = f_{jk}^c(b_k, c_k, u_k(t)) . \qquad (22)$$

For state feedback control, the control law reads as

$$u_k(t) = - K_P y(t) - K_I \int y dt - K_D \dot{y}(t) \qquad (23)$$

where the control gains are summarized in the matrices K_P, K_I, K_D. However, the dynamical behaviour of the element E_k as well as the phenomena due to digital electronic control devices are not properly modeled by (22) and (23). Therefore, a combined software-hardware simulation is a realistic approach. It turns out that the relative motion of the controlled active element and the forces and torques generated represent the essential interface variables.

The Newton-Euler equations of the global system are summarized in matrix notation as follows. The inertial properties are written in the 6px6p-diagonal matrix $\overline{\overline{M}}$, the 6px1-force vectors $\overline{q}^c$ and $\overline{q}^a$ represent gyroscopic forces and applied forces, respectively, in the following scheme

$$\bar{q} = [f_1^T \ f_2^T \ \ldots \ 1_1^T \ \ldots \ 1_p^T]^T \ . \tag{24}$$

Similar schemes are used for the global 6pxf-matrix $\bar{J}$ and the global 6pxg-matrix $\bar{L}$, respectively, as well as for the global 6px(q+r)-distribution matrix $\bar{Q}$. Then, for holonomic systems from it is obtained

$$\bar{\bar{M}} \ \bar{J} \ \ddot{y}(t) + \bar{q}^c(y,\dot{y},t) = \bar{q}^{am}(y,\dot{y},t) + \bar{Q} \ g(t) + \bar{q}^{ac} \tag{25}$$

and for nonholonomic systems it follows

$$\bar{\bar{M}} \ \bar{L} \ \dot{z}(t) + \bar{q}^c(y,z,t) = \bar{q}^{am}(y,z,t) + \bar{Q} \ g(t) + \bar{q}^{ac} \ . \tag{26}$$

The Newton-Euler equations represent for all systems 6p scalar algebraic and differential equations. The numerical solution of such equations is not straightforward, further mathematical treatment is recommended.

The dynamical principles of D'Alembert and Jourdain result in vanishing virtual work of all constraint forces and vanishing virtual power, respectively. Thus, these principles can be used to separate the Newton-Euler equations into purely differential equations for the application of standard solution techniques. The equations of motion are obtained by premultiplication with the transposed global Jacobian matrix. Then, three advantages are achieved simultaneously: i) symmetrization of the inertia matrix, ii) reduction to minimal order of the differential equation system, iii) elimination of the constraint forces and torques.

Holonomic systems with proportional - differential forces result in ordinary multibody systems. The equations of motion are obtained as

$$M(y,t) \ \ddot{y}(t) + k(y,\dot{y},t) = q^m(y,\dot{y},t) + \bar{J}^T \bar{q}^c \tag{27}$$

where the fxf-symmetric positive definite inertia matrix M and the fx1-vectors k and q of generalized gyroscopic and applied forces appear. Multibody systems are called general iff they are not ordinary. Nonholonomic constraints produce general multibody systems. The complete set of equations read as

$$\dot{y} = \dot{y}(y,z,t),$$
$$M(y,z,t)\dot{z}(t) + k(y,z,t) = q^m(y,z,t) + \bar{L}^T\bar{q}^c . \qquad (28)$$

The number of dynamical equations is further reduced now characterized by the symmetric positive definite gxg-inertia matrix M and the gx1-vectors k and q of the generalized gyroscopic and applied forces.

A main problem in the dynamics of multibody systems is the derivation of the equations of motion. Computer-aided formalisms represent the adequate solution of the problem. The formalism NEWEUL uses formula manipulation for the equations of motion realized by index coding on the basis of FORTRAN 77. This results in an excellent portability of the formalism. The resulting symbolical equations of motion offer easy access to all dependent variables like interface variables.

Intelligent Test Rig

The equations of motion (27) can not be solved by simulation since the generalized applied control forces $\bar{q}^c$ are not specified accurately. However, these forces can be measured in a test rig. For this purpose the hardware controlled element, Fig. 2, is assembled in a test rig. The global system is partitioned in software simulation of the mechanical parts and hardware measurements of the controlled element. The input variables of the test rig are the translational and rotational relative motion b_k, c_k of element E_k of node P_{jk} according to (20). Further, the information of the state $y(t)$, $\dot{y}(t)$ of the system is required for the electronic control device. The node P_{jk} is fixed in the test rig and all the forces $f_{jk}^c(t)$ and torques $l_{jk}^c(t)$ are measured at this boundary of the controlled element.

The relative motion of node P_{jk} can be generated by six position actuators as shown in Fig. 3. If B_k means the nominal 3x1-vector between P_{ik} and P_{jk} and the 3x1-vectors r_{kAm}, R_{kAm} defining attachment points A_m, O_m in the P_{ik}-fixed frame, $m=1(1)6$, then the deviations x_{km} of the actuator length l read as

$$x_{km} = 1 - |b_k - B_k + (C_k - E) r_{kAm} - R_{kAm}| \qquad (29)$$

where (20) is again to be considered.

It has to be mentioned that the computation has to be executed in real time. This means that only very simple and very fast integration codes can be used e.g. the Euler foreward method. Further, simplifications of the model of the mechanical parts may be helpful. For this purpose the influence of the generalized gyroscopic forces has to be checked since these forces are sometimes very small. With the increasing power and speed of computers, the real time computation will be less difficult in the future. The state-of-the-art in real time simulations of a moving platform has been demonstrated by the Daimler-Benz driving Simulator, see Drosdol et al. [6].

Active Vehicle Suspension

As a simple example an active automobile suspension will be treated. A complete theoretical analysis has been published in Ref. [7]. Now some simulation results will be presented. The system is defined in Fig. 4, the active element is also simulated on the computer. Therefore, only the partitioning of the multibody system is demonstrated. The excitation of the vehicle is due a quasiperiodic road profile.

Figure 5 shows the excitation and the motion of the mechanical parts, Fig. 6 presents the forces of the active controlled element due to the relative motion ($\dot{y}_1 - \dot{y}_2$) and the control feedback $u = -k_6 \dot{y}_1$. In real active elements usually both components of the force are found. In particular, elastic suspensions of a controlled element result always in forces due to the relative motion.

Conclusions

The problem of hardware-software interfaces for dynamical simulations using the multibody system approach has been treated in detail. The interface variables are the relative motion at the one end of the actively controlled element and the forces generated at the other end of this element. In addition the total information of the system's state is necessary for feeding the controller device. An essential problem remains the real time simulation of the motion of the mechanical parts. Very simple integration codes like Euler's foreward method offer today a chance for real time computation. It is excepted that the increasing power of computers will improve this situation.

References

1. Magnus, K. (ed.): Dynamics of multibody systems. Berlin/...: Springer-Verlag 1978.

2. Bianchi, G.; Schiehlen, W. (eds.): Dynamics of multibody systems. Berlin/...: Springer-Verlag 1986.

3. Mansour, M.: Lineare dynamische Systeme. Stuttgart: Teubner 1988.

4. Schweitzer, G.: Mechatronic. Z. angew. Math. Mech., to appear.

5. Schiehlen, W.: Technische Dynamik. Stuttgart: Teubner 1985.

6. Drosdol, J.; Käding, W.; Panik, F.: The Daimler-Benz Driving Simulator - New technologies demand new instruments. In: The Dynamics of Vehicles, O. Nordström (ed). Lisse: Swets & Zeitlinger 1986, S. 44-57.

7. Schiehlen, W.: Optimierung von Radaufhängungen. Z. angew. Math.Mech. 61(1981), S. T56-T58.

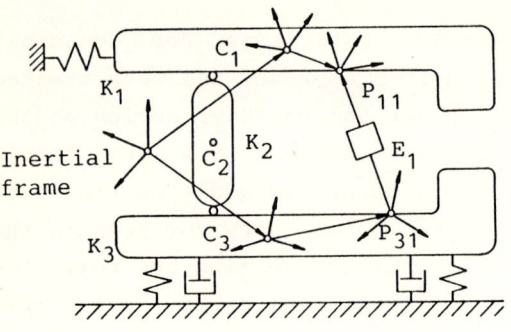

Fig.1. Multibody system

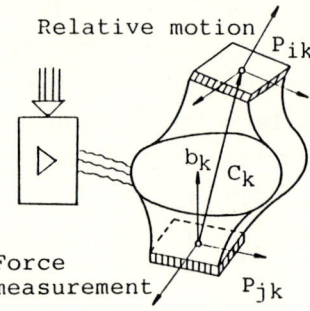

Fig.2. Controlled element

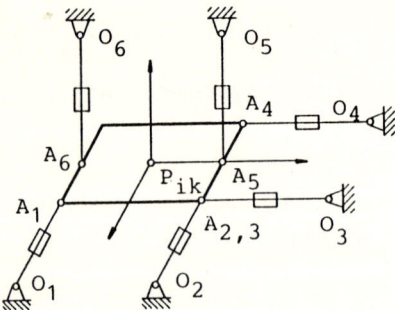

Fig.3. Position actuators

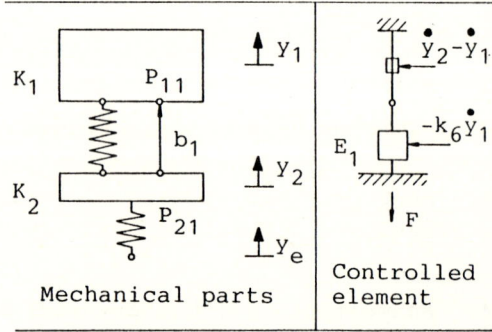

Fig.4. Active suspension

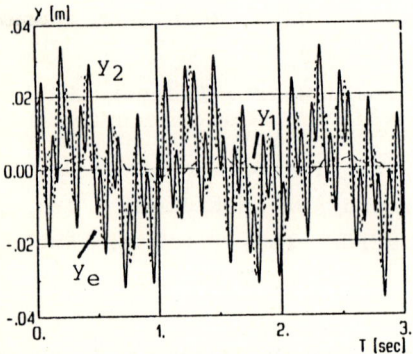

Fig.5. Time history of motion

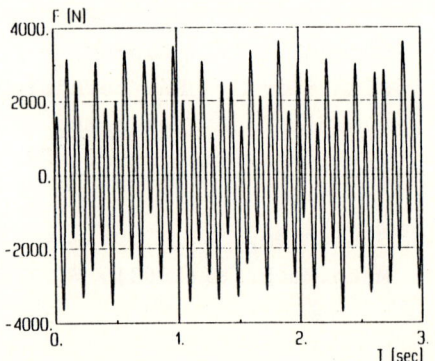

Fig.6. Time history of force

Graphical Tools

Towards Graphical Programming in Control of Mechanical Systems

Dr. Georg E. Maier

Computer Science Department CRBC.2
Asea Brown Boveri Corporate Research
CH-5405 Baden, Switzerland

Abstract

A graphical programming language for real-time programming is presented and discussed with an example. The language is based on well defined interfaces, no side-effects, step-wise refinement. Each program part is represented in a data flow view and a control flow view. A Macintosh-style tool supports direct manipulation of the graphical representations, maintains consistency between the two complementary views, and automatically generates code. An application engineer implementing a complex control system benefits by a reduced need for specific computer science skills (e.g. multi-tasking, synchronization) and by better software documentation, quality, and easier maintainability and reuse. Major computer science aspects are the step from textual to graphical program representation, the way to compose programs by connecting available modules, and the target system independence. A program inherently specifies the most parallel execution but may also be run sequentially.

Introduction

An implementor of a complex, dynamically controlled mechanical system (e.g. a robot) faces computer science problems such as multi-tasking (in order to support several controllers), synchronization (dependent controllers), exception handling (coping with faults in the process to be controlled), low level device control, programming languages, and software engineering. Often, this knowledge is not available and the resulting software is costly, unreliable, slow, of pure functionality, and difficult to maintain, i.e. there are enormous difficulties to achieve the desired performance, to incorporate new sensors, new actuators, or new algorithms, and to reuse software.

With the availability of low cost, high resolution graphical work stations new solutions to these problems become feasible. A control engineer is used to graphically represent his control systems in block diagrams for closed loop control and in state or event diagrams for sequencing control. In computer science similar techniques, e.g. data flow diagrams and flow charts, are used to represent programs. These similarities are a key to application-oriented programming. Graphical tools with modern man machine interfaces supporting graphical programming languages and automatic code generation **are promising** solutions towards allowing engineers with limited knowledge and experience in computer science to produce high-quality, low cost, and efficient software within their application domain. Unfortunately most modern methods - for example SA/SD [1], ProMod, SADT etc. for specification and design or Petri Nets [2] - do not reduce the skills an application engineer needs. Their positive aspect is that ideas and designs are represented in diagrams as it is common practice in all established engineering disciplines. But these diagrams still must be translated into code manually and extra work and the danger of inconsistencies arises when modifications have to be applied to both the diagrams and the code.

A graphical tool [3, 4] for functional programming of programmable logic controllers within large scale, continuously working control systems has already been designed and implemented at the ABB Research Center and is in use (e.g. for power plants, transportation systems, and industrial automation). It has been very well accepted because programs are only represented graphically for programming, debugging, and documentation. It is used for continuous open- and closed-loop control problems with a small depth of connectivity. It mainly supports functional programming by graphical representation of the data flow. The lack of possibilities to handle events and control flow is the reason why it is not suited for complex, mixed continuous and event-driven control.

This paper describes a new project at the ABB Research Center which aims to widen the application domain to general real-time programming. The key idea is to represent each program part in a data flow view and a separate control flow view. The project is further influenced by experience made in a general purpose automation controller project [5] at the IBM Watson Research Center.

The next section discusses our requirements to a graphical programming language. Then the language definition is presented and illustrated by a programming example. A few details on the programming tool are given followed by conclusions.

Requirements for a Graphical Programming Language

A graphical programming language needs a precise, complete semantic to allow code generation. In contrast to this, most diagram techniques only cover specific aspects and do not contain enough semantic to fully represent a program.

Language concepts must be selected carefully to be appropriate for visualization. The attempt to directly visualize concepts used in textual languages would certainly not lead to optimum results.

Data flow representations are suited to visualize data dependencies and functional programming. In case of data triggering these dependencies also specify the order of execution. The language must support the difference between ordinary data dependencies and data triggering to unburden the programmer as far as possible.

Control flow representations are suited to visualize dependencies in the order of execution, events, parallel paths, and synchronization (mutual exclusive blocks, waiting for stimuli).

The language must attempt to hide the target system configuration. Especially the task structure, which gives the amount of parallelism at run-time on a specific target system, should not be shown. A program should inherently include the most possible parallelism which may be mapped differently to different target systems during code generation. (In fact an application engineer is not interested at all in the task structure. He is thinking in terms of - possibly parallel - data dependencies or multiple controllers - which may possibly be executed in parallel.)

Further, the language must be based on commonly accepted computer science concepts as well defined interfaces, no side-effects, step-wise refinement, and well defined behaviour in case of errors (exception-handling).

Tool Requirements

The language must be supported by an interactive, menu-driven tool with a user interface similar to the de-facto standard which has been established by the Macintosh computers.

The basic requirements are on-line syntax check, maintaining consistency between data flow and control flow automatically (appropriate editing operations necessary), automatic code generation, and laser printer support to produce graphical documentation.

Definition of a Graphical Programming Language

Data Types

A data type defines a scalar or structured range of values including a default value. Data types are defined similar to types in Modula-2. If nothing is specified properties of Modula-2 types may be assumed.

A structured data type is either a *RECORD* type (fixed size, ordered collection of named fields of possibly different types) or an *ARRAY* type (fixed size, ordered collection of numbered elements of equal type). A *RECORD* type definition includes a default value for each field, an *ARRAY* type inherits the default value from its base type.

Predefined data types are *BOOLEAN, INTEGER, REAL, CHAR*, and *STRING which* is a variable length, ordered collection of characters terminated by *0C*.

Additional data types may be defined by the user. If a type definition does not include a default value, it is copied from the base type or is automatically determined as stated below.

- The range of values of an enumeration type is defined by a set of named constants. The first constant is taken as default value if none is specified.
- The range of values of a subrange type is defined by a low and high bound within the range of a scalar type. The low bound is taken as default value if none is specified and if the range does not include the default value of the scalar type.
- A data type re-definition may have the purpose to define a new default value only.

Variables

A variable is a possibly named instance of a data type. A variable is part of a program, a function implementation, or a device type implementation.

Attributes

An attribute is a named instance of a data type. An attribute is part of a program, a function implementation, or a device type implementation. Its value does not change during program execution (as a Modula-2 constant). Attributes are visible from everywhere. They are not hidden by interfaces. The attribute mode determines whether its value may be set only locally (mode *local*) or from everywhere (mode *global*).

Attributes allow to parametrize functions, devices, and programs conveniently without any cost at run-time (e.g. coefficients of a controller). If attributes were part of function interfaces, the definitions of functions would become rather clumsy. Nevertheless, the consistent use of attributes is checked automatically.

Functions

A function is a named side-effect-free, reentrant operation. Its interface consists of its inputs and outputs (data flow) and its termination events (control flow). A function instance is part of a program, a function implementation, or a device type implementation.

The interface of a function must be provided before the function can be instantiated and before its implementation, which is separate from the definition, can be defined.

Inputs and outputs are named instances of data types. They are called scalar or structured according to their data type. An input or output is called discrete if one data object is consumed or produced per execution of the function instance. It is called continuous if multiple data objects are consumed or produced per execution of the function instance.

Similar to an IN-OUT parameter of a conventional procedure, a function output may be bi-directional, i.e. be read when execution of a function instance starts.

Figure 2 shows examples of function interfaces.

Devices

A device type is a self-acting *RECORD* type. An instance of a device type is called a device an is part of a program, a function implementation, or a device type implementation.

Similar to an I/O peripheral, a device is an abstraction of a task to be performed independently which is controlled through a data interface. In contrast to a variable, a device is active, i.e. it may further process or modify its interface data autonomously.

The interface of a device type must be provided before it can be instantiated and before its implementation, which is separate from the definition, can be defined.

Figure 2 shows the interface of a device type.

Data Flow View

A data flow view shows data flow aspects of either the implementation of a function or a device type, or a program. It consists of function instances, devices, variables, constants, and data flow connections. Examples are found in figures 1a, 3,4, and 5.

The functions instances must be uniquely named. The function name is used as default name of the instance as long as only one instance of the function appears within the same data flow view.

Data flow connections define the flow of data objects between constants, variables, function instances, and devices. A data flow connection points from a data source to one or several data sinks. A data flow connection is attached to at least one function instance or an input or output of the implementation. Constants, variables, and devices must not be connected with each others directly.

Data triggering is expressed by one or several direct data flow connections between two function instances. The order of execution within a set of data triggered function instances is fully specified by the flow of data. The effective flow of control is determined during code generation automatically. The control flow may be specified explicitly by avoiding data triggering, i.e. by inserting variables between function instances.

A discrete data flow connection models the flow of one data object per execution of the attached function instance. Discrete data flow connections may only be attached to discrete function inputs and outputs. A continuous data flow connection models the flow of multiple data objects per execution of the attached function instance. Continuous data flow connections may only be attached to continuous function inputs and outputs.

Data flow connections must obey data type compatibility. If necessary, a data flow connection of a structured type may be expanded to several connections of the corresponding component types or several data flow connections may be compressed to one connection of the corresponding structured type.

Structured data flow connections are graphically represented by thick lines, continuous data flow connections by double arrows.

Control Flow View

A control flow view shows the control flow aspects of either the implementation of a function or a device type, or a program. It consists of actions, control flow connections, event sources, and blocks with parallel paths. Examples are found in figures 1b, 3,4, and 5.

An action is either a function instance or a set of data triggered function instances. Its name identifies the corresponding function instance(s) in the data flow view. The termination of the execution of an action is marked by an event. An internal event is visible in the control flow interface of the corresponding function, i.e. an internal event comes from the action itself. An external event is a termination condition which cannot be influenced by the action.

Control flow connections define the execution order of actions. A control flow connection typically points from an event to an action to be executed when the event occurs.

An event source creates multiple events (e.g. an interrupt or a periodic timer, see figure 5).

Devices used within a control flow view appear in the "uses"-list. Devices are initialized first. In figure 1b the effective flow of control is obtained by nesting the *xAxis* control flow (see figure 5) into the "use Servo"-block of the *Gripper* control flow, the *yAxis* control flow into the "use Servo"-block of *xAxis*, and so on up to the control flow of *PickAndPlace*.

Standard Functions

A standard expression evaluator function may be used as generic function to perform simple calculations (see function instances *M1* and *M2* in figure 4).

The standard action CASE may be used in control flow views to branch on disjunctive conditions each formulated as a boolean expression (see figure 4).

The standard action WAIT may be used to wait on any external events each formulated as boolean expression (see figure 4).

Programs and Libraries

A program is the entity which may be executed in a run-time environment. A program consists of declarations (library imports, data types, functions, and device types), of a data flow view, and of a control flow view.

A library is a collection of declarations (data types, functions, and device types). A library may be imported by several programs.

Programming Example

The use of the language is illustrated with a example of robotics. A program to pick and place an object is developed starting from servo control and commands as *Move* to move one or more axis of a robot and centering *Grasp* to grasp an object. Many real-world problems (e.g. a reasonable number of axis and kinematics) have been omitted to allow the example to fit into this paper.

System Configuration and Control Concepts

The system consists of an X-Y-table with a two-finger gripper moving in X-direction. Each of the three axis is equipped with an actuator and a absolute position sensor. The two fingers of the gripper are each furnished with a binary touch sensor.

Each axis is to be controlled by a PI-controller which repeatedly compares the actual position from the sensor with the desired position and computes new values to be passed to the actuator.

As long as this desired position is only updated by small increments, it may be assumed that the actual axis position follows continuously and smoothly. Therefore, the desired position may also be used as current position of the axis by high level commands, and the interface from a high level command to an axis consists only of the desired position.

On top of servo control a command *Move* is used for coordinated straight-line motion of one or several axis of a robot from their current positions to given goal positions. A *Move* consists of two steps. First a trajectory of the motion is planned and parametrized in time, and then a set point generator repeatedly adjusts the desired positions of the involved servos to produce a smooth, coordinated movement of all axis.

The purpose of the centering *Grasp* command is to pick up an object with a gripper furnished with touch sensors. These sensors are used to avoid that the object is dropped when one finger of the gripper hits it before the other. A wrist movement to compensate the closing of the fingers is initiated in case that one finger touches the object first. Moving the fingers and moving the wrist is done using the *Move* command.

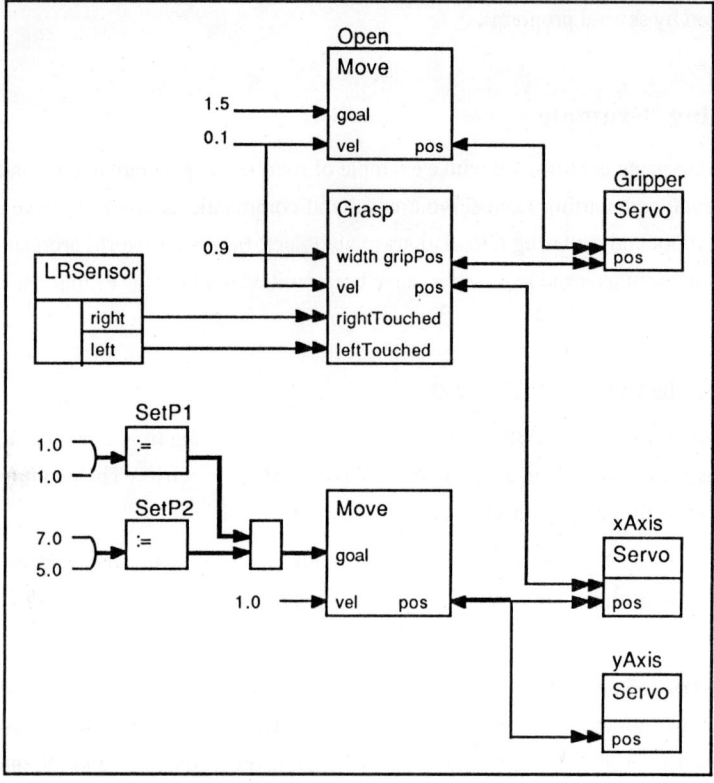

Fig. 1a: Data flow view of main program

Implementation

The main program shown in figure 1a (data flow view) and 1b (control flow view) opens the gripper, moves it to a first point in order to pick up an object, moves to a second point, and opens the gripper to release the object again.

The three axis and the gripper sensors are shown as devices. A device type *Servo* is postulated as an abstraction of the servo controller. Its interface consists of the data field *pos* modelling the desired position of the axis. The device type *LRSensor* models two binary inputs. Two instances of the function *Move* and an instance of *Grasp* access the devices to perform the desired operations.

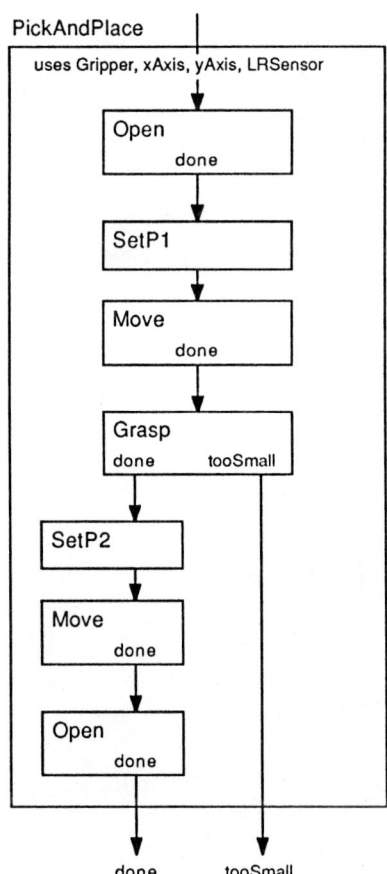

Fig. 1b: Control flow view of main program

Figure 2 shows the interfaces of the functions *Move*, *DMove*, and *Grasp* and of the device type *Servo*. *DMove* is similar to *Move* but the final position is given by an increment *delta* relative to the current position instead of an absolute goal position. The implementations of *Move*, *Grasp*, and *Servo* follow in figures 3-5.

Depending on the target computer configuration the example would be mapped to different task structures. In case of a single processor, one task to be invoked every 20 ms would execute all three instances of *Servo*. On a multi-processor, the servos could be assigned to three tasks running on different processors.

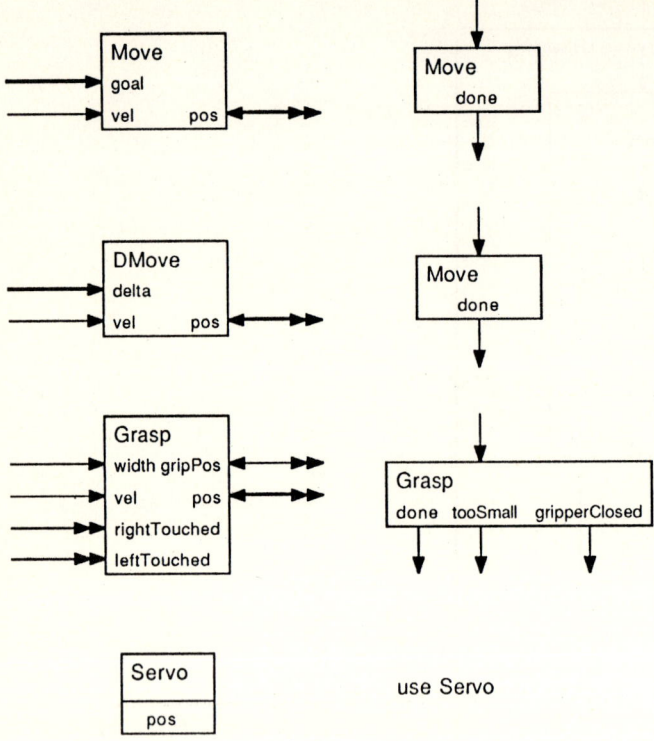

Fig. 2: Interfaces of functions *Move*, *DMove*, *Grasp* and of device type *Servo*

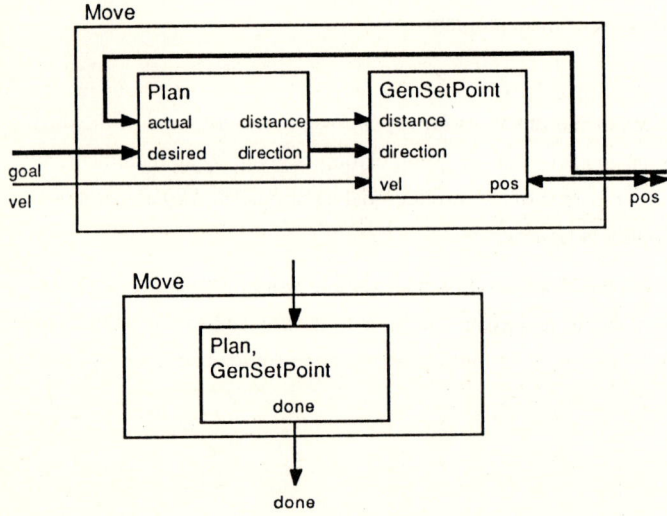

Fig. 3: Implementation of function *Move*

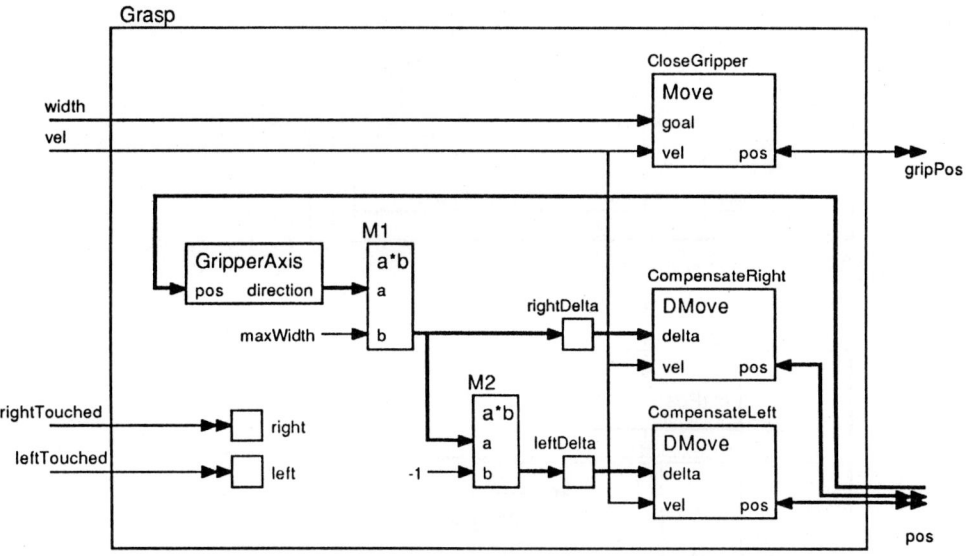
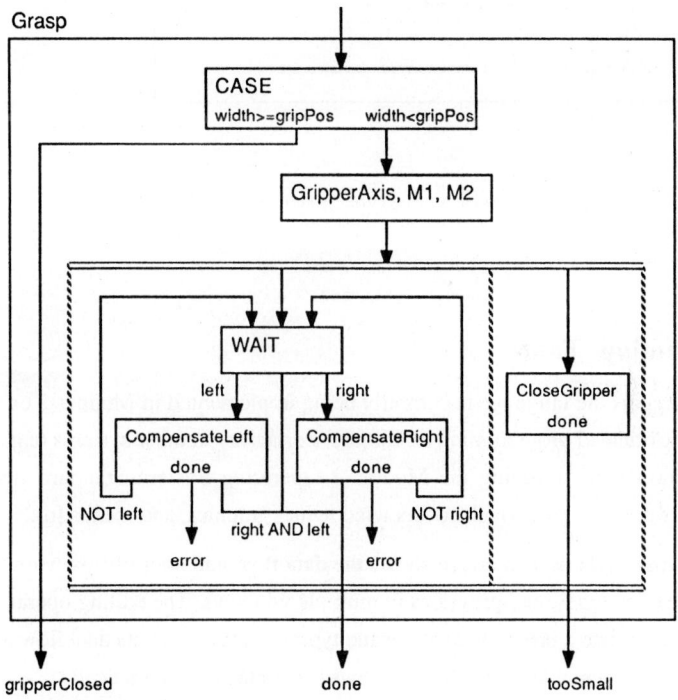

Fig. 4: Implementation of function *Grasp*

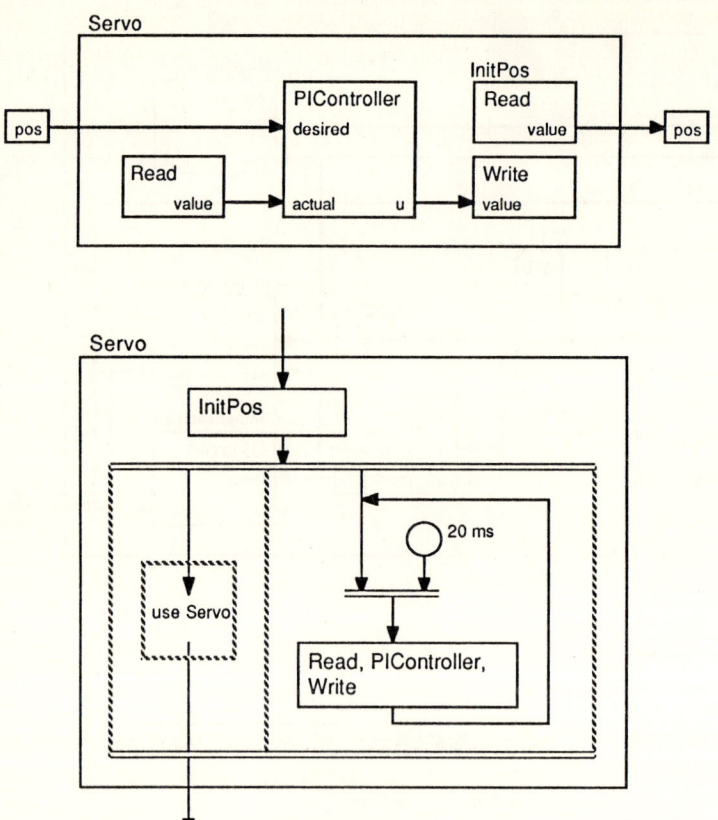

Fig. 5: Implementation of device type *Servo*

Programming Tool

A tool to support the language is currently being implemented in Modula-2 on Macintosh II using object-oriented programming techniques. Portability to other systems (e.g. VAXstation) will be obtained by applying the Modula-2 Operating System Standard Interface OSSI including its optional part which covers windowing, graphics, and menus [6].

The Macintosh-style user interface shows the data flow and control flow view of individual functions, device types, and programs in multiple windows. The editing operations provided in menus manipulate a program logically and typically affect both its data flow and its control flow view. Consistency between the two views is maintained automatically.

The prototype tool will automatically generate Modula-2 code. A rule how to map Modula-2 procedure heads to function interfaces will allow graphical programs to base on existing, conventional code.

Conclusion

A new approach for real-time programming has been presented based on graphical representation of both data flow and control flow in two complementary views.

An implementor of complex control systems may expect the following benefits from such a programming environment: An application engineer will need less computer science knowledge in programming language syntax, multi-tasking, synchronization etc. Problems specific to the target system like the design of a task structure, of synchronized access of shared data, and message exchange to realize data flow between functions will no more appear as part of the program code. Further improvements are expected in software documentation, quality, reuse, and productivity.

From the computer science point of view there are three major aspects which might have drastic long term consequences on programming:

- The use of graphics to represent and manipulate programs instead of text.
- The possibility to compose programs by configuring existing modules which do not know about each other and therefore are fully reusable (programming-in-the-large).
- Independence of target system: Programs defining their most parallel execution are portable from single processors to multi-processors.

Although a simple example has been given in some detail many questions (e.g. exception handling, code generation) concerning the language are not yet answered and it is not yet clear which application areas would benefit most from this rather new way of programming.

References

1. Hatley, D.J.; Pirbhai, I.A.: Strategies for real-time system specification.
   Dorset House Publishing, New York 1987.
2. Peterson, J.L.: Petri net theory and the modelling of systems.
   Prentice-Hall, Englewood Cliffs N.J. 1981.
3. Schillinger, D.: A high level control language based on functional programming.
   IEEE proceedings IECON 85, San Francisco November 1985, pp. 788-793.

4. Schillinger, D.: Programmierung in der prozessnahen Leittechnik. Diss. ETH Nr. 8287, Zürich 1987.

5. Maier, G.E; Taylor, R.H.; Korein, J.U.: A dynamically configurable general purpose automation controller. IFAC/IFIP symposium SOCOCO 86, Graz May 1986.

6. Biagoni, E.; Hinrichs, K.; Heiser, G.; Muller, C.: A portable operating system interface and utility library. IEEE Software, November 1986, pp.18-26.

Graphical Verification of Complex Multibody Motion in Space Applications

P. PUTZ

Department of Robotics and Automatic Control
Space Transportation and Orbital System Division
Dornier System GmbH, Friedrichshafen, F.R. Germany

Summary

Space applications include problems where particularly complex multibody motion needs to be designed, analyzed, and verified: actively controlled satellites with flexible appendages, docking spacecraft, space robots on orbiting platforms. Computer simulation is one of the chief means to support these goals. Two 'traditional' classes of tools are characterized by their capabilities and limitations: nonlinear dynamic simulation software and 3D solid model-based CAD systems with kinematic analysis features. An environment is proposed where the two classes can be integrated in a synergistic fashion to support the complete design and analysis cycle. The benefits of this concept are discussed and realizations at Dornier are introduced together with examples from recent applications and an outlook on further developments.

1. GRAPHICAL SIMULATION IN SPACECRAFT DESIGN

The high complexity and extreme demands on current European space projects result in extraordinarily high importance of pre-mission testing on ground. Yet, some of the most dominant space conditions cannot be satisfactorily reproduced in a laboratory, such as the absence of gravity and the various orbital dynamics effects. As a consequence, the emphasis has to be on kinematic and dynamic simulation for the analysis of performance, operational and functional characteristics.

Besides the obvious importance of computer simulation of controlled electromechanical systems motion such as docking spacecraft or robots on orbiting platforms, the issue of detailed visualization gains increasingly strong impact on their design and evaluation. This has strongly promoted the use of 3D solid model-based interactive computer graphics CAD systems as invaluable tools in all phases of product design and development [1].

Hence there have traditionally been two classes of CAE tools relevant for the verification of complex multibody motion: nonlinear dynamic multibody systems simulation software packages and 3D solid model-based CAD systems with ultra-realistic graphics display features, but essentially restricted to kinematic motion simulation. This paper will expound typical capabilities and representatives of both families and their distinct domains of applicability, show up their potential interfaces, and suggest environments that integrate these capabilities in a highly beneficial synergy. This will be backed by examples from our recent experience and supplemented by an outlook on promising further proceeding.

2. TOOLS FOR DYNAMIC MULTIBODY SYSTEMS SIMULATION

2.1 Typical Capabilities

The required features of state-of-the-art multibody dynamic simulation tools can be classified as follows:

Model Formulation
Definition of the kinematic model; dynamic model including elastic properties, nonlinear sensors, actuators, passive devices; driving inputs; control law and application specific models.

Desired Results
Automatic generation of the overall nonlinear equations of motion; evolution of the overall system state; nonlinear time response simulation; frequency domain analysis and modal analysis of linearized systems; time domain control law synthesis; and

graphical output on paper and on terminal screens (2D time response plots, 3D stick figures).

2.2 Typical Representatives

Here, a few systems will be presented that are being used at Dornier for spacecraft and robotics simulation.

2.2.1 General-Purpose Multibody Systems Simulation

DCAP (Dynamics and Control Analysis Package) [2] which will in the future be called MIDAS (Multibody Interactive Dynamic Analysis System) is a major effort of the European Space Agency ESA to provide an automated design and checking tool for the dynamics and control of rigid and flexible mechanical structures. It has traditionally been applied to actively controlled satellites, but is equally useful for terrestrial systems and robotics. A good evaluation with respect to elastic robots is contained in [3]. The chief advantages of DCAP are its wide scope of applicability (including elastic structures with interfaces to FEM data), high flexibility (user defined models), and widespread use in European space industry. Drawbacks are an inherently low efficiency (Lagrange formalism) and low user friendliness that is only recently being improved.

2.2.2 Robot Dynamics Simulation

For the specific needs of robot dynamics simulation we use ROBSCAD [4] developed at the TH Darmstadt. A multitude of commonly used robotics modules (rigid links, joints, actuators, sensors, control schemes, path planning methods, universal kinematic coordinate transformation) are selected and parameterized in an interactive dialog. Other modules can be added by the user. The robot motion can be commanded by a high level 'robot program'. This makes ROBSCAD very convenient for quick analyses of different concepts. Major drawbacks are the restriction to one rigid robot with no more than 12 joints and the low speed for high model complexity (gearbox elasticities, friction, backlash). DCAP and ADAMS can be used for robotics problems beyond that scope (e.g. mobile/multi-arm/multiple robots).

2.2.3 Specialized Orbital Spacecraft Dynamics Simulation

For the specific problem of AOCS (attitude and orbit control system) design of spacecraft, a dedicated tool AOCSIM [5] was established at Dornier. It mainly offers convenience for modeling the orbital kinematics and dynamics.

2.3 Inherent Limitations

For the purpose of verification of complex system motion, the above mentioned tools have a few limitations in common:

- They all provide responses far from 'real-time'. This is not surprising given their detailed and involved analysis, yet extremely bothersome for the assessment of 'man-in-the-loop' systems such as teleoperated robots.

- They all lack detailed geometric model information and 3D display qualities. This, however, is essential for assessing complex spatial relationships in moving systems that often cannot be anticipated (collision, functional or operational inadequacies) and demand the intuitive information compactness of pictures.

3. 3D SOLID MODEL-BASED KINEMATIC SIMULATION CAD TOOLS

3.1 Typical Capabilities

3D Solid Modeling CAD systems that have traditionally only been viewed in the mechanical design context are becoming increasingly attractive for complex multibody motion verification by virtue of their kinematic and geometrical analysis capabilities. For a good survey of 3D solid geometry modeling, see [6]. Basically, the following features are expected:

<u>Model Formulation</u>
Definition of the 3D solid geometry; the 3D system hierarchy; the relative location of the entities within the system, the grouping of such poses into sequences; and definition of the kinematic model.

<u>Viewing and Display Features</u>
Definition of a parallel/central viewing projection, 3D viewport clipping, a layout of multiple views, a display mode (wire frame, removal of hidden lines, shaded image displays), lighting and shading conditions, and labeling and blanking options.

<u>Graphics Output</u>
Static or animated 3D displays in the selected viewing and display mode, with often extremely high realism, at 'real-time' speeds, and augmented by auxiliary displays (cross sections, exploded views, transparent parts).

<u>Design Analysis</u>
Distance and angle measurements; automatic computation of properties such as volume, mass, surface area, center of mass, moments of inertia of individual entities or the whole system; automatic interference analysis between any two entities or groups; kinematic analysis (trajectories of system variables during prescribed stationary motion, equilibrium forces/torques, traces of points, animation of 3D system motion).

Robotics Specific Features

Some CAD packages have dedicated robotics tools. They should offer standard robot libraries; specific robots kinematics definition; robot programming commands; displays of robot status; animated 3D robot kinematic motion simulation; and conversion into standard robot programming languages.

3.3 Typical Representatives

At Dornier, we mostly use CATIA, CAEDS and CADAM. CADAM still has less importance in 3D systems analysis and will not be described any further.

CATIA:

CATIA (Computer-graphics Aided Three-dimensional Interactive Application) [7] is a major commercial CAD/CAM system consisting of several independent modules for applications such as 2D drafting, advanced 3D curve and surface design, 3D solid geometry design, kinematic analysis, NC machining programming, and robotics. It meets most of the above listed requirements to a high degree, with the notable absence of a FEM pre- and postprocessor (which should become available soon). Especially the Kinematics and Robotics modules offer excellent interactive support. Compared with CAEDS, CATIA is faster, has much better 2D drafting and dimensioning and distinctly superior kinematics and robotics features, but a somehow less systematic internal structure and disadvantages for storing system motions on file for quick re-play. Beyond the standard features, the user may define application specific macros with the IUA (Interactive User Access) capability or use a powerful FORTRAN library of CATIA functions for integration with other systems.

CAEDS/I-DEAS:

CAEDS (Computer-Aided Engineering Design System) [8] which is also marketed as I-DEAS is another major mechanical CAE system and in most aspects a direct competitor with similar functionality as CATIA. Its several modules (Object Modeler, System Modeler, Graphics Finite Element Modeler, System Dynamics Analysis) have a stronger bias towards structural and dynamic analysis and fewer or no features in 2D drafting and NC machining. CAEDS includes a kinematics module ("Mechanism Design" within the System Modeler) and also robotic engineering problems can be solved, yet with less direct support than CATIA. A definite strong point of CAEDS is its Finite Element pre- and postprocessor (mesh generation, solution display on the solid model) which is important for the simulation of flexible structures. CAEDS offers good macro programming capabilities, standard data exchange, and a complete relational database management system for internal or external project data.

Robotics Specific CAD Systems

For robot kinematics design, workspace analysis, workcell layout, off-line programming, motion control, and verification, the capabilities of solid modelers are extremely attractive, especially when coupled with realistic and highly interactive graphics for "real-time" evaluation of task execution. Hence, a number of such dedicated robotics CAD tools have emerged. Surveys are given in [9, 10] and some NASA approaches for space robots are described in [11, 12].

3.4   Inherent Limitations

For our purpose of complex systems motion verification, the discussed CAD-type systems have, for all their advantages, the drawback that no dynamic or control effects are modeled and only "nominal" motions are displayed. In a more general context, their capabilities are not integrated with the dynamic/control simulation capabilities from Chapter 2. Such an integration is the subject of the rest of this paper.

4. AN INTEGRATED ANALYSIS AND SIMULATION ENVIRONMENT

4.1 The Design/Analysis Cycle

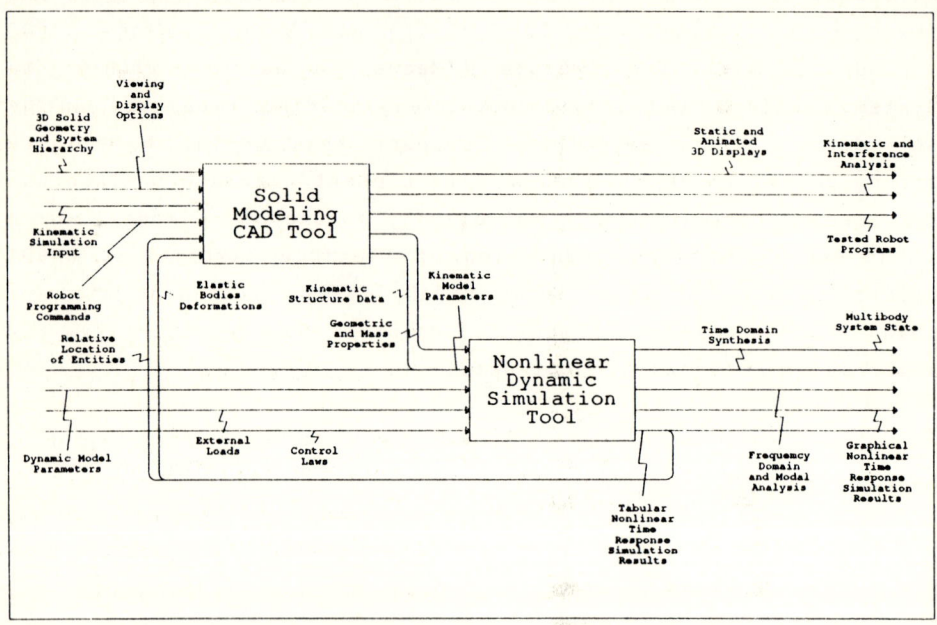

Fig. 1: An Integrated Design/Analysis Environment

After having analyzed the capabilities, benefits, and limitations of both nonlinear dynamic simulation and solid model-based CAD systems, we have enough motivation to investigate a potential integration of the two for a synergistic compound design/analysis environment such as depicted very generically in Fig. 1. A typical design/analysis cycle would then proceed as follows:

1. Design of the mechanical system on the CAD Tool.
2. Kinematic analysis on the CAD Tool, i.e. assessing kinematic and operational functionality (for robots: workspace and dexterity analysis, task programs preparation).

3. If necessary, modification of the kinematic and geometric model on the CAD system (e.g. workcell layout optimization, avoidance of interferences or obstructions).

4. (Automatic) extraction of relevant input data for the dynamic simulation, most notably on kinematic structure and topology, geometric, and mass properties of the single bodies. The more this process can be automated (involving a conversion between the probably different internal representations), the easier it is to guarantee consistency between the CAD and dynamic models - an important issue when modifications tend to arise frequently !

5. Augmentation of the inputs to the dynamic simulation tool by further kinematic model data, all dynamic and control law descriptions and system loads.

6. Detailed dynamic simulation, assessing performance, stability, robustness of the controlled system. The outputs are mostly time response trajectory data of dynamic system variables that can immediately be displayed as conventional 2D plots.

7. (Automatic) feedback of the dynamic time response simulation data to the CAD tool where they give rise to relative motion or deflections of the system's bodies. Again this may involve conversions between internal representations.

8. Analysis of the (dynamic) system motion in all its physical detail on the CAD system via animated displays and exploiting all the viewing and display capabilities. This way, unanticipated behavior and problems due to geometric detail can become immediately obvious (collisions, clearances). The analytical interference checking feature of the solid modeler will not rely on inspection alone to detect malfunctions, which is very important for complex and intricately compact mechanisms or environments.

9. If needed, modifications may be made on the control law and steps 6 - 8 iterated to optimize the dynamic design.

It may even be desired to modify the geometry or the kinematic structure. In any case, this CAE environment offers features to analyze the design in a multitude of facets and the automated coupling relieves the user from tedious housekeeping and permits to concentrate on the actual engineering problems. A side benefit will always be excellent documentation by stunningly realistic images or animations which convey understanding and verification of even very complex spatial system motion in an appealingly compact and intuitive fashion.

4.2 Realization of Integrated Design/Analysis CAE Tools

At Dornier, work along the outlined approach has started in 1985 with a coupling of CAEDS and DCAP for rigid multibody systems [13]. Applications to space robotics have been reported in [14, 15]. A coupling of CAEDS and ROBSCAD for robotics analysis was done in [16]. The reason why CAEDS was used in these projects was that CAEDS offered excellent interface possibilities that only recently are becoming available for CATIA.

As an example, Fig. 2 shows a detail of a dynamic robot motion animation on CAEDS. A small experiment manipulator transports a materials sample from its containment and inserts it into a melting furnace. The dynamic simulation of the controlled system was done on DCAP with inputs from the mechanical design on CAEDS. The critical motion analysis concerns the avoidance of collisions that may result from dynamic overshoot effects.

Fig. 3 illustrates a few steps during a complex motion whereby a robot winds filaments on a Y-shaped workpiece in an automated carbon fiber composite structures manufacturing process. The coordinated motion of robot and workpiece for generation of prescribed windings was computed by an application specific program and the results displayed with CAEDS to study feasibility (avoidance of collisions) and to derive clues for process optimization.

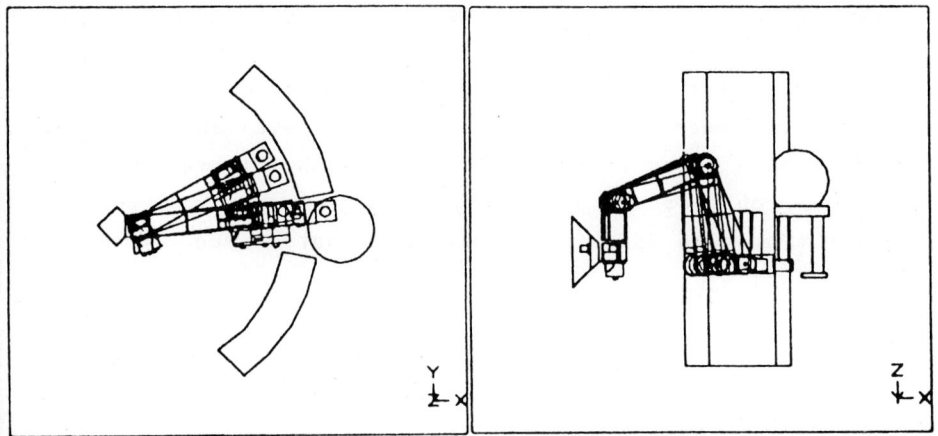

Fig. 2: CAEDS Animation of a Controlled Dynamic Robot Motion

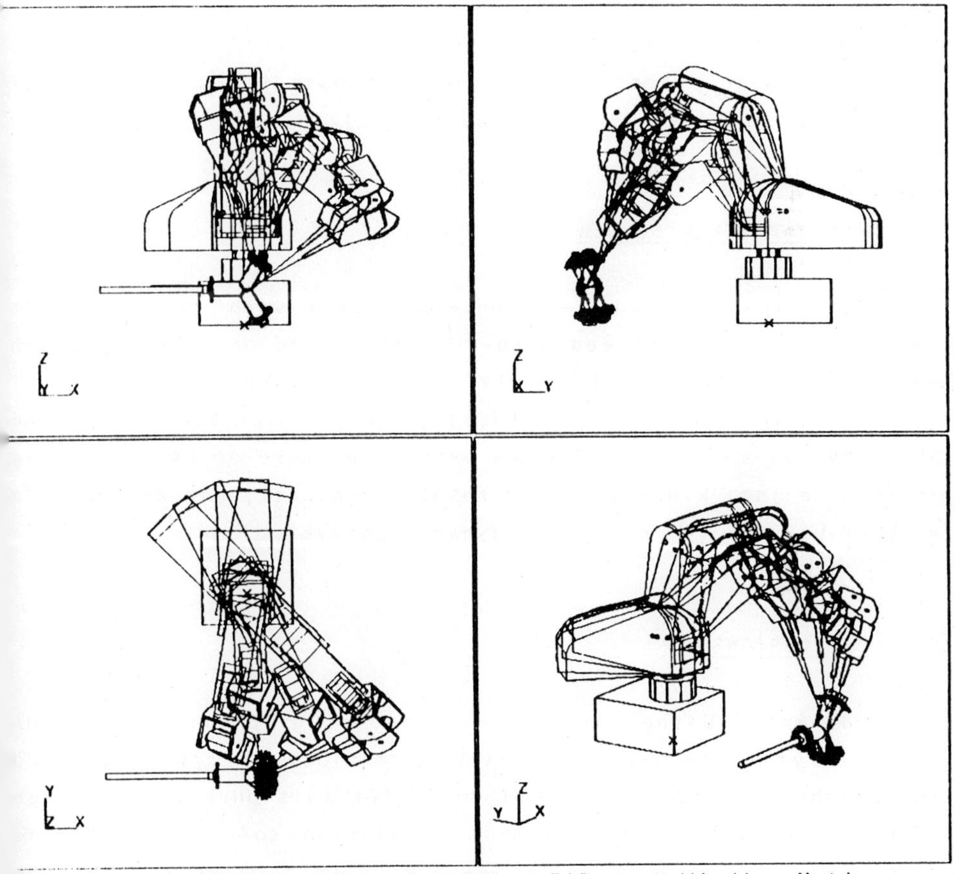

Fig. 3: CAEDS Animation of a Robot Filament Winding Motion

A loose coupling between independently developed and by their nature rather distinct software tools is of course not optimal in view of response time and overall performance, but a reasonable approach when these tools are already available. Recently, new tools have begun to emerge offering such integrated capabilities for robotics in one homogeneous package: a German ROSI [17] and a British ROSI [18] (which does not include solid modeling, though).

On a somewhat wider scope, both NASA [11, 19, 20] and ESA [21] have defined large concepts for space telerobotics simulation facilities incorporating real-time computer graphics and varying degrees of dynamic effects for robot system development, mission and task planning, operator training, and on-line mission support. A NASA system IDEAS[2] integrating the IDEAS Solid Modeling CAE system with spacecraft analysis software for development of the US Space Station is described in [22].

4.3   Plans for the Future

Motivated by the good experience achieved with rather modest means, we plan to proceed with a somehow more unifying approach which is outlined in [23] for robotic engineering. It shall involve an integrated CAE database, the Daimler-Benz 'CAE Data Bus' [24] as a generic exchange mechanism, more of CATIA for the solid modeling, kinematic, and robotic features, and an improved release of DCAP for multibody dynamic analysis.

5.   CONCLUSIONS

After laying out the particular impact of graphical simulation for verifying the complex multibody motions of spacecraft and space robotics, two classes of tools for this purpose have been characterized: nonlinear dynamic simulation software and solid model-based CAD systems. The message of this paper is that the benefits of these classes can be greatly augmented by a synergistic integration to support the complete design and analysis cycle of controlled mechanical systems. Examples were given for more or less complete realizations of this concept and results from recent applications at Dornier were shown.

6. REFERENCES

1. Putz, P.: Using Solid Modeling to Design Space Systems. IBM Seminar on Advanced Engineering Techniques, La Hulpe, Belgium (1987).

2. Aeritalia: Dynamic Control and Analysis Package (DCAP), Rel. 5, Theoretical Manual. D2-MA-AI-002, Issue 2 (1987).

3. Eichberger, A.: Modeling, Control, and Simulation of Elastic Space Robots. Diploma Thesis, TU München, Inst. B f. Mechanik (1987).

4. Ersü, E. et al: ROBSCAD - A Software Package for Dynamic Simulation and Design of Industrial Robots and Control Components. In: VDI-Bericht 598 "Steuerung und Regelung von Robotern", pp. 15 - 26. Düsseldorf, VDI Verlag 1986 (in German).

5. Dornier System GmbH: AOCSIM User's Manual (1988).

6. Requicha, A.A.G. and Voelcker, H.B.: Solid Modeling-Current Status and Research Directions. IEEE Computer Graphics and Applications, Vol. 3, Nr. 7 (1983) 25 - 37.

7. IBM, CATIA Product Description, Program Nr. 5668 - 836 (1986).

8. IBM, CAEDS Product Description, G320-9434-1 (1986).

9. Derby, S.J.: Computer Graphics Robot Simulation Programs - A Comparison. ASME Robotics Research and Advanced Applications. Winter Annual Meeting (1982) 203 - 211.

10. Dombre, E.; Fournier, A.; Quaro, C. and Borrel, P.: Trends in CAD/CAM Systems for Robotics. Proc. 1986 IEEE Int. Conf. on Robotics and Automation (1986) 1913 - 1918.

11. Fernandez, K.: The Use of Computer Graphic Simulation in the Development of Robotic Systems. International Aerospace Conference IAF 86 (1986).

12. Baumann, E.W.: Real-Time Graphic Simulation for Space Telerobotics Applications. Proc. NASA-JPL Workshop on Space Telerobots, JPL Publ. 87 - 13, Vol. 2 (1987) 207 - 217.

13. Putz, P.: An Integrated Software Environment for Simulation of Multibody System Dynamics. Dornier System R + D Report 79 696 (1985).

14. Finsterwalder, R.: Dynamic Simulation and Control of Space Manipulators. Diploma Thesis, Univ. Stuttgart, Inst. A f. Mechanik (1986).

15. Putz, P. and Hilzenbecher, U.: 3D Solid Modeling for Graphical Simulation of Robot Dynamics. In: VDI Bericht 598 'Steuerung und Regelung von Robotern', pp. 39 - 50. Düsseldorf: VDI Verlag 1986 (in German).

16. Mau, K.-D.: Advanced Simulation and Control of a Space Manipulator. Diploma Thesis, Univ. Stuttgart, Inst. A f. Mechanik (1987).

17. Dillmann, R. and Huck, M.: A Software System for the Simulation of Robot Based Manufacturing Processes, Robotics 2 (1986) 3 - 18.

18. Williams, S.J.: The Use of ROSI in Robot Dynamic Simulation. ARI Nuclear Robotic Workshop (1987).

19. Harrison, F.W. and Pennington, J.E.: Systems Simulations Supporting NASA Telerobotics. Proc. NASA-JPL Workshop on Space Telerobots, JPL Publ. 87 - 13, Vol. 2 (1987) 293 - 299.

20. Brown, R. et al: A Space Systems Perspective of Graphics Simulation Integration. Proc. NASA-JPL Workshop on Space Telerobots, JPL Publ. 87 - 13, Vol. 2 (1987) 267 - 272.

21. Pronk, C.N.A. et al.: Definition of the EUROSIM Simulation Subsystem. 1st European In-Orbit Operations Technologies Sympos., ESA SP-272 (1987).

22. Baker, M. et al.: Space Station Multidisciplinary Analysis Capability - IDEAS^2, AIAA Conference (1985).

23. Putz, P.; Mau, K.-D. and Eichberger, A.: Integrated CAE Tools for Robot System Design and Analysis. Dornier System R + D Report 79986 (1987).

24. Haase, E.: An Approach towards Integrated Information Processing in Computer Analysis. In: VDI Tagung Fahrzeugbau: Berechnung im Automobilbau (1984), in German.

Examples for the Dynamics of Controlled Mechanical Systems

Active Vibration Control of a Cantilever Beam by a Piezoelectric Ceramic Actuator

M. Chiba[*], J. Tani[**], G. Liu[**], F. Takahashi[**], S. Kodama[**] & H. Doki[***]

* Dept. Mechanical Engineering, Iwate University, Morioka 020, Japan
** Inst. of High Speed Mechanics, Tohoku University, Sendai 980, Japan
*** Mining College, Akita University, Akita, Japan

Summary

The theoretical analysis is presented for the active vibration control of a cantilever beam using a piezoelectric ceramic actuator. Control forces (moments) are induced by a pair of piezoelectric ceramic actuators partially bonded on the upper and lower side of the beam. The problem is first reduced to a finite degree of freedom system with the Galerkin method, and the control is determined by means of the optimal regulator theory. Numerical calculations are carried out for six degree of freedom system, and the effects of the location and length of the actuator are examined.

1 INTRODUCTION

As structures grow larger and lighter, unexpected vibrations which are caused due to the lack of the structural stiffness, are easily occurred. In these cases, usually, the passive vibration control method, in which the energy of the induced vibrations is absorbed, have been used to suppress the induced vibrations. Though this type of method with passive damper has simple configuration and is very effective for vibration suppressions of the structures indeed, when the vibration characteristics of the system vary with time, the efficiency for vibration suppression decreases considerably. On the other hand, the active vibration control method, in which the energy for vibration suppression is applied by the controller, has been recently studied and applied to various kinds of industrial fields[1, 2].

In this study, an active vibration control for a cantilever beam, which is one of the fundamental element of the structures, is presented by using the optimal regulator theory. A pair of piezoelectric ceramic, partially bonded on the upper and lower side of the beam, is used to induce a control moment as the actuator. In the numerical calculations, the effect of the location and length of the piezoelectric ceramic on the efficiency as a vibration damper are examined.

2 THEORETICAL ANALYSIS

2.1 Dynamic model

Figure 1(a) shows a uniform cantilever beam with length L, cross-section area A, flexural rigidity EI and mass density ρ. At the upper and lower side of the beam with distance x_1 from the clamped edge, a pair of piezoelectric ceramic actuators with length $x_2 - x_1$ is bonded on.

When the electric current flows through the upper and lower piezoelectric ceramics in the opposite direction and with the same magnitude, respectively, the relation between the induced moment M and the added voltage V is given by

$$M(t) = k \frac{d_{31}}{h_p} V(t) \tag{1}$$

where d_{31}, h_p and k are a piezoelectric constant, the thickness of piezoelectric ceramics and the constant determined from Young's modulus and the cross-section geometry of piezoelectric ceramics and beam, respectively (Figure 2), while t is time. Then, the distributions of induced moment along the beam become as shown in Figure 1(b).

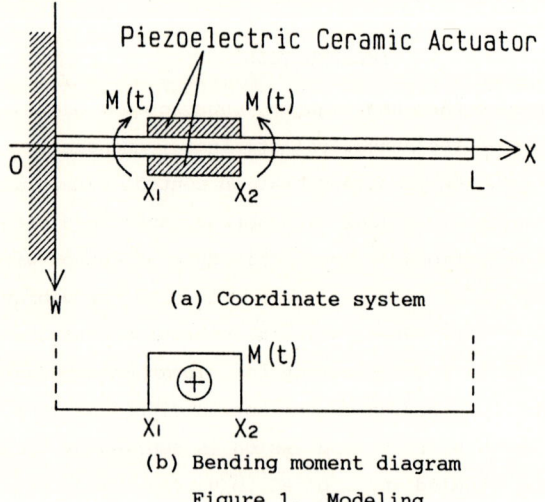

Figure 1. Modeling

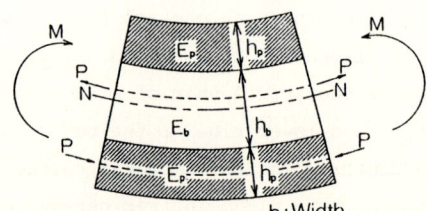

Figure 2. Bending of a combined beam with width b

On the assumption that the mass and the rigidity of the piezoelectric ceramics can be neglected, the equation of motions and the boundary conditions of the cantilever beam are given by

$$\frac{\partial^4 w(\xi,\tau)}{\partial \xi^4} + c_1 \frac{\partial^5 w(\xi,\tau)}{\partial \tau \partial \xi^4} + \frac{\partial^2 w(\xi,\tau)}{\partial \tau^2} = [\frac{\partial \delta(\xi,\xi_1)}{\partial \xi} - \frac{\partial \delta(\xi,\xi_2)}{\partial \xi}]u(\tau) \quad (2)$$

$$w(0,\tau) = \frac{\partial w(0,\tau)}{\partial \xi} = 0, \quad \frac{\partial^2 w(1,\tau)}{\partial \xi^2} = \frac{\partial^3 w(1,\tau)}{\partial \xi^3} = 0 \quad (3)$$

where $\delta(\xi, \xi_i)$ is a Dirac delta function and c_1 is a material damping coefficient, and further, in the foregoing the following non-dimensional parameters are used.

$$u(\tau) = \frac{LM(\tau)}{EI}, \quad \xi = \frac{x}{L}, \quad w = \frac{\bar{w}}{L}, \quad \tau = \frac{t}{L^2\sqrt{\rho A/EI}}, \quad c_1 = \frac{c}{L^2\sqrt{\rho A/EI}} \quad (4)$$

To solve the equation (2), we will apply the Galerkin method to eqn(2) by using the eigen-function of cantilever beam $\phi_n(\xi)$ which satisfy the boundary conditions (3)

$$w(\xi,\tau) = \sum_n a_n(\tau)\phi_n(\xi) \quad (5)$$

and which yields the following equations

$$\sum_{j=1}^{n} \{\delta_{ij} \ddot{a}_j(\tau) + c_1 \delta_{ij} \alpha_j^4 \dot{a}_j(\tau) + \delta_{ij} \alpha_j^4 a_j(\tau)\} = \{\phi_i'(\xi_2) - \phi_i'(\xi_1)\} u(\tau) \quad (6)$$
$$(i = 1, 2, \cdots, n)$$

where $a_j(\tau)$ is unknown time function and the dot stands for differentiation with respect to non-dimensional time τ, while α_j is the non-dimensional eigen frequency of the beam with order j, and δ_{ij} is the kronecker delta.

2.2 Control design

In order to apply the optimal control theory, equation (6) should be better rewritten in a state equation form as

$$\dot{x}(\tau) = Ax(\tau) + Bu(\tau) \quad (7)$$

$$y(\tau) = Cx(\tau) \quad (8)$$

$$x(\tau) = [\ a_1(\tau), a_2(\tau), \cdots, a_n(\tau), \dot{a}_1(\tau), \dot{a}_2(\tau), \cdots, \dot{a}_n(\tau)\ ]' \quad (9)$$

$$A = \begin{bmatrix} 0, & I \\ -\delta_{ij}\alpha_j^4, & -c_1\delta_{ij}\alpha_j^4 \end{bmatrix}, \quad B = \begin{bmatrix} 0 \\ \phi_i'(\xi_2) - \phi_i'(\xi_1) \end{bmatrix}, \quad C = [\phi_i(\xi_T), 0\ ] \quad (10)$$

Then, to conduct the vibration control of the beam, it is necessary to minimize the deflection of the beam as well as the control variable. Thus, this control problem can be formulated as the optimal regulator problem minimizing the performance index as follows:

$$J = \int_0^\infty (x^T Q x + r u^2) \, d\tau \tag{11}$$

where Q and r are weighting matrix and weighting factor, respectively. The optimal control variable $u^*(\tau)$ which minimize the index J is given by

$$u^*(\tau) = -r^{-1} B^T P x(\tau) \tag{12}$$

in which P is a constant matrix which is obtained as the positive definite solution of the following Ricatti matrix equation

$$PA + A^T P - r^{-1} PBB^T P + Q = 0 \tag{13}$$

3 NUMERICAL CALCULATIONS

Based on the preceding analysis, numerical calculations were carried out taking the number $n=6$ in equation (5), and the parameters c_1, Q and r as 0.03, 9.0[I] and 0.5, respectively. The initial conditions of the beam were assumed as $w_0(1, 0)=0.1$, $\dot{w}_0(1, 0)=0$.

At first, as an example of the results for vibration control, a comparison between the controlled and uncontrolled responses of the deflection at the top of the beam, and the change of the control variable with time are shown in Figure 3, when the ceramic actuator is located at $\xi_1=0.26$, $\xi_2=0.46$. In this case, the length of the piezoelectric ceramic was taken $0.2L : \xi_2-\xi_1=0.2$. Concerning the deflection w at the top of the beam, it is normalized by the initial deflection of $w_0=0.1$, and in the figure, thinner line corresponds for the result which is not controlled, while thicker line corresponds for that of controlled. From the figure, it can be seen that owing to the control variable u^* suggested in the lower diagram of the Figure 3, the deflection at the top of the beam can be rapidly damped.

In the example as shown in Figure 3, there was no limitation on the control variable u^*. But in actual problems, when we use piezoelectric ceramic as an actuator, the magnitude of induced moments produced by the actuator may be limited, from various physical reasons. Then, the results are shown in Figure 4, when the control variable u^* has limitted value with $u_L=0.2$ or 0.05. The initial conditions are the same as in Figure 3. In the figure, thinner lines correspond for the results when $u_L=0.2$, while thicker lines correspond for those when $u_L=0.05$, respectively. From the figure, inspite of the existence of the limitation on u^*, excellent vibration damp-

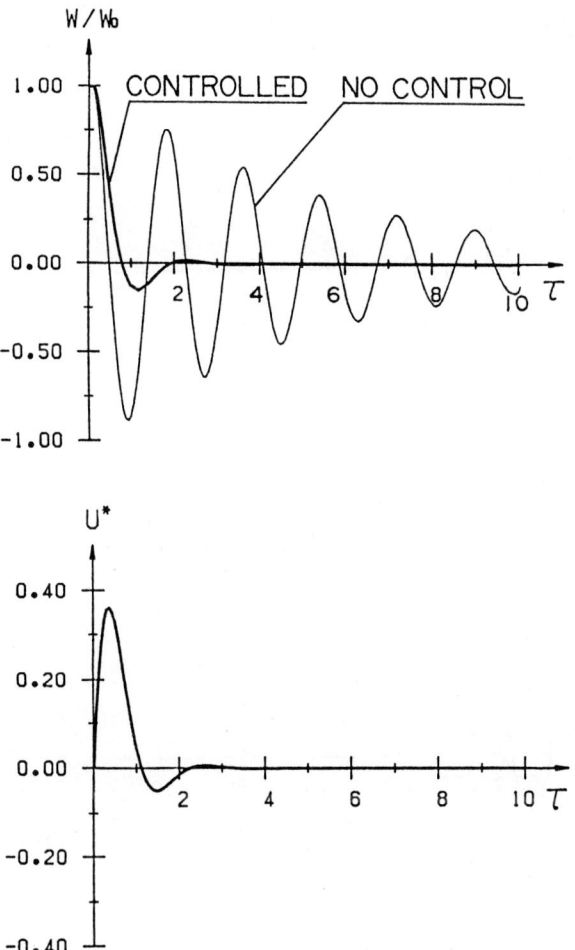

Figure 3. Comparison between controlled and uncontrolled responses of beam at free end, and control variable u*

ing can be seen on the deflection w, except for a little lowering of damping efficiency for lower value with $u_L=0.05$.

In order to get a higher damping efficiency, it is necessary to carry out optimal design of the piezoelectric ceramic as an actuator, concerning on it's location and length. So, at first, we will examine about the location of the piezoelectric ceramic actuator. In Figure 5, the variation of performance index J with the location of the piezoelectric ceramic ξ_1 are

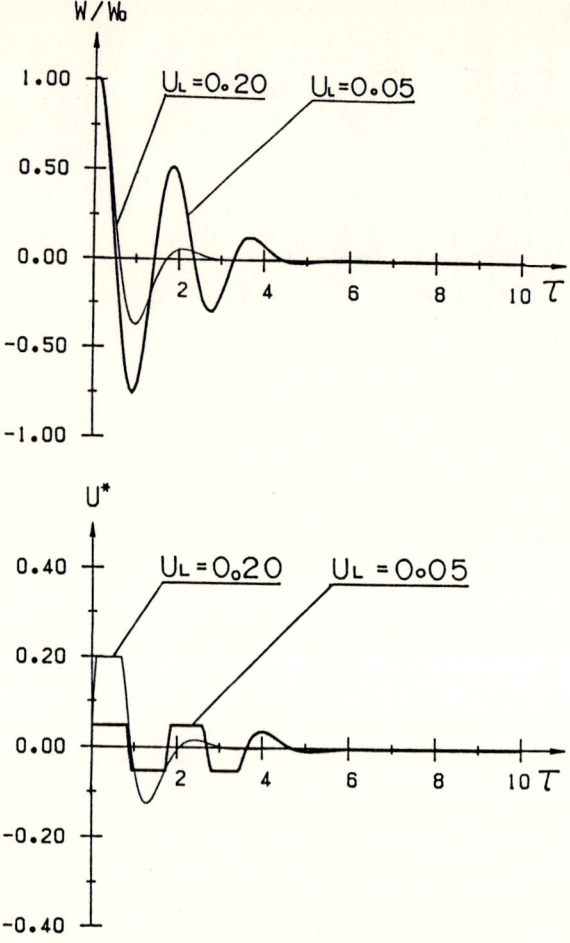

Figure 4. Response of beam at free end and control variable with limitation on u*

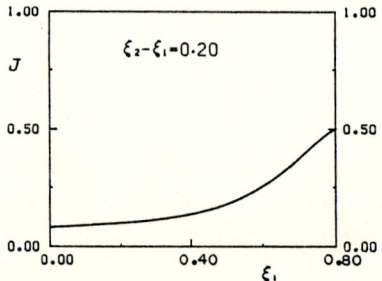

Figure 5. Variation of J with location of actuator ξ_1 : $\xi_2 - \xi_1 = 0.2$

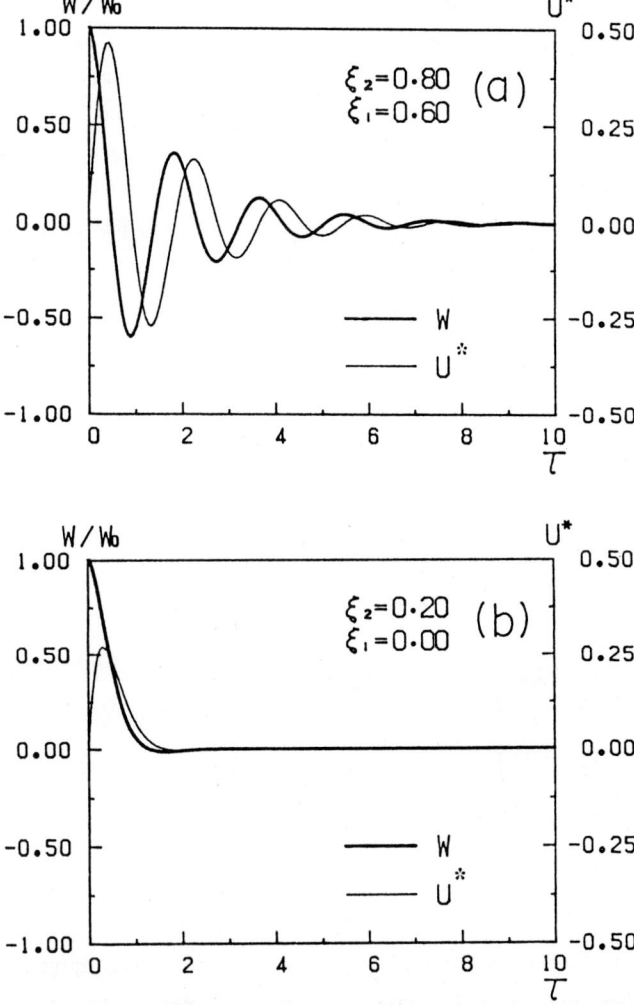

Figure 6. Response of beam at free end and control variable : $\xi_2-\xi_1=0.2$;(a) $\xi_1=0.6$, $\xi_2=0.8$; (b) $\xi_1=0.0$, $\xi_2=0.2$

shown. In this case, the left-hand side of the edge of the ceramic, which has length of $\xi_2-\xi_1=0.2$, is located at $\xi_1(0\leq\xi_1\leq 0.8)$ on the beam. From the figure, it can be seen that when the actuator is located at the clamped end of the beam, J takes minimum, whereas J increases as the actuator moves to the free end.

Here, to make clear the influence of the location of the actuator, on both the time response of the beam and the control variable, the time response of the free end of the beam and the control variable, when the actuator is located at $\xi_1=0.6$ where J takes relatively large value, and at $\xi_1=0.0$ where J takes minimum value, are shown in Figure 6(a) and 6(b). From Figure 6(b), when the actuator is located at the clamped end of the beam, it seems clear that the deflection at the top of the beam, suggested by thicker line, is rapidly damped, and the corresponding control variable, suggested by thinner line, is a little in comparison with the result of Figure 6(a), when the actuator is located at the neighborhood of the free end.

Now, in practically, not all the state variables can be measured. In such a case, we can use state vector y(eqn(8)), and instead of eqn(11), following performance index can be defined

$$J' = \int_0^\infty (y^2 + ru^2)\, d\tau \qquad (14)$$

And here, by using the above index, similar diagram can be obtained on

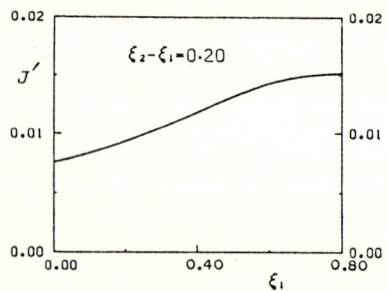

Figure 7. Variation of $J' = \int_0^\infty (y^2 + ru^2)d\tau$ with location of actuator ξ_1 : $\xi_1-\xi_2=0.2$

the variation of J' with the location of the actuator, as Figure 5. The results are shown in Figure 7, when ξ_L in eqn(10) was taken as 1.0. As seems in the figure, J' takes minimum when the actuator is located at $\xi_1=0.0$, and maximum when located at $\xi_1=0.8$ as similar as in Figure 5. The time response of the free end of the beam and of the control variable, at this case, when the actuator is located at $\xi_1=0.6$ and $\xi_1=0.0$ the same location in Figure 6, are shown in Figure 8. Comparing Figure 6 with Figure 8, damping efficiency reduces when we take J' as the performance index. It should be noted that when the actuator is located at the neighborhood of the free end of the beam, Fig. 8(a), there seems few suppression of vibration.

Next, we will examine the effect of the length of the piezoelectric ceramic actuator on the performance of the vibration control. It is a very important factor for economical aspect as well as the efficiency for dam-

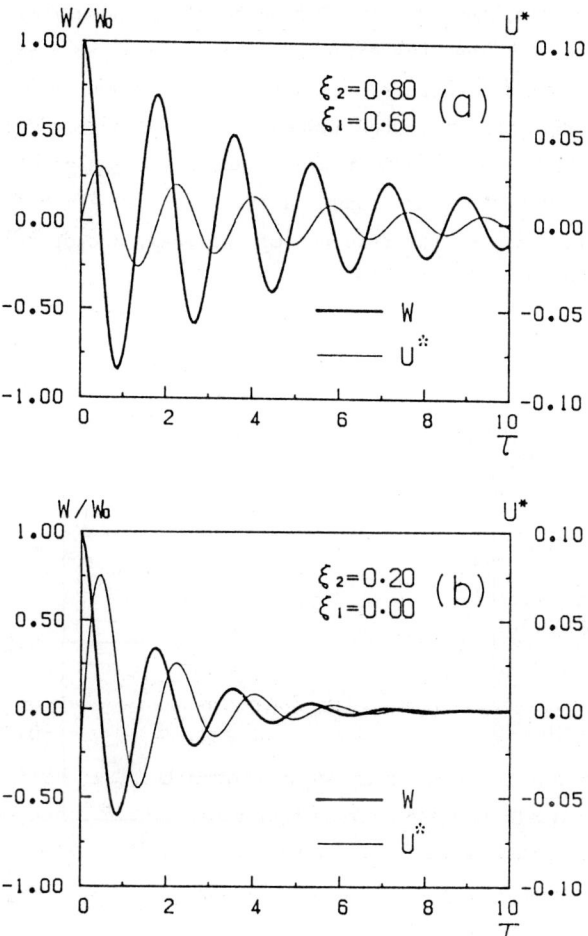

Figure 8. Response of beam at free end and control variable : controlled with J': $\xi_2 - \xi_1 = 0.2$; (a) $\xi_1 = 0.6$, $\xi_2 = 0.8$; (b) $\xi_1 = 0.0$, $\xi_2 = 0.2$

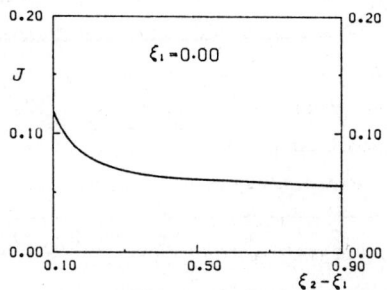

Figure 9. Variation of J with length of piezoelectric ceramic : $\xi_1 = 0.0$

pers. As an example, the variation of performance index J with the length of the ceramic $\xi_2-\xi_1$, when the ceramic is bonded on from clamped end of the beam, are shown in Figure 9. As shown in the figure, J decreases and tends to a saturated value as the increase of the length. This means that the length of the actuator is suitably selected, we can make an optimal vibration control of the beam economically.

The time response of the free end of the beam, and of the control variable, when the length of the actuator $\xi_2-\xi_1$ is 0.1 and 0.5 with $\xi_1=0.0$, are shown in Figure 10. Comparing this figure with Figure 6, the length seems enough with $0.2l$, in practically.

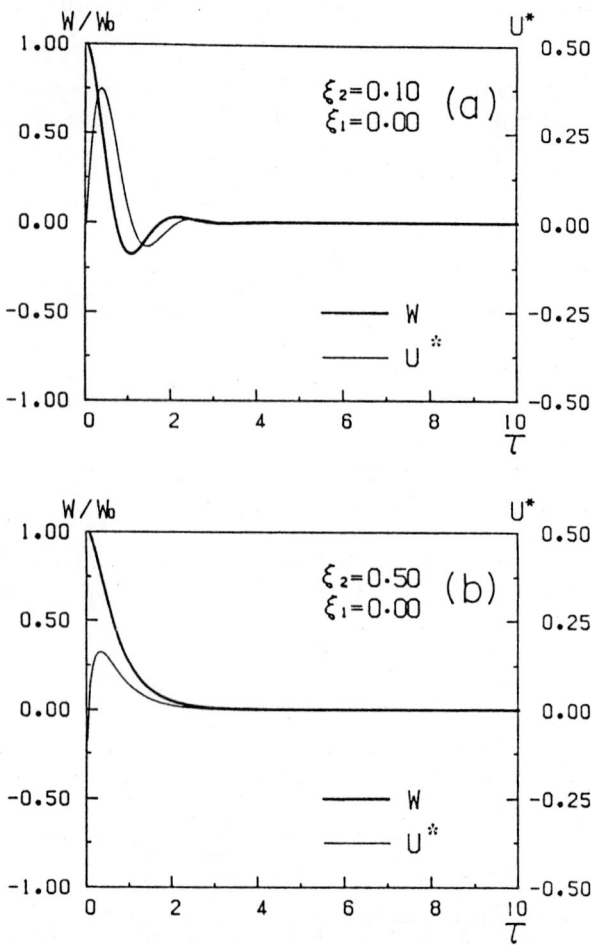

Figure 10. Response of beam at free end and control variable, with different length of piezoelectric ceramic: $\xi_1=0.0$; (a) $\xi_2=0.1$; (b) $\xi_2=0.5$

4 CONCLUSIONS

Theoretical analysis is presented for the active vibration control for a cantilever beam using a piezoelectric ceramic actuator. The problem is solved by using the Galerkin method, and the control is determined by means of the optimal regulator theory. The main results obtained in the range of the present analysis are summarized as follows:

(1) The possibility of the use of a piezoelectric ceramic, as an actuator for active vibration control of the light weight structures, had been confirmed.
(2) The numerical simulations on the active vibration control showed good results, even when there exists the limitation on the performance of the actuator, which seems to happen in practical problem.
(3) The optimal location of piezoelectric ceramic actuator was found at the clamped end of the beam. When the length of the actuator is suitably selected, we can make optimal vibration control of the beam, economically.
(4) When not all the state variables can be measured, vibration control can be also conducted by using a feed back of out-put y, but the efficiency as a damper decreases in this case. To avoid this situation, it is necessary to use the observer to presume the state variables.

REFERENCES

1. Bailey, T.; Hubbard Jr., J. E. : Distributed piezoelectric-polymer active vibration control of a cantilever beam, J. Guidance, Control, Dynamics, 8 (1985) 605-611.
2. Okudaira, et al : Vibration control of flexible space structures, Proc. of JSME, No.870-3 (1987) 49.

Fiber Connected Tug of War

K. Furuta, M. Yamakita
Department of Control Engineering
Tokyo Institute of Technology
Oh-Okayama Meguroku Tokyo Japan

N. Sugiyama, K. Asaka
Showa Electric Wire and Cable Co. Ltd.
4-1-1 Minamihashimoto, Sagamihara-shi, Kanagawa, Japan

ABSTRACT

"Tug of War" is one of the oldest and most popular sports among the world. The game is simple and known that two teams pull a rope each other, and one pulls the tug fixed on the center of rope to its side is the winner. The problem of this game is that all teams should be gathered at a same place. Therefore the game between Zurich and Munich was not possible. The objective of this research is to develop the system to make the game possible for teams in different places. The system consists of two machines and fiber connecting them, and a machine placed at different places is playing a role of the opponent for each team. The positions of ropes at different places are controlled to track output of a model driven by the difference of the measured forces and the sum of the lengths of two ropes is controlled not to be varied.

This paper presents a control system for the above mentioned tug of war. It is implemented in a miniature system, and checked the validity before the realization of a practical system. The miniature system worked as expected. The practical system is to be demonstrated at Aomori-Hakodate Exposition.

1. Introduction

The tug of war was once an official game in Olympiad and is one of the most popular games in the world. The game is known that two teams pull a single rope each other, and the team bringing tug put on the center of the rope to its side wins, where a team consists of eight players. Thus the game could be held only when all teams gathered at one place, and the game between Zurich and Munich could not be held.

This paper is concerned with the development of the system which makes the tug of war between distant places possible. The hardware of system is consisting of two machines located at each place and a fiber connecting them. The machine pulls the rope according to the reference given by the control system. The control system has two layers, and the upper layer generates the reference position based on a model according to the difference of the forces measured at each rope. The secondary layer is for the tracking control of the rope position to the output of the model. The control system is devised based on the idea of the virtual internal model following control (K.Kosuge, K.Furuta, T. Yokoyama,1987). The control system of the second layer is designed based on the model following servo controller (K. Furuta, K.Komiya,1982), which has been developed from Davison's servo control and Kreindler's model following control, and the design algorithm is derived taking use of the idea of K.Furuta (1987). Moreover, an adaptive controller with siding mode proposed by J.E.Slotine and W.Li (1987) for a manipulator control is used as a servo controller in the second layer and the performance is compared to that of the model following servo controller.

The control system is designed and implemented by a personal computer (NEC PC) for the miniature tug of war system. It is constructed for experimental analysis and evaluation. For the actual implementation saturation of input devices and winding up of integrators are considered and a reset method (K.Furuta,K. Kosuge,M.Yamakita,1985) is employed for the control systems to prevent such problems. Since this experimental system works satisfactorily, the practical system has been developed based on the results as in Photo 1.

2. Problem Formulation

First of all, we will consider desired properties for the tug of war machine explained in the previous section. They should be as follows:
1) The exerted tension at each side of the rope is the same
2) Movement of the rope is completely complement, which means that an absolute value of the movement from the initial position of one side is the same as that of opponent side and the signs of the movements are opposite.

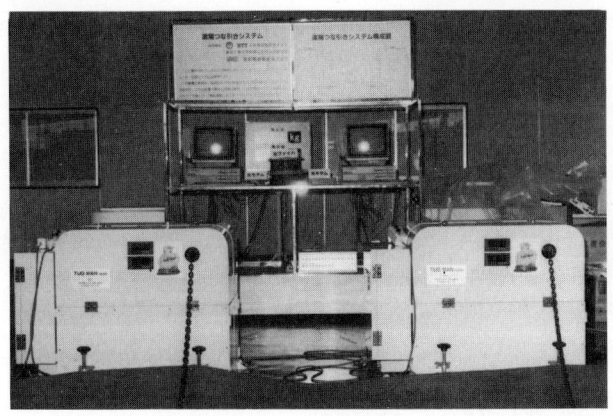

Photo 1 Practical tug of war system : Tug Man

It is, however, impossible to make the system satisfy both properties. Therefore, our machine is designed to control the tug position with satisfying the property 1) because victory or defeat is judged by the movement and players will play the game or exert the force to a rope by watching it. Since the movement is not achieved by direct connection of a rope as the real tug game, it must be controlled by an actuator according to the difference of forces given to each rope.

In order to realize such a machine, Model Following Servo (MFS) mechanism will be employed for each side of machine and it is divided to two layers concerning to their designed function. The upper layer generates the reference position based on a model using the difference of the forces. Usually the model is designed to simulate the dynamics of the rope. If, however, the game should be done with allowing a certain handicap to a team, it will be easily realized by modifying the model. The lower layer achieves the position control to track the reference position given from the upper layer. Therefore the problem can be divided to two subproblems as follows:

I) How to design the model driven by the difference of the forces.
II) How to design the tracking system.

In our system, the model is assumed to be a linear time invariant system like a mechanical one whose parameters are characterized by a mass, a viscosity and a spring constant. After determing such parameters, the remained problem is how to design a tracking system which will be explained in the next section.

3. Design of Control System I

In the following discussion, we will refer an actuator (i. e. a motor) and a rope to a plant and assign a number 1 or 2 to each plant for convention. The dynamic equation for each plant can be represented by

$$\dot{x}_{pi} = A_i x_{pi} + B_{pi} u_{pi} + d_{i1} \tag{1.a}$$

$$y_{pi} = C_{pi} x_{pi} + d_{i2}, \qquad (i=1, 2) \tag{1.b}$$

where $u_{pi} \in R$ is input, $x_{pi} \in R^2$ is state, $y_{pi} \in R$ is output, $d_{i1} \in R^2$ is state disturbance and $d_{i2} \in R$ is observation noise. A linear time invariant model for an upper layer can be represented by

$$r = f_1 - f_2 \tag{2.a}$$

$$\frac{d}{dt} x_m = A_m x_m + B_m r \tag{2.b}$$

$$y_{m1} = C_m x_m \tag{2.c}$$

$$y_{m2} = -C_m x_m, \tag{2.d}$$

where $f_i \in R$ is a measured force, A_m, B_m, C_m are matrices having proper dimensions and y_{mi} is desired movement for each rope. In order to achieve quick response for the tracking system, the difference of the forces is assumed to be output of a system as follow:

$$\frac{d}{dt} x_r = A_r x_r \tag{3.a}$$

$$r = C_r x_r. \tag{3.b}$$

Combining (2.b) and (3.a), the following augmented state model is obtained.

$$\frac{d}{dt} \hat{x}_m = \hat{A}_m \hat{x},$$

where

$$\hat{x}_m := \begin{bmatrix} x_m \\ x_r \end{bmatrix}, \qquad \hat{A}_m := \begin{bmatrix} A_m & B_m C_r \\ 0 & A_r \end{bmatrix}.$$

The purpose of the servo controller is to keep the error defined by the following equation as small as possible under several disturbances.

$$e_i = y_{pi} - y_{mi}. \quad (i=1,2) \tag{4}$$

In order to design the robust servo controller to disturbance or noise, the following operator is introduced,

$$\Phi_d(s)d_{ij}=0 \quad (i,j = 1,2) \tag{5}$$

and it is called a disturbance rejection operator, which is the same operator defined in (K.Furuta,1987). Using the operator, above equations can be rewritten as follows:

$$\Phi_d \dot{x}_{pi} = A_{pi}\Phi_d x_{pi} + B_{pi}\Phi_d u_{pi} \tag{6.a}$$

$$\Phi_d e_i = C_{pi}\Phi_d x_{pi} - C_m \Phi_d x_m \tag{6.b}$$

$$\Phi_d \dot{\hat{x}} = \hat{A}_m \Phi_d \hat{x}_m. \tag{6.c}$$

Combining (6.a), (6.b) and (6.c), we have a next augment system.

$$\frac{d}{dt}\begin{bmatrix}\Phi_d x_{pi} \\ e_i \\ \Phi_d \hat{x}_m\end{bmatrix} = \begin{bmatrix}A_i & 0 & 0 \\ C_i & 0 & -C_m \\ 0 & 0 & \hat{A}_m\end{bmatrix}\begin{bmatrix}\Phi_d x_{pi} \\ e_i \\ \Phi_d \hat{x}_m\end{bmatrix} + \begin{bmatrix}B_{pi} \\ 0 \\ 0\end{bmatrix}\Phi_d u_{pi}. \tag{7}$$

For this augment system the following criterion is minimized.

$$J(u_{pi}) = \int_0^\infty (\|e_i\|_Q^2 + \|\Phi_d u_{pi}\|_R^2)dt, \tag{8}$$

where Q is a semi positive definite matrix and R is a positive definite one. The optimal control to the above criterion can be obtained by state feedback under come condition as follow : (Kreindler,1969)

$$\Phi_d u_{pi} = F_{1i}\Phi_d x_{pi} + F_{2i}e_i + F_{3i}\Phi_d \hat{x}_m \tag{9}$$

Therefore actual input is given by

$$u_{pi} = F_{1i}x_{pi} + F_{2i}\Phi_d^{-1}e_i + F_{3i}\hat{x}_m. \tag{9'}$$

4. Design of Control system II

In this section an adaptive controller with sliding mode (J.E.Slotine and W.Li,1987) for a simple mechanical system is illustrated. A mechanical system given by

$$Ya := m\ddot{x} + c\dot{x} + kx = u \qquad (10)$$

where $Y := \begin{bmatrix} \ddot{x} & \dot{x} & x \end{bmatrix}$, $a := \begin{bmatrix} m & c & k \end{bmatrix}^T$,

is considered, where the parameters are not known but m is not zero. The aim of the control system is to make the output of the plant track the output of a model given by

$$m_m \ddot{x}_m + c_m \dot{x}_m + k_m x = u_m. \qquad (11)$$

For the above systems a sliding mode is defined as follows:

$$S := \dot{\tilde{x}} + c_0 \tilde{x}, \quad c_0 > 0. \qquad (12.a)$$

where $\tilde{x} := x - x_m$. $\qquad (12.b)$

In the original paper, the sliding mode S has been introduced to prevent from using $\ddot{x}$ in the control loop. Here, however, the sliding mode S has rather meaning of specifying a mode in which error converges.

The control input of the adaptive controller with a sliding mode is given by

$$u = \begin{bmatrix} \ddot{x} - \dot{S} & \dot{x} & x \end{bmatrix} \begin{bmatrix} \hat{m} \\ \hat{c} \\ \hat{k} \end{bmatrix} - k_d S$$

$$:= Y_c \hat{a} - k_d S, \qquad (13)$$

where $Y_c := \begin{bmatrix} \ddot{x} - \dot{S} & \dot{x} & x \end{bmatrix}$, $\hat{a} := \begin{bmatrix} \hat{m} & \hat{c} & \hat{k} \end{bmatrix}^T$,

and the adaptation law is give by

$$\dot{\hat{a}} = -\Gamma^{-1} Y_c S, \qquad (14)$$

where $\hat{a}$ contains the estimated parameters and Γ is a positive definite matrix. The stability of the closed loop system can be proved by taking the following functional as a Lyapunov function.

$$V(t) := \frac{1}{2}(SmS + \tilde{a}^T \Gamma \tilde{a}), \qquad (15)$$

where $\tilde{a} := a - \hat{a}$.

Taking a derivative of eqn. (15),

$$\dot{V}(t) = Sm\dot{S} + \tilde{a}^T \Gamma \dot{\tilde{a}}$$
$$= S\{m\dot{S} + \begin{bmatrix} \ddot{x}-\dot{S} & \dot{x} & x \end{bmatrix} a - Y_c \hat{a}\}$$
$$= S\{u - Y_c \hat{a}\} = -SK_d S \leq 0. \tag{16}$$

This inequality shows the global stability of the system.

5. Experimental Apparatus

In our experiments the plant is composed of a servo motor, a pulley and a rope as in Photo 2. Its input and output relationship can be represented by a following transfer function

$$H_{pi}(s) = \frac{K_{pi}}{s(T_{pi}s+1)}, \tag{17}$$

where input is consumed voltage and output is rotational angle. In (17) the dynamics due to electric servo between input voltage and electric current have been ignored because its response is very fast comparing to that of other parts. Eqn. (17) can be also represented by a state space equation as follows :

Photo 2   Miniature tug of war system

$$\frac{d}{dt}x_{pi} = \begin{bmatrix} 0 & 1 \\ 0 & \frac{-1}{T} \end{bmatrix} x_{pi} + \begin{bmatrix} 0 \\ \frac{K}{T} \end{bmatrix} u_{pi} \qquad (18.a)$$

$$y_{pi} = \begin{bmatrix} 1 & 0 \end{bmatrix} x_{pi} \qquad (18.b)$$

This equation is corresponding to (1.a) of the previous section. The parameters of the plant 1 are as follows:

K = 1.56 [rad/sec/V], T = 1.8 [sec].

Parameters of the plant 2 were assumed to be similar to those of the plant 1, where they have been determined by an individual identification procedure. Using such parameters a state space model of plant 1 is presented by

$$\frac{d}{dt}x_{p1} = \begin{bmatrix} 0.0 & 1.0 \\ 0.0 & -0.16 \end{bmatrix} x_{p1} + \begin{bmatrix} 0.0 \\ 0.98 \end{bmatrix} u_{p1} \qquad (19.a)$$

$$y_{p1} = \begin{bmatrix} 1.0 & 0.0 \end{bmatrix} x_{p1} \qquad (19.b)$$

The model driven by difference of forces is a mechanical system given by the following equation.

$$m\ddot{y}_m + c\dot{y}_m + ky_m = Gr,$$

where m is a mass, c is a viscosity, k is a spring constant and G is a constant gain. Eqn. (13) can be also represented by state space equation as follows :

$$\frac{d}{dt}x_m = \begin{bmatrix} 0 & 1 \\ \frac{-k}{m} & \frac{-c}{m} \end{bmatrix} x_m + \begin{bmatrix} 0 \\ \frac{G}{m} \end{bmatrix} r \qquad (20.a)$$

$$y_m = \begin{bmatrix} 1 & 0 \end{bmatrix} x_m \qquad (20.b)$$

In our experiment the spring constant have been set to zero to simulate the dynamics of a rope. This equation is corresponding to (2.b) or (11). Since in our experiment we tried to simulate a real rope as exactly as possible, the model generating reference position to the tracking system have been chosen as follows :

$$\frac{d}{dt}x_m = \begin{bmatrix} 0.0 & 1.0 \\ 0.0 & -4.0 \end{bmatrix} x_m + \begin{bmatrix} 0.0 \\ 4.0 \end{bmatrix} r \qquad (21.a)$$

$$y_{m1} = \begin{bmatrix} 40.0 & 0.0 \end{bmatrix} x_m \qquad (21.b)$$

The pole of the model was chosen by experiments so that the movement to a force is natural and excessive input was not required to achieve the movement. And the output gain 40 has been determined so that the movement of the rope to the exerted force was long enough.

5. Experimental Result I

Since in our design of the control system state disturbance and observation noise are assumed to be constant, the disturbance rejection operator can be chosen as follow :

$$\Phi_d(s) = s. \qquad (22)$$

Corresponding to the selection of the operator, the control is determined by

$$u_{pi} = F_{1i} x_{pi} + F_{2i} \int_0^t e_i \, dt + F_{3i} \hat{x}_m. \qquad (23)$$

The input r to the model was assumed to be output of an integrator and its state space model was represented by

$$\frac{d}{dt} x_r = 0, \qquad (24.a)$$

$$r = x_r. \qquad (24.b)$$

Using state space models described above, feedforward or feedback gain can be determined if Q and R are specified. For experiments of the miniature model Q and R were determined as follows :

$$Q = 1.0e+6, \quad R = 1.0 \qquad (25)$$

Feedforward and feedback gains for these weighting parameters could be calculated using a CAD system DPACS, which was developed by Furuta Lab., and they were obtained as

$$F_{p1} = \begin{bmatrix} -97.0 \\ -14.0 \end{bmatrix}^T, \quad F_{p2} = -316.0, \quad F_{p3} = \begin{bmatrix} 3.89e3 \\ 4.27e3 \\ 1.70e2 \end{bmatrix}^T \qquad (26)$$

Note that the third gain in F_{p3} is a feedforward gain from input of the model to the plant, which will be shown later to be significant for the stability of the servo system. Fig. 1 shows the block diagram of the control system I for each machine.

Fig. 2 shows the experimental result of the miniature model exerted some forces to a rope. The actual controller has been realized by a digital computer and its sampling interval was 4 [msec]. In order to avoid the problems of saturation of input devices and winding up of integrators the reset algorithm of integrators (K.Furuta etc.1985) has been employed in the actual

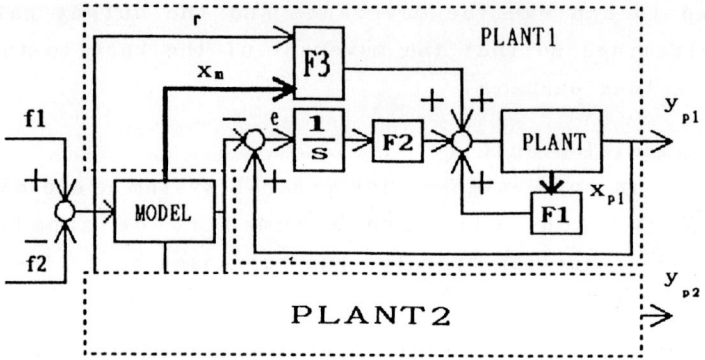

Fig. 1　Block diagram of control system I

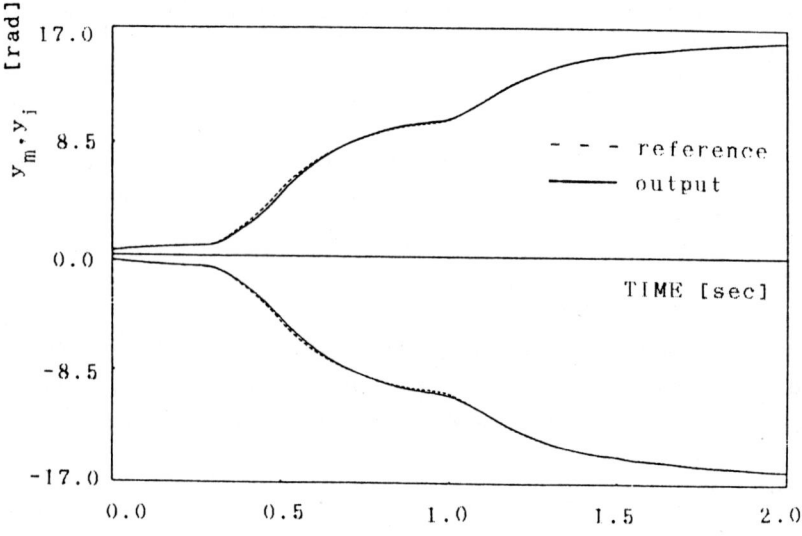

Fig. 2　Experimental result I : reference and output

controller. Fig. 2 shows that the model following servo controller gives good tracking performance to the reference output from the model. It also shows that the servo controller is insensitive for parameter variation of the plant since plant 2 has a good response to the reference even if parameters of the system are not identified and are assumed to be similar to those of plant 1.

6. Experimental Result II

The adaptive controller with a sliding mode explained in the section 4 was also applied to the same plant (18). The parameters of the control system were as follows :

$$k_d = 10000, \quad c_0 = 100 \text{ and } \Gamma = 0.1. \tag{27}$$

Fig. 3 shows the block diagram of the adaptive controller and Fig. 4 shows the experimental result controlled with those parameters, where the estimated parameters were set to zero in advance. The controller was realized by the same digital computer and the sampling interval was 1 [msec]. In the actual implementation of the controller the adaptation law was modified so that the adaptation of the parameters is stopped if control input exceeds over allowed values, otherwise they were to diverge. Fig. 4 shows that the controller gives a good response except small deviation even even if parameters of plants are completely unknown. Therefore, the controller will be a very powerful controller if it is difficult to identify parameters of plant in advance.

7. Analysis of Stability

Sometimes a servo controller controlling a position which affects the input force to the controlled system leads to instability of the system. In this section the stability of the system controlled by the model following servo controller

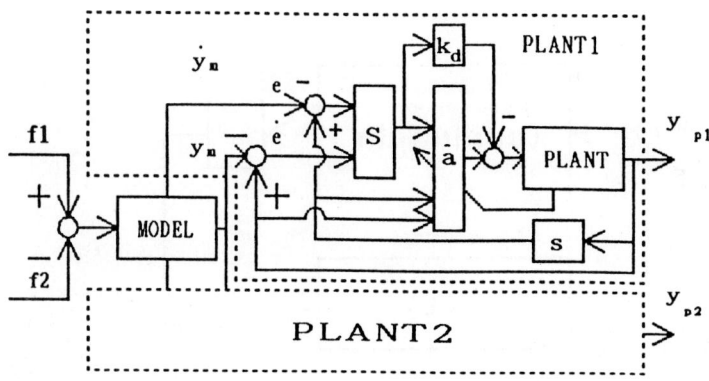

Fig. 3   Block diagram of control system II

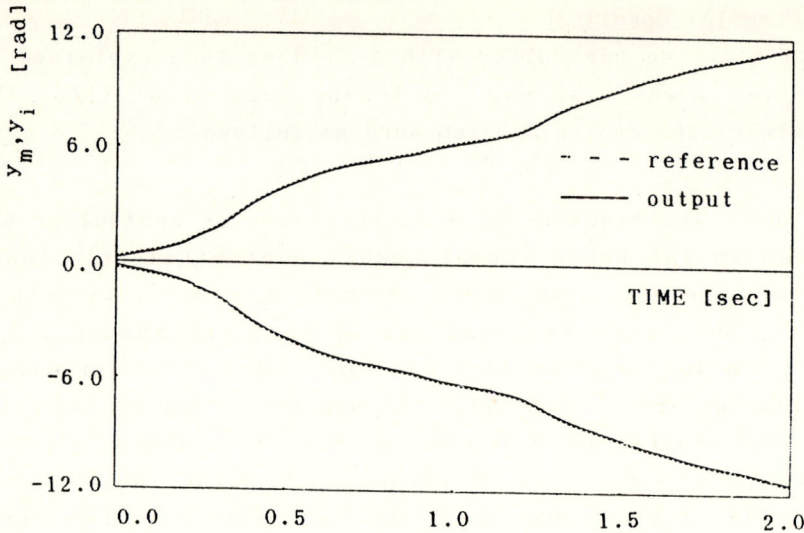

Fig. 4 Experimental result II : reference and output

proposed in the paper is investigated and it will show that the stability of the system is maintained for vast characteristics of the environment which is a model of a player. For simplicity it is assumed that force exerted to a rope in a plant 2 is kept constant, and the stability of the plant 1 is considered. The controlled system which consists of a model and an actuator is considered as a system whose input is force and whose output is movement of a rope, and players of a team can be

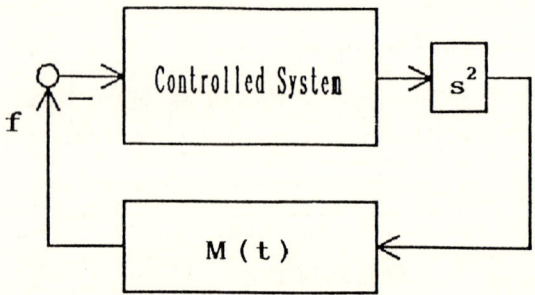

Fig. 5 Block diagram of equivalent closed loop system

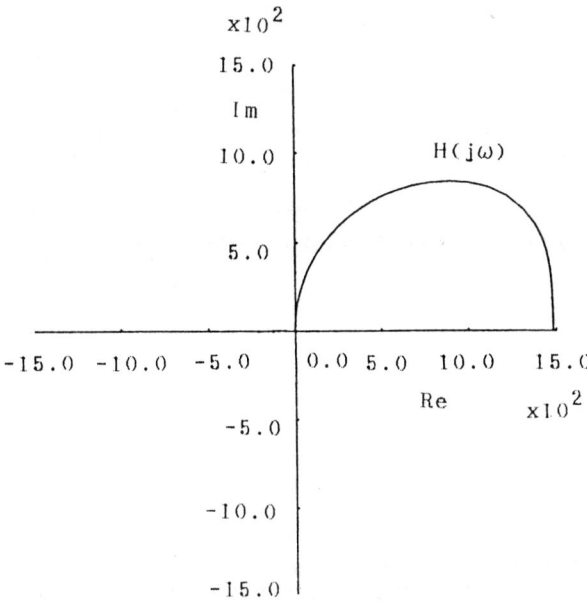

Fig. 6   Nyquist plot of H(s)

considered as an environment which gives a negative force feedback to the system if an actuator pulls the rope against players. (See Fig. 5) It will be valid that the environment can be modeled as a mechanical system whose parameters depend on time and position for small change of movement. First we will assume that the environment is a pure mass whose dynamics is given by

$$M(t)\ddot{y}_{p1} = -f \quad \text{if } y_{p1} \leqq 0 \tag{28.a}$$
$$0 = -f \quad \text{if } y_{p1} > 0, \tag{28.b}$$

where y_{p1} is measured from an equilibrium and it has negative sign if the actuator pulls. The relationship between force and acceleration can be represented as linear time invariant and the corresponding transfer function H(s) for the experimental system is given by

$$H(s) = \frac{s(148s^3+2078s^2+13500s+44000)}{s^4+16.7s^3+135s^2+612s+1100} \tag{29}$$

and its Nyquist plot is given in Fig. 6. As shown that the linear part satisfies the condition of positive realness, we can apply hyper stability theorem to the closed loop system. As denoted in (28), the feedback loop is non-negative, which ensures the stability of the system. It should be noted that it is crucial for H(s) to be positive real that the model has a direct part.

Though the environment has been assumed to be a pure mass in the preceding discussion, the analysis gives a good inference of the stability for general cases. Concerned about the shape of the Nyquist plot in Fig. 6 the stability will be kept if the environment has damping factors. If the environment contains a spring constant, high gain control may lead instability, but the control system has large stability margin since high frequency signals are reduced by unmodeled damping factors.

7. Conclusion

One of realization techniques of a fiber connected tug of war machine has been proposed and the validity of the control system was checked by a miniature model. Since the control system was designed using the idea of virtual internal model following control, it can be realized that the game is done under several conditions, i.e. allowing handicaps to a team or without an opponent team for a training. It has been also studied that the servo system in the control system has good stability under variations of an environment.

The model following servo controller has been implemented as a controller in a practical tug of war machine based on the experiments of the miniature machine. Implementation of the adaptive controller to a practical system however, is under consideration, but it will be implemented in near future because it has been convinced that it has similar performance as the model following servo controller from the experiments and it is expected that the adaptive controller is to be more robust for a change of load.

Finally, we would like to stress that it is very interesting that one of the oldest sports has been reconstructed as a 'High Tech.' sports, a fiber connected tug of war, using modern control theory. There should be other such applications for which modern technologies can be applied.

Authors acknowledze that the idea of the fiber connected tug ofwar is given by NTT, and appreciate their support.

Reference

1] E.J.Davison, H.W.Smith,"Pole assignment in linear time-invariant multivariable systems with constant disturbances," Automatica, vol.7, pp.489-498, 1971

2] E.J.Davison,"The robust control of a servomechanism problem for linear time-invariant multivariable systems," IEEE Trans. Auto. Contr., vol.,ac-21, pp.25-34,1976

3] K. Furuta,"Alternate robust servo-control system and its digital control", Int. J. Contr., 1987, vol. 45, no.1, pp. 183-194

4] E. Kreindler,"On the linear optimal servo problem", Int. J. Control, vol.9, pp.465-472, 1969.

5] K.Furuta, K.Komiya," Design of Model-Following Servo Controller", IEEE Trans. on Aut. Contr., vol. ac-27, no. 3, pp.725-727, 1982.

6] K.Kosuge, K.Furuta, T.Yokoyama, "Mechanical Impedance Control of a Robot Arm by Virtual Internal Model Following Controller", IFAC 10th World Congress on Automatic Control, vol. 4, pp.250-255, 1987

7] K.Furuta, K.Kosuge, M.Yamakita, "Trajectory Tracking Control of Robot Arms Using ORBIX",J. R. Systems, vol.2, pp. 89-112,1985

8] E.E.J.Slotine, W.Li, "ADAPTIVE MANIPULATOR CONTROL A Case Study",IEEE int. Conf. Robotic and Automation,Raleigh,NC,pp. 1392-1400,1987

Structure of Magnetic Bearing Control System for Compensating Unbalance Force

Takeshi Mizuno
Faculty of Engineering, Saitama University, Urawa

Toshiro Higuchi
Institute of Industrial Science, University of Tokyo, Tokyo

summary
 A magnetic bearing control system is constructed in which an unbalanced rotor can be suspended without whirling even if certain system parameters vary from their nominal values. The concept of designing the control system is that unbalance forces are estimated by an observer and cancelled by the electromagnetic forces of the bearing. It is shown theoretically that the original controller can suspend the rotor without whirling for the parameter variations of the controlled object but loses the property for a change of the rotational speed. The structure of a new controller is presented which holds the property even if the rotational speed is perturbed. The controller is obtained by modifying the original controller based upon a formula of Laplace transform.

1 Introduction

The magnetic bearing can suspend a rotor without any mechanical contact and lubrication. For these inherent advantages it has been applied in a various fields: vacuum techniques, space technology, machine tools and so on. An active-type magnetic bearing has another advantage that the bearing force acting on the rotor can be controlled actively according to the states of the rotor. For this property it can have additional functions which have not been achieved by conventional bearings. Making the most use of this feature, the authors have developed a control system in which an unbalanced rotor can be suspended without whirling[1]. The control system was designed based upon the theory of output regulation with internal stability[2]. The concept of design is that the effects of unbalance are estimated by an observer and are cancelled by the electromagnetic forces of the bearing according to the estimation.
 In practical application, sensitivity considerations of

the designed control system are important because there is usually a discrepancy between the physical reality and mathematical model. If the designed control system were proved to be very sensitive to parameter changes, the control method would be useless in practice. This paper shows that the controller, which was constructed according to the theory, holds the property of the output regulation under perturbations of the nominal models of the controlled object but loses the property when the rotational speed varies from the nominal value. In order to overcome this problem the controller is modified to generate compensation signals by using exogenous signals synchronized with actual rotation.

2 Equations of Motion for Magnetic Bearing System

A model, which is used for investigation of a typical totally active magnetic bearing system dynamics, is shown in Fig.1. Since the rotor is treated as a rigid body in this paper, it has six degrees of freedom of motion. In order to keep the rotor rotating about an fixed axis, the magnetic bearing has to control five degrees of freedom of motion. Eight electromagnets, which are numbered as ①,..., ⑧ in Fig.1, are used to control two translational motions and two rotational motions in the radial directions. Two electromagnets

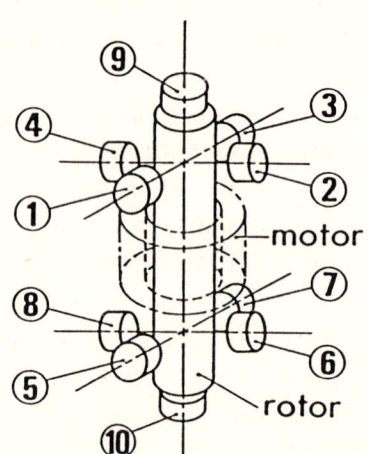

Fig.1 Basic Model of a totally active magnetic bearing (①-⑩ : electromagnets)

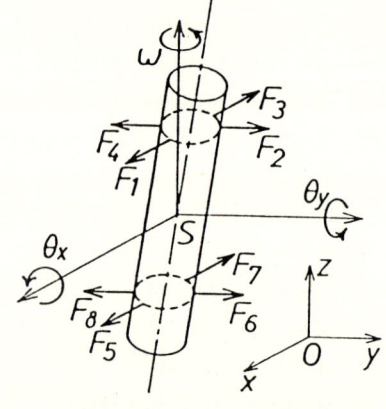

Fig.2 Coordinate axes and forces acting on the rotor

which are numbered as ⑨, ⑩ are used to control one translational motion in the axial direction.

To derive the equations of motion a coordinate frame O-xyz fixed in space is defined as shown in Fig.2; the origin O corresponds to the center of the rotor S in the desired position and z-axis corresponds to its rotating axis.

The attractive force of the magnet numbered as ⓝ is represented as F_n. The directions of $F_1,\ldots,F_8$ are also shown in Fig.2. For small motions about the stationary, F_n can be approximated by a linear relation:

$$F_n = F_0 - Gd_n + Hi_n \qquad n=1,\ldots,8 \qquad (1)$$

where
$\quad F_0$:stationary force
$\quad G,H$:coefficients of the linearised model of the magnet
$\quad i_n$:incremental current flowing through the winding
$\quad d_n$:incremental gap between the rotor and the magnet

Each d_n is determined by the translational and rotational displacements of the rotor.

When the rotor is driven to rotate at a constant speed ω, the equations of motion in the radial directions are given by[1)]

$$m\ddot{x}_s - 4Gx_s = H(i_1 - i_3 + i_5 - i_7) + m\varepsilon\omega^2 \cos(\omega t + \alpha) \qquad (2)$$
$$m\ddot{y}_s - 4Gy_s = H(i_2 - i_4 + i_6 - i_8) + m\varepsilon\omega^2 \sin(\omega t + \alpha) \qquad (3)$$
$$I_r\ddot{\theta}_x + I_a\omega\dot{\theta}_y - 4Gl^2\theta_x$$
$$= H(-i_2 + i_4 + i_6 - i_8)l + (I_r - I_a)\tau\omega^2 \cos(\omega t + \beta) \qquad (4)$$
$$I_r\ddot{\theta}_y - I_a\omega\dot{\theta}_x - 4Gl^2\theta_y$$
$$= H(i_1 - i_3 - i_5 + i_7)l + (I_r - I_a)\tau\omega^2 \sin(\omega t + \beta) \qquad (5)$$

where
$\quad m$:mass of the rotor
I_a, I_r:polar and transverse mass moments of inertia of the rotor
$\quad l$:distance between the center of the rotor and the magnets
$\quad \alpha,\beta$:parameters on angular location of static and dynamic unbalance
$\quad \varepsilon$:eccentricity of the rotor(amount of static unbalance)
$\quad \tau$:angle between the rotational axis and the principal axis (amount of dynamic unbalance)

x_s, y_s: displacements of the rotor center S in x and y directions

θ_x, θ_y: angular displacements of rotor axis about x and y axes

From eqs.(2),...,(5) the dynamics of the magnetic bearing system is expressed by a set of equations of the type:

$$\dot{x}(t) = Ax(t) + Bu(t) + Dw(t) \tag{6}$$

$$\dot{w}(t) = Ew(t) \tag{7}$$

where

$$x = [x_1\ \dot{x}_1\ x_2\ \dot{x}_2]^t, \quad u = [u_1\ u_2]^t, \quad w = [w_1\ w_2]^t$$

$$A = \begin{bmatrix} 0 & 1 & 0 & 0 \\ a & 0 & 0 & c\omega \\ 0 & 0 & 0 & 1 \\ 0 & -c\omega & a & 0 \end{bmatrix} \quad B = \begin{bmatrix} 0 & 0 \\ b & 0 \\ 0 & 0 \\ 0 & b \end{bmatrix} \quad D = \begin{bmatrix} 0 & 0 \\ 1 & 0 \\ 0 & 0 \\ 0 & 1 \end{bmatrix} \quad E = \begin{bmatrix} 0 & -\omega \\ \omega & 0 \end{bmatrix}$$

The variable and coefficients are defined as shown in Table 1. It is remarked that the effects of unbalance are considered to be exogenous disturbance to the system

$$\dot{x}(t) = Ax(t) + Bu(t) \tag{8}$$

and the dynamics of the disturbance can be described by a linear constant-coefficient equation (7).

3 Control System Design

This chapter shows the procedure of designing the control system based upon the theory of output regulation with internal

Table 1
Variables and coefficients in each subsystem

symbol	meaning in subsystem related to translation	meaning in subsystem related to rotation
x_1	x_s	θ_x
x_2	y_s	θ_y
u_1	$i_1 - i_3 + i_5 - i_7$	$-i_2 + i_4 + i_6 - i_8$
u_2	$i_2 - i_4 + i_6 - i_8$	$i_1 - i_3 - i_5 + i_7$
w_1	$\varepsilon\omega^2 \cos(\omega t + \alpha)$	$(1-c)\tau\omega^2 \cos(\omega t + \beta)$
w_2	$\varepsilon\omega^2 \sin(\omega t + \alpha)$	$(1-c)\tau\omega^2 \sin(\omega t + \beta)$
a	$4G/m$	$4Gl^2/I_r$
b	H/m	Hl/I_r
c	0	I_a/I_r

stability. The ways of designing are described as follows.

First, define a combined system as

$$\dot{x}_c(t) = A_c x_c(t) + B_c u(t) \quad (9)$$
$$y_c(t) = C_c x_c(t) \quad (=x(t)) \quad (10)$$

where

$$x_c = \begin{bmatrix} x \\ w \end{bmatrix}, \quad A_c = \begin{bmatrix} A & D \\ 0 & E \end{bmatrix}, \quad B_c = \begin{bmatrix} B \\ 0 \end{bmatrix}, \quad C_c = [I_4 \ 0]$$

I_4: identity matrix

Output regulation with internal stability is achieved by a control

$$u(t) = P x_c(t) \quad (11)$$
$$= P_1 x(t) + P_2 w(t) \quad (12)$$

such that

(i) a closed-loop system

$$\dot{x}(t) = (A + B P_1) x(t) \quad (13)$$

is stable (internal stability) and

(ii) $\quad C_c \exp[(A_c + B_c P) t] \to 0$ as $t \to \infty \quad (14)$

(output regulation)

Second, construct a feedback matrix P which satisfies conditions (i) and (ii). Considering the internal symmetry of the controlled object, P_1 is given in the form[3]

$$P_1 = -\begin{bmatrix} p_d & p_v & -p_c & 0 \\ p_c & 0 & p_d & p_v \end{bmatrix} \quad (15)$$

The elements are selected to satisfy the stability conditions:

(s1) $(bp_d - a) b p_v + b p_c c \omega > 0 \quad (16)$

(s2) $(bp_d - a)(bp_v)^2 - (bp_c)^2 + b^2 p_v p_c c \omega > 0 \quad (17)$

The matrix P_2 is given in the form

$$P_2 = -\frac{1}{b}\begin{bmatrix} 1 & 0 \\ 0 & 1 \end{bmatrix} \quad (18)$$

so that the unbalance forces are cancelled by the magnetic forces.

Third, an observer which estimates $w(t)$ will be constructed since it is difficult to detect the instant value of $w(t)$ directly during rotor running. According to the observer theory a second-order observer can be constructed as

$$\dot{z}(t) = (E - VD) z(t) + (-VA + EV - VDV) x(t) - VBu(t) \quad (19)$$

$$\hat{w}(t)=z(t)+Vx(t) \tag{20}$$

where $\hat{w}(t)(=[\hat{w}_1,\hat{w}_2]^t)$ denotes an estimator of $w(t)$ and

$$z=\begin{bmatrix}z_1\\z_2\end{bmatrix}, \qquad V=\begin{bmatrix}0 & \sigma & 0 & \nu\\0 & -\nu & 0 & \sigma\end{bmatrix}$$

For convergence the parameter σ must satisfy

$$\sigma > 0 \tag{21}$$

Consequently the control input $u(t)$ is determined as

$$u(t)=P_1 x(t)-\hat{w}(t)/b \tag{22}$$

Substituting eqs.(20) and (22) into eq.(19), the state equation of the observer is transformed as

$$\dot{z}(t)=Ez(t)+Rx(t) \tag{23}$$

where

$$R=\begin{bmatrix}r_{11} & r_{12} & -r_{21} & -r_{22}\\r_{21} & r_{22} & r_{11} & r_{12}\end{bmatrix}$$

$$r_{11}=-\sigma(a-bp_d)-\nu bp_c, \qquad r_{12}=\sigma bp_v+\nu(1-c)\omega$$
$$r_{21}=\nu(a-bp_d)-\sigma bp_c, \qquad r_{22}=-\nu bp_v+\sigma(1-c)\omega$$

The block diagram of the magnetic bearing system with the compensator for unbalance is shown in Fig.3. The obtained dynamic compensator has an internal model of the disturbance, that is to say, a generator of two-phase alternating signals whose frequency is equal to the rotational frequency.

4 Sensitivity Analysis

It has been confirmed theoretically and experimentally that the constructed compensator can remove whirling motion due to unbalance completely for the nominal model[1]. In this chapter the influences of deviations of the system parameters on the property of output regulation are analysed.

In the sequel a parameter p ($p=a$, b, c or ω) will be represented as a sum of a nominal value p^0 and a perturbation Δp.

Define complex variables

$$x=x_1+jx_2, \quad u=u_1+ju_2, \quad w=w_1+jw_2, \quad z=z_1+jz_2, \quad \hat{w}=\hat{w}_1+j\hat{w}_2 \tag{24}$$

and denote each Laplace-transformed variable by the corresponding capital. By using these variables the transfer function representation of the system is written as

$$X(s) = \frac{1}{t(s)} \left(W(s) - \frac{b}{b^0} \hat{W}(s) \right) \tag{25}$$

$$Z(s) = \frac{(r_{11}^0 + r_{12}^0 s) + j(r_{21}^0 + r_{22}^0 s)}{s - j\omega^0} X(s) \tag{26}$$

$$\hat{W}(s) = Z(s) + (\sigma - j\nu) s X(s) \tag{27}$$

$$W(s) = \frac{w(0)}{s - j\omega} \tag{28}$$

where the initial values of the variables but $w(t)$ are set to be zero for simplicity; $t(s)$ is defined as

$$t(s) = s^2 + (bp_v - jc\omega)s + (bp_d - a - jbp_c) \tag{29}$$

and r_{ij}^0 denotes the value of r_{ij} for nominal values of the

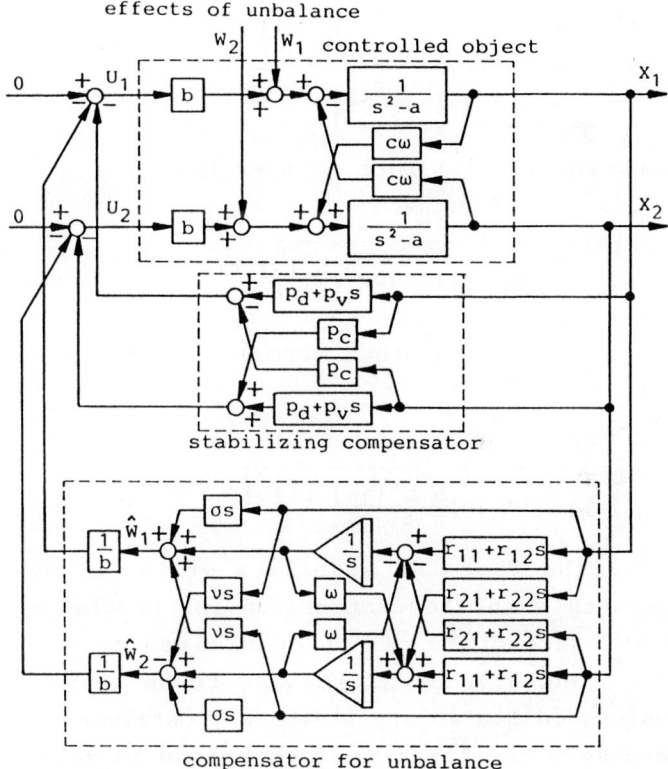

Fig.3 Block diagram of the control system with a compensator for unbalance which has an internal model

parameters. From eqs.(25),...,(27) the estimation by the observer is obtained as follows.

$$\hat{W}(s) = \frac{\sigma - j\nu}{(s - j\omega^0)t(s) + (\sigma - j\nu)t^0(s)\frac{b}{b^0}} t^0(s)W(s) \quad (30)$$

where $t^0(s)$ is defined as

$$t^0(s) = s^2 + (b^0 p_v - jc^0\omega^0)s + (b^0 p_d - a^0 - jb^0 p_c) \quad (31)$$

The estimation error of the observer is given by

$$W(s) - \hat{W}(s) = \frac{s - j\omega^0 + (\sigma - j\nu)\frac{t^0(s)}{t(s)}(\frac{b}{b^0} - 1)}{s - j\omega^0 + (\sigma - j\nu)\frac{b\, t^0(s)}{b^0 t(s)}} W(s) \quad (32)$$

$$= \frac{c_1}{s - j\omega} w(0) + \text{(other terms)} \quad (33)$$

where

$$c_1 = \frac{j\Delta\omega + (\sigma - j\nu)\frac{t^0(j\omega)}{t(j\omega)}(\frac{b}{b^0} - 1)}{j\Delta\omega + (\sigma - j\nu)\frac{b\, t^0(j\omega)}{b^0 t(j\omega)}} \quad (34)$$

Substituting eq.(30) into (25) we have

$$X(s) = \frac{s - j\omega^0}{(s - j\omega^0) + (\sigma - j\nu)\frac{b\, t^0(s)}{b^0 t(s)}} \cdot \frac{W(s)}{t(s)} \quad (35)$$

$$= \frac{c_2}{s - j\omega} w(0) + \text{(other terms)} \quad (36)$$

where

$$c_2 = \frac{j\Delta\omega}{j\Delta\omega + (\sigma - j\nu)\frac{b\, t^0(j\omega)}{b^0 t(j\omega)}} \cdot \frac{1}{t(j\omega)} \quad (37)$$

Assuming that the stability of the closed-loop system incorporated with the compensator for unbalance is preserved, the stationary state can be determined by the first term in eq.(33) or eq.(36). By estimating these terms the following conclusions on stationary states are obtained.

(1) When only the parameters contained in matrix A vary from their nominal values, the output of observer converges to the exact state asymptotically.

(2) When b varies and the rotational speed is kept a nominal value, the error of the observer does not converge to zero but the property of output regulation is hold.

(3) For a change of ω, which means that the rotor speed varies from the nominal value, output regulation fails; for a small change, the amplitude of whirling motion of the rotor is proportional to the amplitude of Δω.

5 Modification of the Compensator

Equation (36) implies that if the value of parameter ω in the internal model is set to the actual value, the whirling motion will disappear. One of the methods by which this property is obtained is that the parameter ω in the controller is changed adaptively according to the output of a sensor which detects the angular frequency of the rotor. In this chapter another method will be presented.

The concept of designing is to construct a model of disturbance dynamics by using exogenous signals synchronized with actual rotation. As is mentioned in Chapter 3, the compensator has an internal model which generates two-phase alternating signals whose frequency is set to that of rotation. Instead of generating the signals, exogenous signals whose frequency is truly equal to the rotational frequency will be used.

The dynamics of the compensator, which is shown in Chapter 3, can be described as

$$Z(s) = F(s)R(s)X(s) \tag{38}$$

where

$$F(s) = \frac{1}{s - j\omega} \tag{39}$$

$$R(s) = (r_{11} + r_{12}s) + j(r_{21} + r_{22}s) \tag{40}$$

The inverse transformed functions of F(s) and R(s)X(s) are

$$f(t) = \exp(j\omega t) \tag{41}$$

$$(r*x)(t) = ((r_{11} + r_{12}\tfrac{d}{dt}) + j(r_{21} + r_{22}\tfrac{d}{dt}))x(t) \tag{42}$$

where x(0) is assumed to be zero. A formula of the Laplace transform says that when the response function for a system is given by the product of two function of s, the corresponding time function of the system can be found by convolving the corresponding two time functions. Applying this formula to eq.(38), z(t) can be represented as

$$z(t) = \int_0^t g(t-\tau)(r*x)(\tau)d\tau$$

$$= \exp(j\omega t)\int_0^t ((r_{11}+r_{12}\frac{d}{d\tau})+j(r_{21}+r_{22}\frac{d}{d\tau}))x(\tau)\exp(-j\omega\tau)d\tau$$

(43)

In eq.(43) the terms to be integrated are composed by a measurable variable x(t) and sinusoidal signals whose angular frequency is equal to the rotational frequency. When sinusoidal signals synchronized with actual rotation are used to calculate the integration, the critical parameter ω contained in the compensator is automatically set to the exact value. As a result the property of output regulation is preserved even if the rotational speed varies from the nominal value. The block diagram of the modified controller is shown in Fig.4.

6 Simulation

To confirm the effectiveness of the modified controller, numerical simulations are performed. The values of parameters used in the simulations are listed in Table 2. In the following

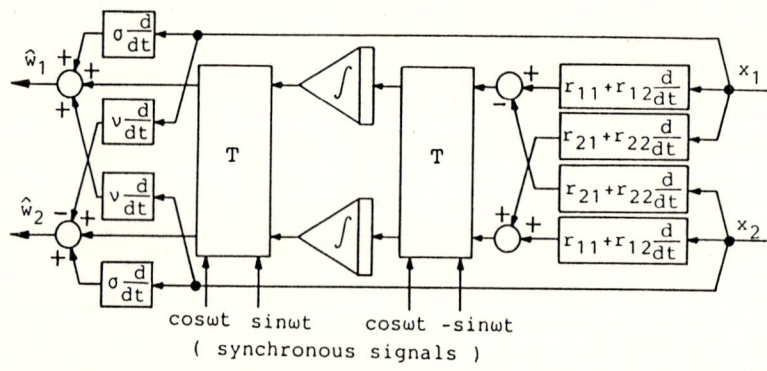

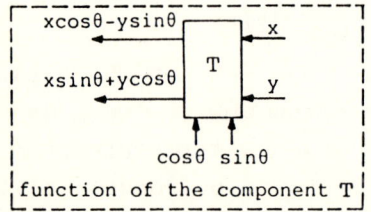

Fig.4
Block diagram of a compensator for unbalance using exogenous synchronized signals

Table 2 Parameters and initial conditions used for simulations

nominal values				values in perturbed systems		
					case(a)	case(b)
a^0	1.00	c^0	0.549	a, b, c	nominal	nominal
b^0	1.00	ω^0	3.46	ω	3.11	3.80
values in the desiged controller				$x_1(0)$	-0.0487	-0.0723
p_d	3.26	r_{11}	2.89	$\dot{x}_1(0)$	0.399	0.471
p_v	2.55	r_{12}	3.26	$x_2(0)$	-0.128	-0.124
p_c	2.42	r_{21}	-3.10	$\dot{x}_2(0)$	-0.152	-0.275
σ	1.28	r_{22}	1.99	$w_1(0)$	0.810	1.21
ν	0.00	$z_1(0)$	0.00	$w_2(0)$	0.00	0.00
		$z_2(0)$	0.00			

simulations, ω is assumed to decrease (case (a)) or increase (case(b)) by 10 per cent from its nominal value; all the other parameters are assumed to be nominal.

Figure 5 and 6 show the responses when the original and modified compensator for unbalance are used; in the figures broken lines show the stationary motion of the rotor before the the compensation for unbalance starts at t=0. When the compensator with an internal model is used, residual whirling motion of the rotor is observed. As contrasted with the original compensator, the modified compensator can completely eliminate the effects of unbalance on the rotor motion even if the rotating speed is perturbed. These results show the modification is effective.

7 Conclusion

This paper presents the design, sensitivity analysis and modification of a magnetic bearing control system with compensation for unbalance. The modified control system, which uses external signals synchronized with actual rotation, can remove completely the whirling motion due to rotor unbalance even if the rotational speed varies from the its nominal value.

References

1. Mizuno, T.; and Higuchi, T.:Compensation for Unbalance in Magnetic Bearing System (in Japanese). Trans. Society of

Instrument and Control Engineers, Vol.20, No.12 (1984)1095-1101.
2. Wonham, W.M.; Pearson, P.G.: Regulation and Internal Stabilization in Linear Multivariable Systems. SIAM J. on Control, Vol.12, No.1(1974)5-18.
3. Mizuno, T.; Higuchi, T.: Design of the Control System of Totally Active Magnetic Bearings ---Structures of the Optimal Regulator---. Proc. 1st International Symposium on Design and Synthesis, Tokyo (1984)534-539.

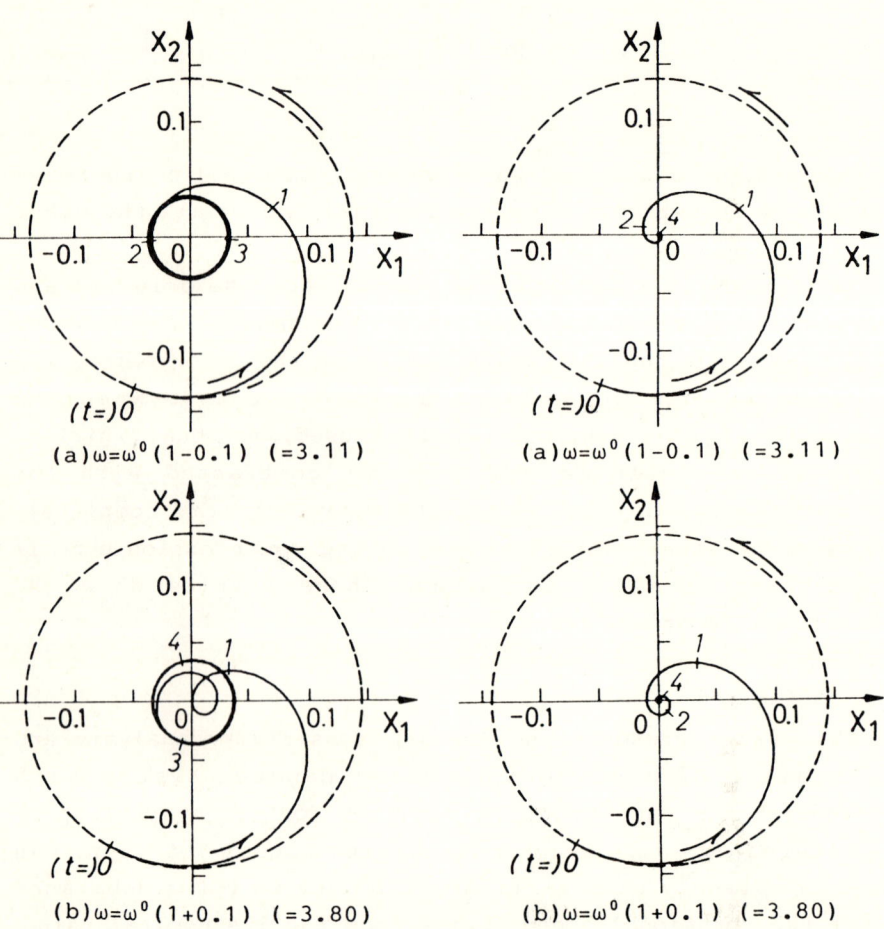

Fig.5
Residual whirling motions in the system with a fixed internal model

Fig.6
Responses of the system with a compensator for unbalance using exogenous synchronized signals

Sensors and Actuators

Placing Dynamic Sensors and Actuators on Flexible Space Structures

Gregory A. Norris and Robert E. Skelton
School of Aeronautics and Astronautics
Purdue University
West Lafayette, IN 47907

ABSTRACT

This paper selects sensors and actuators (location, type and number) from an admissible set. We seek an approximate solution to this integer programming problem. Given the optimal use of the entire admissible set of sensors and actuators, it is possible to decompose the quadratic cost function into contributions from each stochastic input and each weighted output. In the past, these suboptimal cost decomposition methods of sensor and actuator selection have been used to locate perfect (infinite bandwidth) sensors and actuators on large scale systems. This paper extends these ideas to the more practical case of imperfect actuators and sensors with dynamics of their own. Secondly, the old cost decomposition methods are discarded for improved formulas for sensor and actuator deletion (from the admissible set). These results show that there exists an optimal number of actuators (it is possible to use too few and too many). Preliminary attempts to solve this new research question are described. It is also shown that there exists optimal *dynamics* of the actuators. NASA's SCOLE example demonstrates the concepts.

1.0 INTRODUCTION

The objective of this paper is to develop and evaluate a method for the selection of sensors and actuators in the control of finite-dimensional linear systems using imperfect sensors and actuators -- devices which do not provide instantaneous responses, but have dynamics of their own. In addition, the actuator and sensor noise may be correlated. This important case allows the use of accelerometer sensors (this always yields correlated plant and measurement noise). Correlation of the noise and the presence of dynamics in the actuator and sensor devices can significantly affect the optimal selection of both the number and location of sensors and actuators. Also, the dynamics of these devices can be tuned for better system performance. (Actuators can be too slow or too fast). Hence, the algorithms herein produce *design requirements* (time constants and noise levels permitted) for actuator devices.

Consider the series connection of three dynamical elements, the actuators described by the dynamics $y_a = C_a(sI-A_a)^{-1}B_a(u+w_{in})$, the plant described by the dynamics $y_p = C_p(sI-A_p)^{-1}B_p(y_a+w_{out})$, the sensors described by the dynamics $z = C_s(sI-A_s)^{-1}B_s(y_p+v_{in})$, where

$$w^T(t) \stackrel{\Delta}{=} [w_{in}^T(t), w_{out}^T(t), v_{in}^T,] \qquad (1.1)$$
$$v(t) \stackrel{\Delta}{=} v_{out}(t)$$

are correlated white noise processes

$$E \begin{bmatrix} w(t) \\ v(t) \end{bmatrix} [w^T(\tau), v^T(\tau)] = \begin{bmatrix} W & U \\ U & V \end{bmatrix} \delta(t-\tau) \qquad (1.2)$$

and the dimensions of the vectors are

$$x_a \in \mathbb{R}^{n_a}, \; x_p \in \mathbb{R}^{n_p}, \; x_s \in \mathbb{R}^{n_s}, \; u \in \mathbb{R}^{n_u},$$

$$y_a \in \mathbb{R}^a, \; y_p \in \mathbb{R}^p, \; z \in \mathbb{R}^{n_z}, \; v \in \mathbb{R}^{n_z}.$$

In the control of large space structures, the locations of sensors and actuators becomes a critically significant "degree of freedom" in control design [14-19]. Among over 60 recent contributions to the sensor and actuator selection (SAS) problem, only [4], [7], [10], [11], and [12] consider noisy actuators (W, V nonzero). In all cases, the disturbances are modelled as Gaussian, white, and *uncorrelated* (W, V diagonal, U = 0). Most of the SAS literature takes no account of *actuator or sensor dynamics*. Two exceptions are McClamrock [19], and Howell and Baxter, [6]. In [1] the authors extend the cost decomposition approach [2] to accommodate noise correlation between sensor and actuator noise sources (W, V not diagonal, U ≠ 0). A key conclusion in [1] is that the proper sensor/actuator selection and placement can be drastically affected by noise correlation. For example, the deletion of a noise source (by making an actuator or sensor noise free) may *degrade* performance contrary to the usual expectations when noise sources are uncorrelated. However, [1] does not handle sensor and actuator dynamics. That is the contribution of this paper.

A discussion of the effect of *actuator dynamics* is given by Goh and Caughey [8]. The analysis of [8] and [9] demonstrates that plant frequencies occurring above the actuator bandwidth can lead to closed loop instability, even for co-located sensors and actuators. Goh and Caughey do not address the problem of *selection* of dynamic actuators nor sensors. That is the goal of this paper. The tools of cost decomposition [2-4] have to be modified substantially to handle this case.

This paper is organized as follows. First the augmented system model including sensor and actuator dynamics is examined for controllability, observability. These dynamics are used to define expressions which reflect the effectiveness of each *dynamic* actuator or sensor in minimizing the cost function. Finally, the method is illustrated by application both to small scale numerical examples and to NASA's SCOLE flexible space structure model.

2.0 System Dynamics

The system described by (1.1) has the structure

$$\dot{x} = Ax + Bu + Dw$$
$$y = Cx \quad (2.1a)$$
$$z = Mx + v$$

where

$$A = \begin{bmatrix} A_a & 0 & 0 \\ B_p C_a & A_p & 0 \\ 0 & B_s C_p & A_s \end{bmatrix}, \quad B = \begin{bmatrix} B_a \\ 0 \\ 0 \end{bmatrix}, \quad D = \begin{bmatrix} B_a & 0 & 0 \\ 0 & B_p & 0 \\ 0 & 0 & B_s \end{bmatrix} \quad (2.1b)$$

$$C = \begin{bmatrix} C_a & 0 & 0 \\ 0 & C_p & 0 \end{bmatrix}, \quad x = \begin{bmatrix} x_a \\ x_p \\ x_s \end{bmatrix}, \quad w = \begin{bmatrix} w_{in} \\ w_{out} \\ v_{in} \end{bmatrix} \quad (2.1c)$$

$$M = [0\ 0\ M_s], \quad x \in \mathbb{R}^n, \quad n \triangleq n_a + n_p + n_s$$

2.1 Controllability, Observability Properties

Suppose the plant (A_p, B_p) is controllable. It is of interest to know whether the addition of dynamic sensors and actuators will render the system uncontrollable. This question is resolved by the following result.

Theorem 1:

Suppose A_a, A_p, A_s have no common eigenvalues, and $G_a(s) \triangleq C_a(sI-A_a)^{-1}B_a$, $G_p(s) \triangleq C_p(sI-A_p)^{-1}B_p$, $G_s(s) \triangleq C_s(sI-A_s)^{-1}B_s$. Then (A, B) in (2.1b) is a controllable pair if and only if

i) (A_a, B_a) is a controllable pair.
ii) $rank\ [\lambda I-A_p, B_p G_a(\lambda)] = n_p$ for all $\lambda = \lambda_i[A_p]$,
iii) $rank\ [\lambda I-A_s, B_s G_p(\lambda) G_a(\lambda)] = n_s$ for all $\lambda = \lambda_i[A_s]$.

The proof is an extension of the results in [20] where it is proved that the tandem connection of two dynamic system (A_1, B_1, C_1) and (A_2, B_2, C_2) with transfer functions $G_1(s)$ and $G_2(s)$ respectively, is controllable if and only if (A_1, B_1) is controllable and rank $[\lambda I-A_2, B_2 G_1(\lambda)] = \dim(A_2)$, where $\lambda = \lambda_i[A_1]$, assuming $\lambda_i[A_1] \neq \lambda_j[A_2]$ for any i, j. To prove theorem 1, we need only to apply these results to the tandem connection of *three* dynamic systems $G_a(s), G_p(s), G_s(s)$. The reader will recognize i) and ii) as the necessary and sufficient conditions for controllability of the tandem connection of the actuators and plant $G_a(s)$ and $G_p(s)$. Condition (iii) readily follows by grouping the plant and the sensors as the "second" system $G_2(s) = G_p(s)G_a(s)$ and applying the theorem [20] using $G_1(s) = G_a(s)$.   #

Corollary to Theorem 1:

(A, B) is controllable if (A_a, B_a), (A_p, B_p), (A_s, B_s) are all controllable pairs and
a) rank $[G_a(\lambda)] = a$ for all $\lambda = \lambda_i[A_p]$.
b) rank $[G_p(\lambda)G_a(\lambda)] = p$ for all $\lambda = \lambda_i[A_s]$.

Proof:

Conditions a) and b) will not decrease the column rank of $B_p G_a(\lambda)$ or $B_s G_p(\lambda)G_a(\lambda)$ below that of B_p, B_s respectively. Theorem 1, and full row rank of $[\lambda I - A_p, B_p]$, $[\lambda I - A_s, B_s]$ is equivalent to controllability of (A_p, B_p) and (A_s, B_s).   #

Quite often complete controllability is not required. The following conditions allow the plant x_p to be controllable even if the entire system is not controllable.

Theorem 2

The plant vector x_p is completely controllable if and only if (A_p, B_{pa}) is a controllable pair where $B_{pa} \triangleq B_p C_a$.

Proof:

By definition the plant state x_p is controllable if x_p, taken as an output of (2.1), is "output controllable." This requires

$$\text{rank}[C_x B, C_x AB, C_x A^2 B, \cdots C_x A^{n-1} B] = n_p \qquad (2.2)$$

where

$$A = \begin{bmatrix} A_a & 0 & 0 \\ B_p C_a & A_p & 0 \\ 0 & B_s C_p & A_s \end{bmatrix}, \quad B = \begin{bmatrix} B_a \\ 0 \\ 0 \end{bmatrix} \qquad (2.3)$$

$$C_x = [0 \; I \; 0].$$

Now (2.2) becomes, upon substitutions,

$$\text{rank}[B_p C_a, A_p B_p C_a, A_p^2 B_p C_a, ..., A_p^{n-1} B_p C_a].$$

This concludes the proof.   #

Remark: Note that the plant x_p is controllable if (A_p, B_p) is controllable and C_a has full row rank.

2.2 Observability

The sensor and actuator dynamics can also hinder our ability to observe the entire system from the outputs of the sensors. The following result sorts out these circumstances.

Theorem 3:

Suppose A_a, A_p, A_s have no common eigenvalues. Then (A, M) in (2.1) is an observable pair if and only if

i) (A_s, C_s) is an observable pair

ii) $\text{rank} \begin{bmatrix} \lambda I - A_a \\ G_s(\lambda)G_p(\lambda)C_a \end{bmatrix} = n_a \quad \text{for all} \quad \lambda = \lambda_i[A_p], \lambda_i[A_a],$

iii) $\text{rank} \begin{bmatrix} \lambda I - A_p \\ G_s(\lambda)C_p \end{bmatrix} = n_p, \quad \text{for all} \quad \lambda = \lambda_i[A_s].$

The proof is based upon this result from [ ]: If A_1 and A_2 have no common eigenvalues then the tandem connection of (A_1, B_1, C_1) and (A_2, B_2, C_2) is observable if and only if 1) (A_2, C_2) is observable and, 2) $\text{rank} \begin{bmatrix} \lambda I - A_1 \\ G_2(\lambda)C_1 \end{bmatrix} = \dim A_1$ for all $\lambda = \lambda_i[A_2]$. Now let (A_a, B_a, C_a) represent (A_1, B_1, C_1) and the randem connection of (A_p, B_p, C_p) and (A_s, B_s, C_s) represent system (A_2, B_2, C_2). Hence, $G_2(s) = G_s(s)G_p(s)$, and $G_1(s) = G_a(s)$. It follows that the observability of $G_1(s)$ and $G_2(s)$ requires 1) observability of $G_2(s)$ [written as (A_{ps}, C_{ps})] and the observability of $G_2(s)$ requires i) and iii). #

Remark: (A, M) is observable if (A_a, C_a), (A_p, C_p) and (A_s, C_s) are observable pairs and $G_s(\lambda)$ has linearly independent columns for $\lambda = \lambda_i[A_s]$, $i = 1, 2, ..., n_s$, and $G_s(\beta)G_p(\beta)$ has linearly independent columns for $\beta = \lambda_i[A_p]$, $i = 1, 2, ..., n_p$ and $\beta = \lambda_i[A_s]$, $i = 1, 2, \cdots n_s$.

2.3 Defining the Cost Function

With the properties of the augmented system established, optimal control design for the augmented system is now considered. In the augmented system (2.1), the actuator *command* is given by u(t), the actuator *response* $y_a(t)$ (contained in the augmented output vector y) is distinct from u(t) due to actuator dynamics. We wish to weight both the input and the output of the actuators. For this reason, and in view of the relation of $y_a(t)$ to the design goals as discussed above, minimization of cost functions of the form

$$V = E_\infty [\|y(t)\|_Q^2 + \|u(t)\|_R^2] \qquad (2.4a)$$

and

$$Q = \text{diag}[Q_a, Q_p], \quad Q > 0 \tag{2.4b}$$

provides a stable optimal closed-loop solution, provided (A, C) (A, M) are detectable, (A, D) and (A, B) are stabilizable.

3.0 SELECTION OF DYNAMIC SENSORS AND ACTUATORS
3.1 Closed-Loop Input/Output Cost Analysis

In order to write the expressions for the closed-loop input and output costs, it is first necessary to put the fully augmented system, under closed loop steady-state optimal state-estimate feedback control, in the following state space form:

$$\dot{x}(t) = Ax(t) + Dw(t) \tag{3.1a}$$

$$y(t) = Cx(t) \tag{3.1b}$$

$$V = E_\infty V_o(t), \quad V_o(t) = y^*(t) Q y(t), \tag{3.1c}$$

where

$$x^T = [x^T, x_c^T], \quad y^T = [y]^T \quad w^T = [w^T, v^T] \tag{3.1d}$$

$$A = \begin{bmatrix} A & BG \\ FM & A+BG-FM \end{bmatrix}, D = \begin{bmatrix} D & 0 \\ 0 & F \end{bmatrix}, C = \begin{bmatrix} C & 0 \\ 0 & G \end{bmatrix}, Q = \begin{bmatrix} Q & 0 \\ 0 & R \end{bmatrix}, W = \begin{bmatrix} W & U \\ U^* & V \end{bmatrix} \tag{3.1e}$$

$$G = -R^{-1} B^T K, \quad 0 = KA + A^T K - KBR^{-1} B^T K + C^T Q C \tag{3.1f}$$

$$F = [PM^T + DU]V^{-1}, \quad 0 = [A - DUV^{-1}M]P + P[A - DUV^{-1}M]^T \tag{3.1g}$$

$$- PM^T V^{-1} MP + DWD^T - DUV^{-1} U^T D^T$$

Equation (3.1a) describes a linear system driven by zero mean white noise. The contribution of the ith input $w_i(t)$ in the total cost function (3.1c) is called the "input cost". The contribution of the ith output $y_i(t)$ in (3.1b) in the total cost function (3.1c) is called the "output cost." Formulas for the input and output costs were first derived in [2] and we shall cite the essential results that will be needed here.

For the system (3.1) the "output costs" V_i^y, defined by

$$V_i^y \triangleq (1/2)\{E_\infty (\partial V_o / \partial y_i) y_i\} \tag{3.2a}$$

are calculated as follows [2]

$$V_i^y = [CXC^T Q]_{ii} \tag{3.2b}$$

where X is the steady state covariance satisfying

$$0 = AX + XA^T + DWD^T \tag{3.2c}$$

and where the output costs satisfy the cost decomposition property

$$\sum_{i=1}^{n_y} V_i^y = V . \tag{3.2d}$$

The "input costs" are defined by

$$V_i^w \triangleq (1/2)\{E_\infty(\partial V_o/\partial w_i)w_i\} \tag{3.3a}$$

and are found from [2]

$$V_i^w = [D^T SDW]_{ii} \tag{3.3b}$$

where S satisfies

$$0 = A^T S + SA + C^T QC \tag{3.3c}$$

and where the input costs also satisfy the cost decomposition property

$$\sum_{i=1}^{n_w} V_i^w = V . \tag{3.3d}$$

The input and output costs represent the *in situ* contributions that the noise inputs and the system outputs make in the cost function. We may also wish to know the amount by which the cost function will be reduced if a noise input is eliminated. This amount, ΔV_i^w, is defined as

$$\Delta V_i^w = V - V_{Ri} \tag{3.4}$$

where V_{Ri} is the value of the cost function after the i_{th} *noise input* is eliminated, (but the controller is not redesigned) and ΔV_i^w is the cost reduction due to eliminating w_i. A positive value for ΔV_i^w indicates that elimination of the i_{th} input will *reduce* the cost, while negative ΔV_i^w indicates that a cost *increase* will follow noise elimination. It was shown in [1] that the ΔV_i^w may be positive or negative in the presence of noise correlation. (Hence, the concept of "beneficial noise" in linear systems).

Partitioning the matrices W and D facilitates direct solution for the cost reduction, yielding

$$\Delta V_i^w = 2V_i^w - d_i^* S d_i W_{ii} . \tag{3.5}$$

The closed-loop covariance X may be written

$$X = \begin{bmatrix} P+N & N \\ N & N \end{bmatrix} \tag{3.6}$$

where P satisfies eqn (3.1g) and where N satisfies:

$$0 = N(A+BG)^T + (A+BG)N + FVF^T \qquad (3.7)$$

Also, S has the following form

$$S = \begin{bmatrix} K+L & -L \\ -L & L \end{bmatrix} \qquad (3.8)$$

where K satisfies eqn (3.1f) and where L satisfies

$$0 = L(A-FM) + (A-FM)^T L + G^T RG \qquad (3.9)$$

For notational convenience the steady state covariance X is partitioned as follows:

$$X = [P+N] = \begin{bmatrix} X_p & X_{12} & X_{13} \\ X_{12}^T & X_a & X_{23} \\ X_{13}^T & X_{23}^T & X_s \end{bmatrix} \qquad (3.10)$$

Using the notation of (3.10) and the special structure of the closed-loop system matrices in eqn (3.13) the following expressions may be derived [2] for the output costs

$$V_i^{y_p} = [C_p X_p C_p^T Q_p]_{ii} \quad i = 1, \cdots n_{y_p} \qquad (3.11a)$$

$$V_i^{y_a} = [C_a X_a C_a^T Q_a]_{ii} \quad i = 1, \cdots n_u \qquad (3.11b)$$

$$V_i^u = [GNG^T R]_{ii} \quad i = 1, n_u \qquad (3.11c)$$

and for the input costs

$$V_i^w = [D^T(K+L)DW]_{ii} \qquad i = 1, \cdots n_w \qquad (3.12a)$$

$$V_i^{v_{in}} = [D^T(K+L)DW]_{n_w+i, n_w+i} \quad i = 1, \cdots n_z \qquad (3.12b)$$

$$V_i^{v_{out}} = [F^T LFV]_{ii} \qquad i = 1, \cdots n_z \qquad (3.12c)$$

and the input cost reductions

$$\Delta V_i^w = [D^T(K+L)DW - D^T LFU^T]_{ii} \qquad i = 1, \cdots n_w \qquad (3.13a)$$

$$\Delta V_i^{v_{in}} = [D^T(K+L)DW - D^T LFU^T]_{n_w+i, n_w+i} \quad i = 1, \cdots n_z \qquad (3.13b)$$

$$\Delta V_i^{v_{out}} = [F^T LFV - F^T LFV - F^T LBU]_{ii}. \qquad i = 1, \cdots n_z \qquad (3.13c)$$

A straightforward approach to the selection of sensors and actuators leads to integer programming [23]. Due to the numerical intensity of this approach, we seek a suboptimal alternative. Equations (3.1)-(3.13) provide the ingredients to a "cost decomposition" approach which motivates our approach. However we shall not use the cost decomposition of [2], since it does not lead itself to the inclusion of sensor and actuator dynamics. We shall also modify the basic formulas of [2], since [2] does not utilize the ΔV information available in (3.13).

3.2 Dynamic Actuator Effectiveness Values

Now that the closed-loop input and output costs have been determined for systems with dynamic sensors and actuators, it remains to use the cost decomposition results (3.11-3.13) to define expressions which reflect the effectiveness of each sensor and actuator in the cost function. This section defines the effectiveness values for dynamic actuators. The approach we recommend for non-dynamic actuators is to subtract the contribution the i_{th} actuator's *noise* in the cost function from the contribution of its *control* signal, and to label this difference the "effectiveness" of the i_{th} actuator, V_i^{act}. That is,

$$V_i^{act} = V_i^u - \Delta V_i^w \qquad (3.14)$$

This subtracts the "bad" from the "good" contributions of the actuator to measure its effectiveness. This approach is different from [2] due to the ΔV_i^w term. In [2] only the V_i^w term from (3.3a) was used in (3.14). The results of applying (3.14) to sensor and actuator selection for a range of small and large scale examples have demonstrated the improvement of this approach.

Extending the definition (3.14) for applicability to systems with dynamic actuators, we proceed as follows. In (3.1) there are two noise sources associated with each actuator: command noise, w_u, which is filtered by the actuator dynamics; and output noise, w_f, which is additive with the actuator output. Thus, the noise contribution associated with the i_{th} actuator is given by the sum of $\Delta V_i^{w_u}$ and $\Delta V_i^{w_f}$.

The beneficial control cost for each actuator is not immediately evident. First, recall that *it is the actuator output* $y_a(t)$, *not its input* $u(t)$, *which drives the system.* Next, note that the contribution of the i_{th} actuator's *output* in the cost function, $V_i^{y_a}$, includes the effects of noise w_{ui}. That is, even in the open loop ($u \equiv 0$), $V_i^{y_a} \neq 0$ for $[W_u]_{ii} > 0$ with dynamics. Hence, to define the *beneficial* (control) portion of $V_i^{y_a}$ it is necessary to subtract the portion of $V_i^{y_a}$ which is due to noise. This can not be accomplished exactly, since the actuator command $u(t)$ and the command noise $w_u(\tau)$ are correlated for $t > \tau$. An approximation is obtained, however, by solving for $V_i^{y_a}$ when $u \equiv 0$ (that is, in the open loop). We define the contribution of w_{ui} to $V_i^{y_a}$ and the contribution of u_i to $V_i^{y_a}$ as follows, using the open loop covariance of the actuator states $\underline{X}_a$:

$$[V_i^{y_a}]^w = [C_a \underline{X}_a C_a^T Q_a]_{ii} \qquad (3.15a)$$

and

$$[V_i^{y_a}]^u = V_i^{y_a} - [V_i^{y_a}]^w = [C_a(X_a - \underline{X}_a) C_a^T Q_a]_{ii} \qquad (3.15b)$$

where $\underline{X}_a$ solves

$$0 = A_a \underline{X}_a + \underline{X}_a A_a^T + B_a W_u B_a^T . \tag{3.15c}$$

Finally, the input costs and the decomposition of the output cost $V_i^{y_\bullet}$ are combined in an effectiveness formula for dynamic actuators

$$V_i^{act} = [V_i^{y_\bullet}]^u - \Delta V_i^{Wout} - \Delta V_i^{Win} . \tag{3.16}$$

Note that in the absence of command input noise, $[V_i^{y_\bullet}]^W$ and V_i^{Win} are both zero. Also, in the absence of actuator dynamics, $y_{a_i}(t)$ is equivalent to $u_i(t)$. Thus the expression (3.16) reduces to the original effectiveness formula of [2] in the absence of actuator dynamics. Note also that (3.16) is applicable whether or not the actuator noise signals are correlated with other noise sources, and it is applicable to systems with actuator dynamics of arbitrary order.

3.3 Dynamic Sensor Effectiveness Values

Unlike the actuator noise, (which has a direct path to the output, independently of the controllers influence) the noise associated with sensors reaches the system only through the controller. Since the gains in the Kalman filter of the LQG controller represent an optimal trade-off of each sensor's (beneficial) measurement information versus the (performance degrading) impact of its noise, then a ΔV_i^v of large magnitude is indicative of a *highly effective sensor*. That is, the fact that a sensor's noise is being allowed to heavily affect the cost means that its measurement information is even more critical to performance. For this reason, the following effectiveness formula for non-dynamic sensors generalized to accommodate the possibility of noise correlation, was presented in [1]:

$$V_i^{sen} \triangleq |\Delta V_i^v| . \tag{3.17}$$

For dynamic sensors there are two possible noise inputs associated with each sensor. As in the non-dynamic case, both noise inputs reach the system dynamics through the controller dynamics. Thus a straightforward extension of (3.17) to dynamic sensors is

$$V_i^{sen} = |\Delta V_i^{Vin}| + |\Delta V_i^{Vout}| . \tag{3.18}$$

Note that this formula is applicable in the presence of sensor dynamics of arbitrary order, and applies whether or not any of the noise sources are correlated with one another.

These are new formulas and are quite different from the sensor and actuator effectiveness criteria suggested in [2]. Ref. [2] did not use ΔV information nor could [2] handle dynamic devices.

This section concludes with the suggestion that (3.16) and (3.18) provide effective measures of the contribution of each actuator and sensor in a closed loop optimal LQG control (with sensor and actuator dynamics properly included).

CONCLUSIONS

A new method of sensor and actuator selection (SAS) has been derived for application to systems with dynamic sensors and actuators -- that is, systems in which the response of the sensors and actuators to their inputs is not instantaneous but governed by dynamics. The extended SAS method is applicable to systems in which the sensor and actuator dynamics are of arbitrary order. Application to simple numerical examples in [18] demonstrates that there usually exists optimal dynamics (an actuator can be too fast and too slow). This raises new research questions on the optimum *component design* in large scale systems.

Application of the actuator selection method in detail to NASA's SCOLE space structure demonstrated that even uniform actuator dynamics can affect the optimal selection of actuators.

LIST OF REFERENCES

[1] Skelton, R.E., and Norris, G.A., "Selection of Noisy Sensors and Actuators in the Presence of Correlated Noise," *Journal of Control Theory and Advanced Technology,* to appear.

[2] Skelton, R.E., and DeLorenzo, M.L., "Selection of Noisy Actuators and Sensors in Linear Stochastic Systems," *Journal of Large Scale Systems, Theory and Applications,* Vol. 4, April 1983, pp. 109-136.

[3] Skelton, R.E., and DeLorenzo, M.L., "Space Structure Control Design by Variance Assignment," *Journal of Guidance and Control,* Vol. 8, July-August 1985, pp. 454-462.

[4] DeLorenzo, M.L., "Selection of Noisy Sensors and Actuators for Regulation of Linear Systems," Ph.D. Thesis, School of Aeronautics and Astronautics, Purdue University, West Lafayette, IN, May 1983.

[5] Skelton, R.E., and Hughes, P.C., "Modal Cost Analysis for Linear Matrix-Second-Order Systems," *Journal of Dynamic Systems, Measurement, and Control,* Vol. 102, Sept. 1980.

[6] Howell, K.C., and Baxter, M.J., "Some Considerations of Actuator Dynamics in the Attitude Control of a Flexible Beam," (AIAA/AAS Paper 86-2124).

[7] Chiu, J.D. and Skelton, R.E., "Optimal Selection of Inputs and Outputs in Linear Stochastic Systems," *J. Astronautical Sciences,* Vol. XXXI, No. 3, pp. 399-414, July-Sept. 1983.

[8] Goh, C.J., and Caughey, T.K., "On the Stability Problem Caused by Finite Actuator Dynamics in the Collocated Control of Large Space Structures," *Int. J. Control,* Vol. 41, No. 3, 1985, pp. 787-802.

[9] Balas, M.J., "Feedback Control of Flexible Systems," *IEEE Transactions on Automatic Control,* Vol. AC-23, No. 4, 1978, pp. 673-679.

[10] Malandrakis, C.G., "Optimal Sensor and Controller Allocation for a Class of Distributed Parameter Systems," *Int. J. Syst. Sci.,* Vol. 10, No. 5, pp. 463-480, Sept. 1980.

[11] Ichikawa, A. and Ryan, E.P., "Filtering and Control of Distributed Parameter Systems with Point Observations and Inputs," *Proc. of the 2nd IFAC Symp. on Control of D.P.S.*, pp. 347-357, Coventry, Jul. 1977.

[12] Ichikawa, A. and Ryan, E.P., "Sensor and Controller Location Problems for Distributed Parameter Systems," *Automatica*, Vol. 15, No. 3, pp. 347-352, May 1979.

[13] Hughes, P.C., "Space Structure Vibration Modes: How Many Exist? Which ones are Important?" *IEEE Control Systems Magazine*, February 1987, pp. 22-28.

[14] Schaechter, D.B., "Control Technology Development," *NASA Langley Research Center LSS Tech.*, Mar 1982, pp. 297-311.

[15] Taylor, L.W., and Balakrishnan, A.V., "A Mathematical Problem and a Spacecraft Control Laboratory Experiment (SCOLE) Used to Evaluate Control Laws for Flexible Spacecraft ... NASA/IEEE Design Challenge," January 1984, unpublished, available from Lawrence W. Taylor, Jr., Spacecraft Control Branch, NASA Langley Research Center, Hampton, VA, 23665.

[16] Hotz, A.F., Collins, E., and Skelton, R.E., "Linearized Dynamic Model for the NASA/IEEE SCOLE Configuration," NASA Contractor Report 172394, Langley Research Center, Hampton, VA, Sept. 1984.

[17] King, A.M., Norris, G.A., and Skelton, R.E., "Controller Design for Vibration and Shape Control of an Offset Reflector Satellite," contractor report to SPARTA, Inc., May 1986.

[18] Norris, G.A., "Selection of Non-Ideal Noisy Actuators and Sensors in the Control of Linear Systems," Master's Thesis, School of Aeronautics and Astronautics, Purdue University, West Lafayette, IN, May 1987.

[19] McClamrock, H. "Control of Large Space Structures Using Electro-Mechanical Actuators," CSDL-P-1607, The Charles Starke Draper Lab., Inc., Cambridge, Mass., July 1982.

[20] Hautus, M.L.J., "Input Regularity of Cascaded Systems," *IEEE Trans. Auto. Control*, Vol. AC-20, No. 1, Feb. 1975, pp. 120-123.

[21] Hughes, P.C., and R.E. Skelton, "Controllability and Observability of Linear Matrix Second Order Systems," *J. Applied Mechanics*, Vol. 47, June 1980, pp. 452-459.

[22] Laskin, R.A., R.W. Longman, and P.W. Likins, "Actuator Placement in Modal Systems Using Bounded-Time Fuel-Optimal Degree of Controllability," *Proceedings 20th Annual Allerton Conference on Comm. Control and Computing*, Oct. 6-8, 1982, pp. 813-822.

[23] Chen, W.H., and J.H. Seinfeld, "Optimal Location of Process Measurements." *Int. J. Control*, 21, 6, 1003-1014, 1975.

Aerospace

A Simple Active Controller to Supress Helicopter Air Resonance in Hover and Forward Flight

P. P. Friedmann and M. D. Takahashi

Mechanical, Aerospace, and Nuclear Engineering Department
University of California, Los Angeles, California, U.S.A.

Summary
A coupled rotor/fuselage helicopter analysis with the important effects of blade torsional flexibility, unsteady aerodynamics, and forward flight is presented. This model is used to illustrate the effect of unsteady aerodynamics, forward flight, and torsional flexibility on air resonance. Next a nominal configuration, which experiences air resonance in forward flight, is selected. A simple multivariable compensator using conventional swashplate inputs and a single body roll rate measurement is then designed. The controller design is based on a linear estimator in conjunction with optimal feedback gains, and the design is done in the frequency domain using the Loop Transfer Recovery method. The controller is shown to suppress the air resonance instability throughout wide range helicopter loading conditions and forward flight speeds.

Nomenclature
Variables with an overbar are dimensional. Unless otherwise stated, variables without an overbar are non-dimensionalized by the blade mass $\overline{M}_B$, rotor radius $\overline{R}$, and the rotor rate $\overline{\Omega}$.

a	Rotor blade lift curve slope
a_T	Horizontal tail lift curve slope
AR	Horizontal tail aspect ratio
A, B, C	First order system, control, and output matrices
b	Blade semi chord
C_{d0}	Blade drag coefficient
C_{d0T}	Horizontal tail drag coefficient
e	Hinge offset
f	Fuselage drag area = $\overline{f}/2\overline{b}\overline{R}$
FF^T, GG^T	State and observation noise covariances
G(s), K(s)	System and compensator matrices
I_b	Blade flap inertia about hinge offset
I_{Cxx}, I_{Cyy}	Fuselage roll and pitch inertias
J_x	Blade pitch inertia
J_y, J_z	Integral of the blade flap and lead-lag bending inertias
K_c, K_f	Feedback and filter gains
K_x, K_y, K_z	Flap, lag, and torsion spring constants
l	Blade length
l_m	Model error function
L(s)	Unstructured multiplicative error matrix
M_F	Fuselage mass
N_b	Number of blades
P_c, P_f	Positive semi-definite solutions to the Riccati equation
q_c	Recovery factor
Q, R	State weight and control weight matrices
R_c	Elastic coupling coefficient
R_{Mx}, R_{My}, R_{Mz}	Translational degrees of freedom of the fuselage

$S(s), T(s)$	Sensitivity and command response transfer matrices
S_T	Horizontal tail area
V	Forward flight speed
w_s, w_o	State and observation noise processes
x_A	Blade aerodynamic center offset from the blade elastic axis
x_b, y_b, z_b	Position of the blade center of mass from the hinge offset
X_{MC}, Z_{MC}	X and Z position of the fuselage center of mass
X_{MH}, Z_{MH}	X and Z position of the rotor hub center from point M
X_{MT}, Z_{MT}	X and Z position of the horizontal tail a.c. from point M
x, u, y	System state, control, and input vectors
$\hat{x}, \hat{y}$	Estimator state and output vectors
α_R	Rotor trim pitch angle
β_p	Blade precone angle
γ	Lock number
$\theta_0, \theta_{1s}, \theta_{1c}$	Collective, sine, and cosine inputs
θ_{pk}	Pitch of k-th blade
σ	Solidity ratio $= 2N_b b/\pi$
μ	Advance ratio $= \overline{V} \cos(\alpha_R)/\overline{R\Omega}$
ψ_k	K-th blade angle $= \psi + (k-1)2\pi/N_b$
ψ	Azimuth angle of blade measure from straight aft position
ω_c	Cross over frequency
ω_L	Inplane lead-lag frequency
$\omega_{F1}, \omega_{L1}, \omega_{T1}$	Rotating first flap, lag, and torsional blade frequencies
$\underline{\sigma}[\bullet], \bar{\sigma}[\bullet]$	Mimimum and Maximum singular values
$(\dot{\bullet})$	Derivative wrt to the azimuth angle
MIMO	Multiple Input/Multiple Output
SISO	Single input/Single output
LTR	Loop Transfer Recovery

Introduction

The need to reduce the mechanical complexity and weight of the rotor hub on helicopters has generated considerable interest in hingeless and bearingless rotors. Though these new rotor configurations are simple and lightweight they can experience other undesirable dynamic problems. One important problem that can arise in soft-in-plane rotor systems is termed "air resonance", and is a condition where the blade lead-lag motions strongly interact with the fuselage pitch or roll motion in flight [1,2]. This aeromechanical phenomenon produces large fuselage oscillations and is clearly undesirable when unstable or weakly stable. The approach to suppressing ground resonance in articulated rotor systems has been through lead-lag dampers for each rotor blade. This approach can also be applied to air resonance of hingeless rotors systems, but this solution tends to destroy the mechanical simplicity and aerodynamic cleanliness inherent in hingeless and bearingless rotors. Another possible means of stabilizing or augmenting stability of air resonance is through an active controller operating with a conventional swashplate. This approach is feasible from a practical point of view only if it is simple to implement since it must compete against the straightforward mechanical solution to this problem based on lag dampers. Such an active controller would need sensing and actuating devices leading to an expensive system. However, with the inevitable introduction of other active control devices such as higher harmonic control (HHC) for vibration suppression [3,4,5] this argument is considerably weakened. Vibration control requires sensors and actuators with bandwidths well above the 1/rev frequency. Since the air resonance instability results in an unstable lead-lag regressing mode (i.e. the mode associated with the $|1 - \omega_L|$ frequency) these devices would also be sufficient for air resonance control. Thus, sensing and actuator hardware, which may be already available, could be used for additional purposes below the frequency range intended for the vibration control objective.

Research in the active control of air and ground resonance has been limited to a few studies [6, 7, 8], where various theoretical active control studies were presented. The helicopter models used in these studies were quite limited since important effects such as torsional flexibility of the rotor blades, forward flight, and unsteady aerodynamic were all neglected. Furthermore, the studies dealing with the active control of air resonance did not adequately demonstrate the ability of the control schemes to operate through the wide range of operating conditions which can normally be encountered. The primary objectives of this paper are:

(1) To illustrate the importance of unsteady aerodynamics, blade torsional flexibility and the role of periodic coefficients (or forward flight) on this problem.

(2) To remove the limitations inherent in previous studies by using a coupled rotor/fuselage model, in which the important effects of forward flight, unsteady aerodynanics, and blade torsional flexibility are included.

(3) To demonstrate the feasibility of designing a simple active controller capable of suppressing air resonance throughout the flight envelope representive of the wide range of operating conditions which may be encountered by a helicopter.

Mathematical Model

The mathematical model of the rotor/fuselage system is that of Ref. 9 and 10, and its salient features are described next. The fuselage is represented as a rigid body with five degrees of freedom, where three of these are linear translations and two are angular positions of pitch and roll (Fig. 1). Yaw is ignored since its effect in the air resonance problem is known to be small. A simple offset hinged spring restrained rigid blade model is used to represent a hingeless rotor blade (Fig. 2). This assumption simplifies the equations of motion, while retaining the essential features of the air resonance problem. In this model, the blade elasticity is concentrated at a single point called the hinge offset point, and torsional springs are used to represent this flexibility. The dynamic behavior of the rotor blade is represented by three degrees of freedom for each blade, which are flap, lag, and torsion motions. The aerodynamic loads of the rotor blades are based on quasi-steady Greenberg's theory, which is a two dimensional potential flow strip theory [11, 12]. Compressibility and dynamic stall effects are neglected, though they could be important at high advance ratios. Greenberg's theory is an extension of Theodorsen theory, which accounts for a time dependent lead-lag motion and constant collective pitch of the blade. Unsteady aerodynamic effects, which are created by the time dependent wake shed by the airfoil as it undergoes arbitrary time dependent motion, are accounted for by using a dynamic inflow model. This simple model uses a third order set of linear differential equations driven by pertubations in the aerodynamic thrust, roll moment, and pitch moment at the rotor hub. The three states of these equations describe the behavior of perturbations in the induced inflow through the rotor plane. The model coefficients used in this paper are those of Ref. 13.

The equations of motion of the coupled rotor/fuselage system are very large and contain geometrically nonlinear terms due to moderate blade deflections in the aerodynamic, inertial, and structural forces. Furthermore, the coupled rotor/fuselage equations have additional complexity due to the presence of the fuselage degrees of freedom. To reduce the equations to a manageable size, an ordering scheme is used in the derivation of the equations of motion to systematically remove the higher order nonlinear terms [14]. The ordering scheme is based on the assumption that

$$1 + O(\varepsilon^2) \simeq 1 \tag{1}$$

which states that terms of order ε^2 are negligible relative to terms of order unity. The term ε is a non-dimensional parameter, which quantifies the meaning of a "small" term. For our purposes, it represents the slopes of the deflections of the blades, which usually are of an order of magnitude which is less than .15. The blade degrees of freedom are assigned an order of ε, while the fuselage degrees of freedom are of order $\varepsilon^{3/2}$. A symbolic manipulation program is then used to generate the nonlinear set of equations of the rotor/fuselage system using the ordering scheme. Five fuselage equations result of which three enforce the fuselage translational equilibrium and two enforce the roll and pitch equilibrium. The three resulting rotor blade equations are associated with the flap, lag, and torsional motions of each blade. Also, the aerodynamic thrust and roll moments at the hub center are determined for the perturbation aerodynamics in the dynamic inflow equation. All of these equations can be found in detail in Ref. 9.

The active control to suppress the air resonance instability is implemented through a conventional swashplate. The pitch of the k-th rotor blade is given by the expression

$$\theta_{pk} = (\theta_0 + \Delta\theta_0) + (\theta_{1c} + \Delta\theta_{1c})\cos(\psi_k) + (\theta_{1s} + \Delta\theta_{1s})\sin(\psi_k) \tag{2}$$

The Δ terms are small and these represent the active control inputs, while those without Δ are the inputs necessary to trim the vehicle.

The stability of the system is determined through the linearization of the equations of motion about a blade equilibrium solution and the helicopter trim solution. The helicopter trim and equilibrium solution are extracted simultaneously using harmonic balance for a straight and level flight condition [10]. After linearization, a multi-blade coordinate transformation is applied, which transforms the set of rotating blade degrees of freedom to a set of hub fixed non-rotating coordinates [15]. This transformation is introduced to take advantage of the favorable properties of the non-rotating coordinate representation. The original representation has periodic coefficients with a fundamental frequency of unity, however, the transformed system has coefficients with a higher fundamental frequency. These higher frequency periodic terms have a reduced influence on the behavior of the system and can be ignored in some analyses at low advance ratios [14]. In hover, the original system has periodic coefficients with a frequency of unity, but the transformed system has constant coefficients.

Once the transformation is carried out, the system is rewritten in first order form.

$$\dot{x} = A(\psi)x + B(\psi)u \tag{3}$$

The fundamental frequency of the coefficient matrices depends on the number of rotor blades. For an odd bladed system the fundamental frequency is N_b per revolution, while for an even bladed system the fundamental frequency is $N_b/2$ per revolution [15]. Stability can now be determined using either an eigenvalue analysis or Floquet theory for the periodic problem in forward flight. An approximate stability analysis in forward flight is also possible by performing an eigen analysis on the constant coefficient portion of the system matrices in Eq. (3).

The mathematical model was carefully tested by comparing results to other investigators' analytical and experimental results. The correlation with these results was good and verified that the effects of torsion, unsteady aerodynamics, and forward flight were accurately represented in the model [9, 10].

Influence of New Modeling Effects on the Helicopter Configuration

The configuration used in this paper is the same as the "Nominal Configuration" in Ref. 10, and the data for the configuration is shown in Table 1. The parameters are selected so as to yield a nominal configuration somewhat similar to the MBB 105 helicopter in size and weight. The nominal configuration differs from the MBB 105 in that it has an unstable air resonance mode, which was induced by adjustments in some rotor and body parameters. The system has 37 states. The five body degrees of freedom and the twelve rotor degrees of freedom (three degrees of freedom for each blade) produce 34 position and rate states. The dynamic inflow model augments the system with three more states giving a total system order of 37. Figure 3 shows the pole locations in the s-plane of the dominant modes of the nominal configuration at $\mu = 0.3$. The lead-lag regressing mode is associated with the air resonance instability and is mildly unstable in this flight condition. It is with the body roll mode that the lead-lag regressing mode interacts. Thus, for this particular configuration, the dominant body motion of the instability is the rolling motion of the fuselage.

TABLE 1
Data of the nominal configuration.

Characteristic Dimensions
    Blade mass = 52 kg
    Rotor radius = 4.9 m
    Rotor rate = 425 RPM

Rotor Data
    $l = .85$ $e = .15$ $\omega_{F1} = 1.15$ at zero pitch
    $x_b = .36$ $\gamma = 5.0$ $\omega_{L1} = .620$
    $I_b = .18$ $C_{d0} = .01$ $\omega_{T1} = 3.00$
    $J_x = .00015$ $a = 5.90$ .5 percent damping
    $J_y = 0.$ $x_A = 0.$ $\sigma = .07$
    $J_z = .00015$ $y_b = 0.$ $R_c = 1.0$
    $\beta_p = 0.$ $b = 0.02749$ $N_b = 4.0$

Fuselage Data
    $M_F = 32.$ $f = .60$
    $I_{Cxx} = 1.0$ $Z_{MH} = .2667$
    $I_{Cyy} = 4.0$ $Z_{MC} = .0333$

Horizontal Tail Data
    $X_{MT} = 1.0$ $a_T = 5.0$
    $S_T = .04$ $C_{d0T} = .007$
    $AR = 5.5$

Figure 4 illustrates the influence of unsteady aerodynamics as well as the the effect of periodic coefficients (or forward flight) on the lead-lag regressing mode damping of the open loop configuration. The two sets of curves represent air resonance damping of the configuration with quasi-steady aerodynamics and with dynamic inflow at various advance ratios. Dynamic inflow captures primarily the low frequency unsteady aerodynamic effect which is known to be important for coupled rotor/fuselage aeromechanical problems such as air resonance. The stabilizing effect of forward flight, which is evident in the figure, is consistent with behavior observed in previous studies [14]. For hover, the system has constant coefficients and thus the constant coefficient approximation and the periodic system produce the same results, as is clearly evident in the figure. It is also evident from the figure that the effect of periodic coefficients is relatively minor. The quasisteady aerodynamic model produces a more stable system than the model which includes the unsteady aerodynamic effects as represented by the dynamic inflow model. It is also worthwhile mentioning that considerable differences between the two models exist particularily at low advance ratios.

Figure 5 shows that neglecting the torsional degree of freedom on the nominal configuration increases the instability of the lead-lag regressing mode. The trend of the two curves also tends to diverge at high advance ratios. The addition of torsion also tends to amplify the effect of the periodic terms. At high values of advance ratio, the flap-lag-torsion model shows a much greater difference between the constant and periodic stability analysis than does the flap-lag analysis. Additional results of other effects can be found in Refs. 9 and 10. Furthermore, in Ref. 10, preliminary control studies were conducted on the configuration at the nominal weight to assess the importance of various modeling effects. In these studies, simple full state feedback from the linear deterministic optimal regulator problem was used [16]. The relevant results that will be used throughout this paper are:

(1) The torsional degree of freedom and unsteady aerodynamics are an important effect in an air resonance controller design model. Significant errors can arise in the closed loop damping if these effects are ignored.

(2) The collective pitch input is not important in controlling the air resonance instability in forward flight up to $\mu = .4$.

(3) The periodic coefficients of the linearized model have a small effect on the open and closed loop damping of the air resonance mode for advance ratios up to 0.4. Thus, the constant coefficient approximation of the model should be sufficient for the initial control design.

The feasibility of using a simple controller to suppress the air resonance instability throughout a wide range of operating conditions is one of the primary objectives of this paper. To accomplish this, parameters must be varied and the stability of the closed loop system must be evaluated. The parameter variations considered in this paper are limited to those that change during the normal operation of the helicopter. Thus, the significant parameters are the advance ratio μ, fuselage mass M_F, fuselage inertias I_{Cxx}, I_{Cyy}, and the fuselage center of gravity position X_{MC} and Z_{MC}. Checking the stability for every combination of these parameters would require an excessive amount of labor. A more convenient approach consists of introducing approximate relations which govern the variations of I_{Cxx}, I_{Cyy}, Z_{MC}, and X_{MC}, resulting from practical combinations of fuel, cargo and passenger mass which may be encountered during the normal operation of the aircraft. These relations can be found in Ref. 9.

Since the preliminary studies revealed that the periodic terms are negligible, the stability analyses presented are based on the constant coefficient model (i.e. the constant portion of A and B of Eq. (3)), unless otherwise indicated. The open loop lead-lag regressing damping of the helicopter configuration throughout its the flight regime is shown in Fig. 6. The horizontal axis is the advance ratio, while the vertical axis is the fuselage mass non-dimensionalized by the blade mass of 52 kg. A non-dimensional fuselage mass of 32 plus four blades corresponds to the nominal total mass of 1872 kg. The figure indicates the system experiences an air resonance instability throughout most of the flight regime. Marginal stability exists at an advance ratio greater than .35 and the point of the deepest instability is at $M_F = 30$ and in the vicinity of hover.

Compensator Design Method

The controller aimed at suppressing air resonance in the flight envelope of the helicopter is based on an optimal state estimator in conjunction with optimal feedback gains [16]. A constant coefficient model is assumed since the results of the preliminary control studies [10] indicated a periodic model was unnecesary. Summarizing, we assume a linear system of the form

$$\dot{x} = Ax + Bu + w_s \qquad x \in R^n \qquad u \in R^m \qquad (4)$$

$$y = Cx + w_o \qquad y \in R^l \qquad (5)$$

where w_s and w_o are the state and observation noise processes. A few measurements y are used to drive the estimator

$$\dot{\hat{x}} = A\hat{x} + Bu + K_f(y - \hat{y}) \qquad (6)$$

$$\hat{y} = C\hat{x} \qquad (7)$$

$$K_f = P_f C^T (GG^T)^{-1} \qquad (8)$$

The optimal filter gains K_f come from the steady state Riccati equation

$$0 = AP_f + P_f A^T + FF^T - P_f C^T (GG^T)^{-1} CP_f \qquad (9)$$

where the state and observation noise processes are uncorrelated zero mean white noise processes with state and observation noise covariances FF^T and GG^T. The estimator states are then used to form the control law

$$u = -K_c \hat{x} = -R^{-1} B^T P_c \hat{x} \qquad (10)$$

The feedback gains are determined from the linear quadratic Guassian (LQG) optimal control problem which minimizes

$$J = E\{\lim_{T_f \to \infty} \frac{1}{T_f} \int_0^{T_f} [x^T Q x + u^T R u] \, dt\} \qquad (11)$$

The matrix Q is the positive semi-definite state weight matrix and R is the positive definite input weight matrix. The gains result from selecting these weight matrices and solving for the positive semi-definite solution of

$$0 = A^T P_c + P_c A + Q - P_c B R^{-1} B^T P_c \qquad (12)$$

which is the dual of the filter Riccati equation. In the s-plane, the estimator and optimal feedback gains form a compensator

$$K(s) = K_c(sI - A + BK_c + K_fC)^{-1}K_f \tag{13}$$

The approach outlined above is a powerful approach to feedback design, however, if the design model differs from the actual plant to be controlled, as is the case of any real system, poor performance and even instability can occur. The possibility of a controller lacking "robustness" is not surprising since no provision is made to account for uncertainty in the design process. In all applications, the design model and the actual plant to be controlled will have unavoidable differences due to the limitations associated with formulating models of physical systems. Our objective in this paper is to design a controller at an operating condition and require it to function adequately at the off design conditions. Thus, the differences between the design model and the actual plant to be controlled will be exacerbated. An additional drawback of the design method described above is that there are many possible choices of design variables in the covariance, state weight, and input weight matrices. This selection process is difficult without the use of important concepts (e.g. bandwidth) that have proved so useful in SISO time invariant linear control design [17]. To overcome these difficulties, the multivariable frequency domain design methods of Refs. 18, 19, and 20 are used. This will allow interpretation of the design process using frequency domain concepts and account for the possibility high frequency modeling error. Furthermore, this can be done while retaining the structure of the state space approach previously described.

With these points in mind, it is now necessary to discuss the design process for MIMO systems in the frequency domain. The general problem is one of designing a compensator K(s) to control the MIMO system $(G(s) + \Delta G(s))$ as shown in Fig. 7. How the compensator is selected is not important for this brief discussion, but for this paper it will be accomplished through the state estimator and optimal regulator by selecting the filter covariance and regulator weight matrices. In addition to meeting a given performance specification is the requirement that the controller do so in the presence of modeling errors represented by $\Delta G(s)$ in the figure. The specific representation of this error is in the form of an unstructured multiplicative uncertainty at the model output.

$$G(s) + \Delta G(s) = [1 + L(s)]G(s) \tag{14}$$

Other unstructured uncertainty models are available depending on the type of modeling errors one encounters, but for the objectives of this paper (14) is quite sufficient. What is of particular interest is the singular values of the uncertainty matrix L. In particular, the maximum singular values, which define the error function.

$$l_m(\omega) = \bar{\sigma}[L(j\omega)] \tag{15}$$

The error function $l_m(\omega)$ characterizes the magnitude of the modeling error at all frequencies. The maximum (minimum) singular value of any matrix A is the square root of the maximum (minimum) eigenvalue of the matrix product AA^H. The singular values are quantities used to characterize matrices as either "large" or "small". A small matrix is one with a small maximum singular value, while a large matrix is one with a large minimum singular value. For a system, a typical curve of $l_m(\omega)$ might look like Fig. 8a. A high modeling fidelity is shown at the low frequencies, but this fidelity gives way to large errors when the frequency becomes sufficiently large. Figure 8b indicates where the singular values of this system might be restricted for the error bound shown.

With this specific characterization of the error bound in the model, one can examine two fundamental aspects of control design, performance and stability. For the closed loop system in Fig. 7, a general statement of performance can be found by using the output sensitivity matrix given by

$$S(j\omega) = (I + G(j\omega)K(j\omega))^{-1} \quad (16)$$

Adjustment of the size of this function through K(s) affects the closed loop performance of the system. High gain results in small S, which gives a closed loop system that is less sensitive to the disturbance inputs and command inputs shown in Fig. 7. The closed-loop stability of the system is determined from the MIMO Nyquist stability criteria, which requires

$$\bar{\sigma}[T(j\omega)] = \bar{\sigma}[G(s)K(s)(I + G(s)K(s))^{-1}] < \frac{1}{l_m(\omega)} \quad (17)$$

Equation (17) guarantees stability, of course, if the error in the model is precisely as in Eq. (14). This stability requirement demands a low gain at frequencies where the model uncertainty is high. This is so, since a large error function necessitates small T(s), which in turn requires a small loop gain. Thus, meeting both performance and stability requirements is a task of adjusting the singular values of T to small values when modeling uncertainty is high and adjusting the singular values of S to small values to give good closed loop performance. These two criteria cannot be met simultaneously since both T and S cannot be both made arbitrarily small. This is easily seen when one considers that $S + T = I$ and a decrease in one always requires a increase in the other.

The design process can be carried out using plots of the singular values of T and S to adjust the size of each in the appropriate frequency range [19]. Alternately, this process can be visualized by using the singular values of the open loop feedforward cascade GK [18]. Figure 9 shows an example of the singular values of GK that have been placed between the bounds representing performance and stability requirements. The low frequency requirements are to make the lower singular values clear the performance requirements (high gain to produce small S). The frequency where the error function nears unity is where the loop cross over (i.e. where $\sigma_i[T] \simeq 1$) needs to be placed in order to avoid an instability due to modeling error, which can be seen from Eq. (17). When $\bar{\sigma}[T(j\omega)] << 1$, then $\bar{\sigma}[T(j\omega)] \simeq \bar{\sigma}[G(j\omega)K(j\omega)]$ and the maximum singular value of the open loop transfer function GK must be less than the inverse of l_m for stability to be maintained (low gain for small T).

A convenient means of achieving this loop shape selection is through Loop Transfer Recovery (LTR), which is outlined in Refs. 18 and 20. This method can be considered an optimal balancing of the contradictory requirements of good performance (high gain, small S(s)) and maintaining stability in the face of uncertainty (low gain, small T(s)). To discuss the method, the input weight matrix is chosen as $R = \rho_c^2 I$ and the state weight matrix is $Q = q_c^2 H H^T + Q_o$ in the optimal regulator. In the filter, the observation noise covariance is $E[w_o w_o^T] = \rho_f^2 I$ and the state noise covariance is $E[w_s w_s^T] = q_f^2 F F^T$. The first step in the method is the filter design, where ρ_f, q_f, and F are selected to give an optimal filter gain K_f. These values are selected to give a desired loop shape defined by the maximum and minimum singular values of the matrix $C(Is - A)^{-1} K_f$. The guidelines for selecting this loop shape are those which were previously described for selecting the loop shape G(s)K(s). Once the desired loop shape is determined, the second step is to recover it through the regulator by setting $H = C$ and letting q_c approach a large enough value. This second step is based on the result

$$G(s)K(s) \to C(sI - A)^{-1}K_f \text{ as } q_c \to \infty \tag{18}$$

for minimal phase $G(s)$ (i.e. all transmission zeroes in the left half of the s-plane) [18, 21]. With this result in mind, the statement " q_c large enough" simply means large enough to recover the desired loop shape that was selected in the filter design. The requirement that $G(s)$ be minimal phase is necessary since the recovery process of Eq. (18) inverts the plant dynamics making the zeroes of $G(s)$ the poles of the compensator $K(s)$. If the zeroes are in the right half of the s-plane, then with a large enough gain the compensator poles eventually become unstable. The procedure previously discussed is sometimes referred to as "Sensitivity Recovery" [16].

An important result of interpreting the loop recovery method as an optimal balancing of T and S is that the frequency regions where $C(sI - A)^{-1}K_f$ is large are regions of high penalty on the size of S. Choosing a loop shape that is large in a given frequency range results in small S, which is a region where good performance is expected. Thus, choosing a loop shape shape entails placing the peaks of the loop in the frequency region where "tight" control is desired. Two other properties of the LTR method are that the loop shape is guaranteed to have a high frequency runoff proportional to $1/\omega$ and it has guaranteed robustness properties at the loop cross over. Regarding the last item, if the error is precisely as in Eq. (14), then a 50 percent modeling error at the loop cross over is possible without destabilizing the closed loop system. These two properties are also useful in selecting the loop shapes of the control design. Since the model is expected to have higher errors in the high frequency regions, the loop runoff can be used to attenuate the effects of these errors by proper placement of the loop cross over frequency.

Controller Design

The design approach of this paper is to select an operating point to design a constant gain controller, and use this controller throughout the operating range of the helicopter. The design point is chosen to be in hover ($\mu = 0$) with the nominal weight ($M_F = 32$), which is a point near the region of worst instability for the configuration.

A single roll rate measurement of the fuselage and the sine and cosine swashplate inputs are chosen to control the instability. The selection of the inputs is based on the previous control studies, which demonstrated the ineffectiveness of the collective swashplate input in controlling the air resonance instability in forward flight [10]. The roll rate is selected as the measurement since it is this motion that the lead-lag motion of the blades interacts with during the air resonance instability. Examination of the eigenvectors of the unstable mode confirm this statement showing that the roll motion is dominant when compared to the pitch motion. The lead-lag degrees of freedom of the blades also could serve as measurement. However, it is preferrable to use measurements taken from a non-rotating reference frame (i.e. the frame of the fuselage). This avoids the problem of transmiting signals across the rotor head.

The full model with the given set of inputs and output is not minimal phase, which is a requirement of the loop recovery method for selecting design loop shapes. However, a reasonable minimal phase reduced model can be formulated that closely ressembles the full model input/output characteristics in the frequency range of interest. This is a perfectly acceptable practice provided that the the design model errors are considered during the loop shaping process [20]. The reduced model is formed by removing modes from the full model. This is accomplished by transforming the full system to block diagonal form and then striking out the states from the model that are associated with the undesirable modes. The design model that meets the minimal phase requirement is one consisting of the body roll, body pitch, lead-lag regressing, and the lead-lag progressing mode. The open loop poles of the design model are given in Table 2 and their order is the order of the modes in the model. The collective and differential lead-lag modes are near the frequency range of the air resonance instability (Fig. 3), but are not retained since they are uncontrollable and unobservable in the hover condition. The low frequency body modes are also not retained in the model since control over these modes is not the objective of this paper. These modes are in the frequency range of order .01/rev, which is a full decade below the frequency range of interest. Thus, it is assumed a high pass filter can be used if necessary to leave these modes unaffected by the air resonance controller. This would also prevent any interaction of the controller with pilot inputs or any Stability Augmentation System (SAS) on the vehicle for controlling the low frequency body modes [22].

TABLE 2

Open loop poles of the design model.

-.05231	$\pm$ j .16818	Body Pitch
.00439	$\pm$ j .37095	Lead-Lag Regressing
-.09223	$\pm$ j .40111	Body Roll
-.00634	$\pm$ j 1.8335	Lead-Lag Progressing

The design model is eighth order and closely ressembles the full model, as can be seen in the singular value plot of Fig. 10.. The model has only one output, so the maximum and minimum singular values of the system are the same. The reduced model is very close to the full model in the frequency range of interest capturing the peak due to the lead-lag regressing mode, which is the unstable mode that is to be controlled. The model also captures the sharp peak of the other dominant mode of the system, which is the lead-lag progressing mode. Naturally, removal of the higher and lower frequency modes produces the errors in these regions. The gain and phase plots for each input/output combination were also compared for the full and reduced models and they too showed good agreement up to a frequency near 1/rev. A reduced model is being used in the design and care must be exercised in the placement of the bandwidth of the controller. The reduced model is valid in the frequency range below 1/rev, so the crossover of the loop shape should not greatly exceed this value.

With the design model chosen, the next step might be to generate the error function by generating models at various operating points. However, this approach is of dubious value. In theory, once l_m is defined the closed loop stability can be checked through Eq. (17) without regenerating the models at the full range of operating conditions. Unfortunately, as stated before, Eq. (17) is only true if the errors in the system are precisely as indicated in Eq. (14) (i.e. a multiplicative unstructured uncertainty at the model output), which is not necessarily the case. Because of this, a check of the closed loop stability using the controller on the full model throughout the operating range is still necessary. In addition to the stability problem is the problem of evaluating the performance of the controller. The performance as indicated by the size of S (Eq. (16)) or by the singular value boundary (Fig. 9) are both useful for discussion purposes, but they are not a practical means of evaluating the performance of the problem of this paper. What is of real interest is the amount of damping in the the lead-lag regressing mode, which also requires that the design be checked at all of the operating points. Thus, instead of calculating the error function directly, an assumed error fuction is used to guide the design process through the ideas of loop cross over and loop run off. Stability and performance of the controller is then checked directly at all of the operating points. The assumed error function is to be of the form as in Fig. 8 with good model fidelity at low frequency and poor fidelity at the higher frequencies. The eventual cross over frequency of the loop GK is determined by the location at which the error function of the system becomes too large (i.e. $l_m \geq 1$). Obviously, it is assumed that the air resonance mode is adequately modeled, so a lower bound on the cross over frequency is at the instability frequency of .37. The upper limit on the crossover is limited to 1/rev due to the existence of the unavoidable 1/rev noise and the limitations in the reduced design model.

TABLE 3

Closed loop poles of controller A.

$-.44445 \pm j\, 2.0286$
$-.07980 \pm j\, 1.8333$ Lead-Lag Progressing
$-.30337 \pm j\, .37971$
$-.03659 \pm j\, .37472$
$-.00780 \pm j\, .37441$ Lead-Lag Regressing
$-.10205 \pm j\, .17774$
$-.06748 \pm j\, .17178$
-1.0290
-152.52

The first controller is designated controller A, and is chosen with the filter noise covariances given by $E[w_s w_s^T] = (.001)^2 I$ and $E[w_o w_o^T] = (.001)^2$. Since the state noise covariance is diagonal, all of the states have equal state noise disturbances entering into them. The regulator is chosen with weight matrices $R = I$ and $Q = .3^2 I + q_c^2 C^T C$. A recovery factor of $q_c = 10{,}000$ is sufficient to recover the loop shape show in Fig. 11. Examination of this loop shape shows two peaks near the lead-lag regressing frequency and the lead-lag progressing frequency. The cross over frequency of the first peak is near .73, which is well below the one per revolution requirement. The closed loop poles of the controller applied to the design model are given in Table 3. The lead-lag regressing mode is stabilized from an open loop damping of .00439 to a closed loop damping of -.00780, and the lead-lag progressing mode is shifted from -.0063 to -.080, which is beneficial though not necessary, since this mode was stable before the application of the controller. This shift in the lead-lag progressing mode is the result of the large gain seen in the peak near 1.8 in Fig. 11. Applying this controller to the full model yields a stable lead-lag regressing and progressing damping. Unfortunately, the flap progressing damping is strongly destabilized throughout most of the operating range of the vehicle. Figure 12 shows this mode is only marginally stable at a very high loading condition, and below $M_F = 37$ the mode is unstable for all advance ratios. The reason for this higher mode destabilization is the peak in the loop gain at 1.8, which is not necessary since the lead-lag progressing mode is not what needs to be stabilized. This particular loop shape places a high gain near the lead-lag progressing mode frequency, which is near a region where the design model begins to significantly deviate from the full model. From the discussion in the previous section on modeling uncertainty, it is clear that this choice of loop shape is a poor one.

The next controller, designated controller B, uses the same weight functions as before except with a different state noise covariance given by $E[w_s w_s^T] = (.001)^2 \, \mathrm{diag}[1,1,1,1,1,1,0,0]$. This choice gives input noises into all modes except the lead-lag progressing mode and effectively filters this mode out of the loop shape. The loop shape of this controller is also shown in Fig 11 and it only has one cross over frequency near .73. The peak near 1.8 is eliminated from this loop shape, and the only region of high gain is near the instability. The closed loop poles of this controller are the same as those of Table 3 except for the lead-lag progressing mode, which is not moved from its open loop position of $-.00634 \pm j\, 1.83$. The application of this controller on the full model gives stable lead-lag regressing damping as shown in Fig. 13. The damping is stable throughout the flight regime being the weakest in the vicinity of $M_F = 23$ and $\mu = .11$.

Controller B was checked to verify that the periodic terms in the full model do not significantly alter the stability results. The controller was also checked to show that excessively large control inputs are not necessary to suppress the air resonance instability. A time domain simulation showed that the closed loop system could suppress an angular roll rates as large as 6.5 deg/sec with less than two degrees of swashplate input. Addtional results on the other controller designs can be found in Refs. 9 and 23 .

Concluding Remarks

A coupled rotor/fuselage helicopter model which accounts for the effects of blade torsional flexibility, unsteady aerodynamics, and forward flight was developed. Results obtained from using this model indicated that the role of torsional flexibility and unsteady aerodynamics is important, while the effects of forward flight (or periodic coefficients) is fairly small. Subsequently, the model was used to demonstrate the effectiveness of using an active control system to stabilize air resonance. The helicopter configuration considered was selected to be unstable in the whole flight envelope, thus this paper also demostrates the practical feasibility of using an active controller to augment the stability of the lead-lag regressing mode, which is known to play a key role in helicopter air resonance. The controller was designed using multivariable frequency domain techniques with the optimal estimator and optimal regulator structure. The technique, which is based on transfer function singular values, proved to be particular effective resolving problems that would not be obvious if only the covariance and weight matrices were used in the design process. To select the design loop shapes, Loop Transfer Recovery was used, which can be interpreted as an optimization balancing system performance requirements and the requirement of stability in the presence of modeling errors.

The controller used a single roll rate measurement and both the sine and cosine swashplate inputs. This configuration is particularily simple since the measurement is taken from a non-rotating (frame of the fuselage) reference avoiding the need to send signals across the rotor head. Using sine and cosine inputs is also simple and can be accomplished through a conventional swashplate mechanism. A constant four mode design model consisting of the body roll and pitch modes and the lead-lag regressing and progressing modes was found to be quite practical for control design. The controller was shown to stabilize the system throughout a wide range of loading conditions and forward flight speeds and it required small inputs of the order of three degrees or less.

Acknowledgement

This research was funded by NASA Ames Research Center, Moffett Field, CA under Grants NAG 2-209 and NAG-477.

References

1. Burkam, J.E., and Miao, W., "Exploration of Aeroelastic Stability Boundaries with a Soft-in-plane Hingeless Rotor Model," Journal of the American Helicopter Society, Vol. 17, No. 4, October 1972, pp. 27-35.

2. Donham, R.E., Cardinale, S.V., and Sachs, I.B., "Ground and Air Resonance Characteristics of a Soft Inplane Rigid Rotor System," Journal of the American Helicopter Society, Vol. 14, No. 4, October 1969, pp. 33-44.

3. Shaw, J., and Albion, N., "Active Control of the Helicopter Rotor for Vibration Reduction," 36th Annual Forum of the American Helicopter Society, Washington D.C., May, 1980.

4. Wood, E.R., and Powers, R.W., "Practical Design Considerations for a Flightworthy Higher Harmonic Control system," 36th Annual Forum of the American Helicopter Society, Washington D.C., May, 1980.

5. Wood, R.E., Powers, R.W., Cline, J.H., and Hammond, C.E., "On Developing and Flight Testing a Higher Harmonic Control System," Journal of the American Helicopter Society, Vol. 30, No. 1, January, 1985, pp. 3-20.

6. Straub, F.K., and Warmbrodt, W., "The Use of Active Controls to Augment Rotor/Fuselage Stability," Journal of the American Helicopter Society, Vol. 30, No. 3, July, 1985, pp. 13-22.

7. Straub, F.K., "Optimal Control of Helicopter Aeromechanical Stability," Vertica, Vol. 11, No. 3, 1987, pp. 12-22.

8. Young, M.I., Bailey, D.J. and Hirschbein, M.S., "Open and Closed Loop Stability of Hingeless Rotor Helicopter Air and Ground Resonance," Paper number 20, NASA-SP 352, 1974.

9. Takahashi, M.D., "Active Control of Helicopter Aeromechanical and Aeroelastic Instabilities," Ph.D. Dissertation, University of California, Los Angeles, To Be Published, 1988.

10. Takahashi, M.D., and Friedmann, P.P., "Active Control of Helicopter Air Resonance in Hover and Forward Flight," AIAA paper 88-2407, Proceedings of 29-th AIAA/ASME/ASCE/AHS Structures, Structural Dynamics and Materials Conference, Fort Magruder, VA, April 18-20, 1988, pp. 1521-1532.

11. Greenberg, J.M., "Airfoil in Sinsoidal Motion in a Pulsating Stream," NACA-TN 1326, 1947.

12. Venkatesan C., and Friedmann, P., "Aeroelasic Effects in Multi-Rotor Vehicles with Applications to a Heavy-Lift System," NASA-CR 3822, August 1984.

13. Pitt, D.M., and Peters, D.A., "Theoretical Predictions of Dynamic Inflow Derivatives," Vertica, Vol. 5, 1981, pp.21-34.

14. Friedmann, P.P., "Formulation and Solution of Rotary-Wing Aeroelastic Stability and Response Problems," Vertica, Vol. 7, No. 2, 1983, pp.101-141.

15. Hohenemeser, K.H., and Yin, Sheng-Kuang, "Some Applications of the Method of Multi-Blade Coordinates," Journal of the American Helicopter Society, Vol. 17, No. 4, 1972, pp.1-12.

16. Kwakernaak, H., and Sivan, R., Linear Optimal Control Systems, John Wiley & Sons inc., New York, 1972, Chapters 3 and 5.

17. Horowitz, I.M., and Shaked, U., "Superiority of Transfer Function Over State-Variable Methods in Linear Time-Invariant Feedback System Design," IEEE Transactions on Automatic Control, Vol. AC-20, No. 1, February, 1975, pp. 84-97.

18. Doyle, J.C., and Stein, G., "Multivariable Feedback Design: Concepts for a Classical/Modern Synthesis," IEEE Transactions on Automatic Control, Vol. AC-26, No. 1, February 1981, pp. 4-16.

19. Safonov, M.G., Laub, A.J., and Hartmann, G.L., "Feedback Properties of Multivariable Systems: The Role and Use of the Return Difference Matrix," IEEE Transactions on Automatic Control, Vol. AC-26, No. 1, February, 1981, pp. 47-65.

20. Stein, G., "The LQG/LTR Procedure for Multivariable Feedback Control Design," IEEE Transactions on Automatic Control, Vol. AC-32, No. 2, February 1987, pp. 105-114.

21. Madiwale, A., and Williams, D.E., "Some Extensions of Loop Transfer Recovery," Proceedings of the American Control Conference, Vol. 2, 1985, pp.790-795.

22. Johnson, W., Helicopter Theory, Princeton University Press, 1980.

23. Takahashi, M.D., and Friedmann, P.P., "Design of a Simple Active Controller to Suppress Helicopter Air Resonance," Proceedings of 44-th Annual Forum of the American Helicopter Society, Washington D.C., June 16-18, 1988.

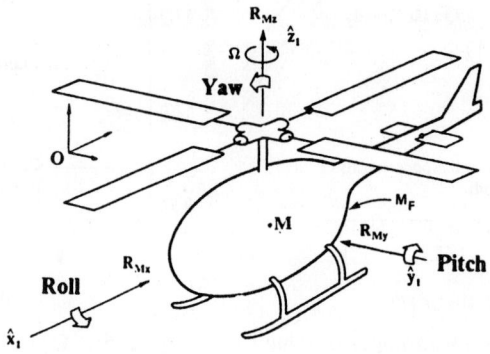

Figure 1:   Rotor/fuselage configuration.

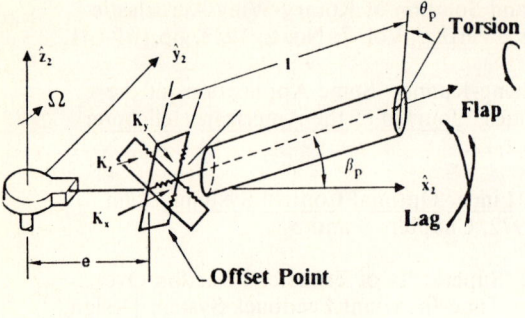

Figure 2: Offset hinged spring restrained rigid rotor blade model.

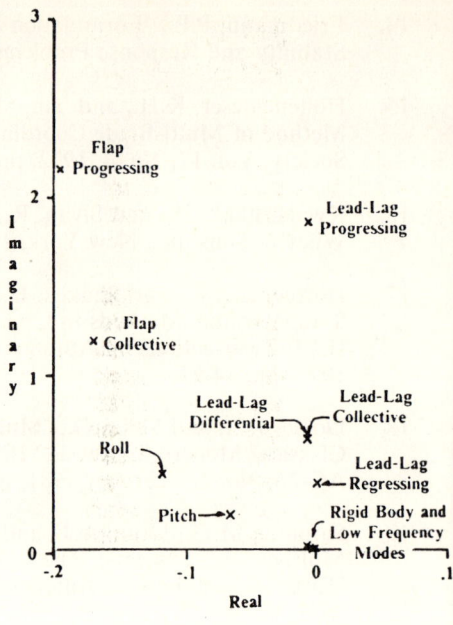

Figure 3: Dominant poles of the full model at advance ratio 0.3.

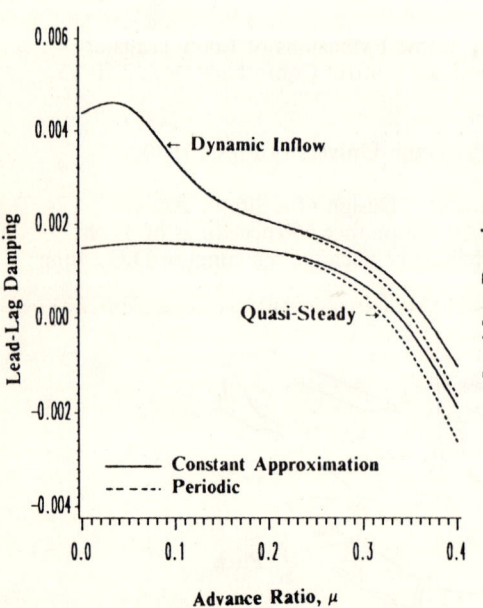

Figure 4: Open loop lead-lag regressing damping of the nominal configuration with and without dynamic inflow.

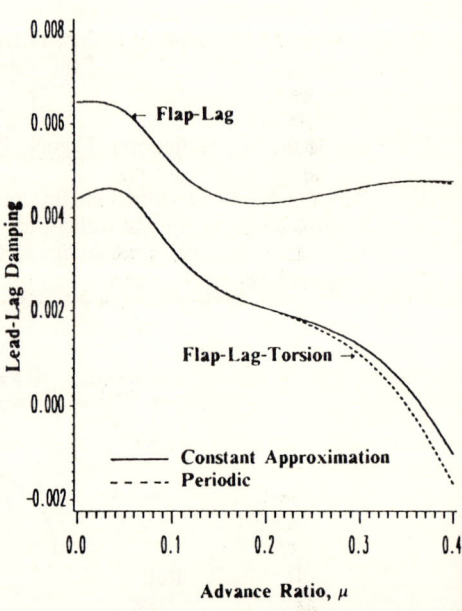

Figure 5: Open loop lead-lag regressing damping of the nominal configuration with and without blade torsional flexibility.

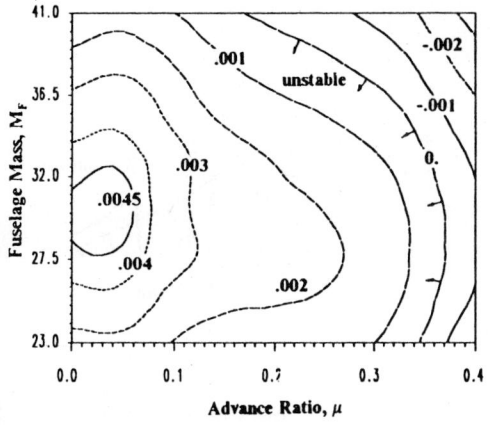

Figure 6: Open loop lead-lag regressing damping at various fuselage weights and advance ratios

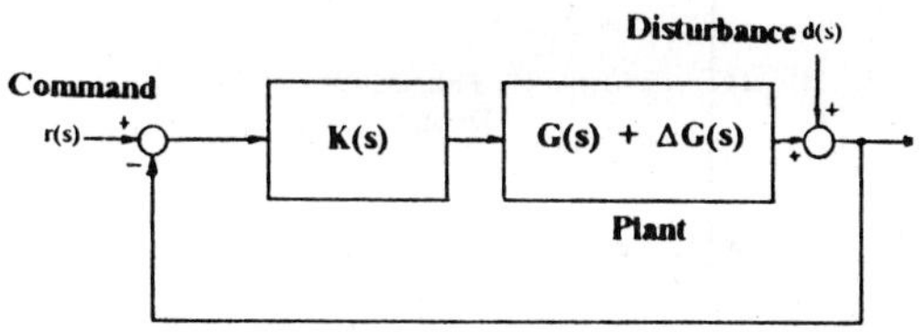

Figure 7: General control configuration.

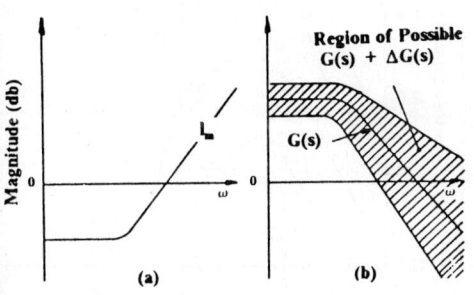

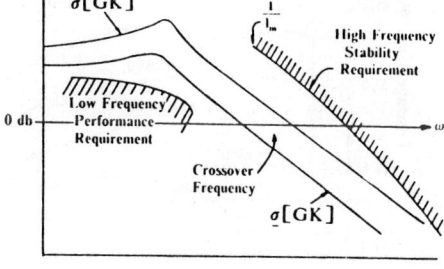

Figure 8: Typical error bound shape.

Figure 9: Hypothetical control design loop shape.

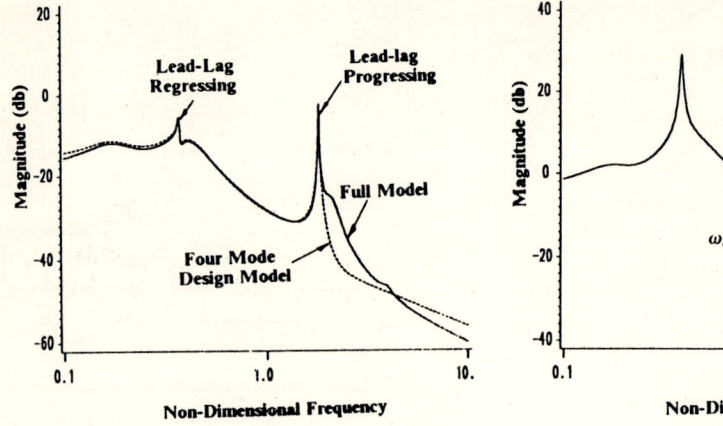

Figure 10: Comparison of the singular values of the full model with the reduced model in hover at the nominal weight.

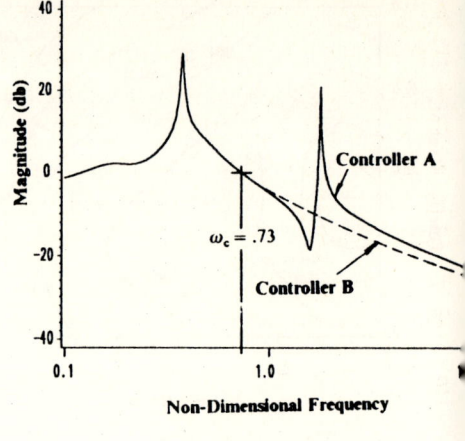

Figure 11: Design loop shapes for controllers A and B.

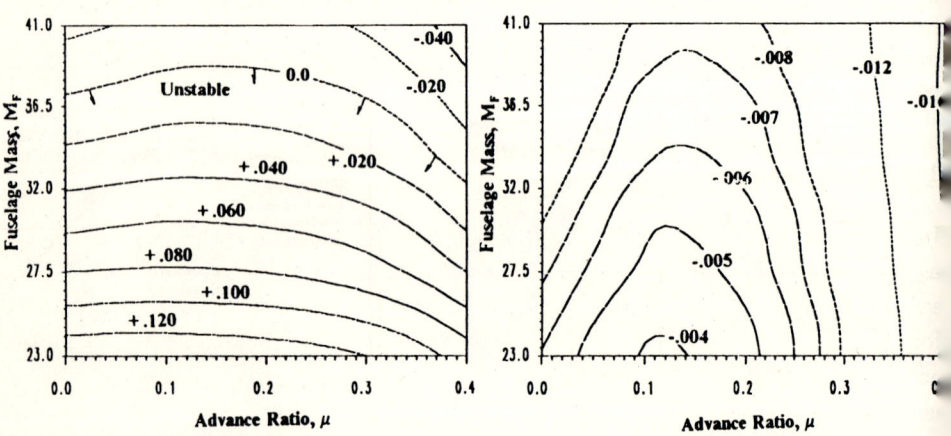

Figure 12: Closed loop flap progressing damping using controller A.

Figure 13: Closed loop lead-lag regressing damping using controller B.

Active Vibration Control for Flexible Space Environment Use Manipulators

T. KOMATSU*, M. UENOHARA*, S. IIKURA*,
H. MIURA**, I. SHIMOYAMA**

* Mechanical Engineering Laboratory
Research and Development Center
Toshiba Corporation, Japan

** Mechanical Engineering
The University of Tokyo, Japan

Summary

A new dynamic control system for flexible space environment use manipulators has been developed from the practical viewpoint. The key concept in the proposed method is that the local position and torque PD feedback loop at each joint should be used for position and structural vibration control. First, the authors derived manipulator dynamics, and then feedback control was developed, using an appropriate potential function. Secondly, an experimental setup using an air suspended SCARA flexible manipulator is described. The effectiveness of this method has been verified by experimental results, adapting it to automatic payload handling.

1. Introduction

In the near future, many robots will be used in space for extravehicular tasks, such as construction of a space station, or periodic repair, cleaning, and maintenance of satellites. Most of these robots must be structurally flexible, reflecting the necessity for their light weight based upon minimum energy consumption and shipping cost, as well as handling of large mass payloads in a no gravity environment. Therefore, it is necessary to control the structural vibration in this flexible arm for quick, precise tracking of the trajectories and accomplishment of tasks.

Recently, there have been a number of studies reported concerning this subject. Sakawa [1] used the optimal control theory selecting the mode amplitude as a control parameter. Cannon [2] used the feedback from the link's end-point position. These methods are effective for a one-link arm. However, it seems difficult to apply these methods to multi-link manipulators from the viewpoint of dynamic model derivation.

The authors developed a new dynamic control method from the application viewpoint. In this method, flexible manipulators are simply controlled by the local position and torque feedback at each joint. An experimental 1.5 m long flexible arm was constructed to investigate the effectiveness of this method. Experimental results obtained through automatic payload handling showed the effectiveness of this method.

2. Mathematical Model Derivation

Let us consider a planar flexible manipulator with n degrees of freedom (d.o.f.). The following assumptions were made for this arm:

(1) Deflection w is small, and any extension is neglected.

(2) Friction and backlash are neglected in the system.

(3) The Euler-Bernoulli model is used for the beam, for which the rotary inertia and shear deformation effects are neglected.

The motion of the arm with a bending vibration is described by the coordinate system shown in the Fig. 1. y_i is defined as follows [2]:

$$y_i(r_i, t) = r_i \theta_i(t) + w_i(r_i, t) \qquad i = 1, \cdots, n. \qquad (1)$$

Also, the local vector $\mathbf{d}_i$ is denoted as follows:

$$\mathbf{d}_i = (r_i, y_i, 1)^t \qquad i = 1, \cdots, n. \qquad (2)$$

Note that superscript t denotes a transpose while the other t denotes time. The position vector for any point at r_i on link i, with respect to the base, is given as:

$$\mathbf{d}_i^0 = \mathbf{R}_i \mathbf{d}_i \qquad i = 1, \cdots, n. \qquad (3)$$

where $\mathbf{R}_i$ denotes a transpose matrix. The system kinetic energy can be written as

$$K = \sum_{i=1}^{n} \int_0^{l_i} k_i \, dr_i \qquad (4)$$

where k_i denotes the kinetic energy per unit length. It can be expressed in a quadratic form as follows:

$$k_i = \frac{\rho_i}{2} trace(\dot{\mathbf{d}}_i^0 \dot{\mathbf{d}}_i^{0t}) \qquad (5)$$

where ρ_i is the mass per unit length and the dot denotes the time derivative. The potential energy V can be written in the form:

$$V = \sum_{i=1}^{n} \int_0^{l_i} v_i \, dr_i \qquad (6)$$

where

$$v_i = \frac{1}{2} E_i I_i \left(\frac{\partial^2 w_i}{\partial r_i^2} \right)^2 = \frac{1}{2} E_i I_i \left(\frac{\partial^2 y_i}{\partial r_i^2} \right)^2.$$

I_i is the area moment of inertia for the link i about the neutral axis. E_i is Young's modulus for the link. The Lagrangian is defined as:

$$L = \sum_{i=1}^{n} \int_0^{l_i} L_i \, dr_i \tag{7}$$

where

$$L_i = k_i - v_i.$$

Using Hamilton's principle[3], the following is obtained:

$$\frac{\partial}{\partial t}\left(\frac{\partial L_i}{\partial \dot{y}_i}\right) - \frac{\partial^2}{\partial r_i^2}\left(\frac{\partial L_i}{\partial y_i''}\right) - \frac{\partial L_i}{\partial y_i} = 0 \qquad 0 < r_i < l_i. \tag{8}$$

The boundary conditions are (Fig. 2):

$$\begin{cases} -\left.\dfrac{\partial L_i}{\partial y_i''}\right|_{r_i=0} = M_i(0) \\ \dfrac{\partial}{\partial r_i}\left(\dfrac{\partial L_i}{\partial y_i''}\right)\bigg|_{r_i=0} = Q_i(0) \\ y_i(0,t) = y_i'(0,t) = 0 \\ -\left.\dfrac{\partial L_i}{\partial y_i''}\right|_{r_i=l_i} = M_i(l_i) \\ \dfrac{\partial}{\partial r_i}\left(\dfrac{\partial L_i}{\partial y_i''}\right)\bigg|_{r_i=l_i} = Q_i(l_i) \end{cases}$$

where the prime denotes the derivative with r_i. Let us define p_i as

$$p_i = \frac{\partial L_i}{\partial \dot{y}_i}. \tag{9}$$

Using this, Eq.(8) is rewritten with boundary conditions as

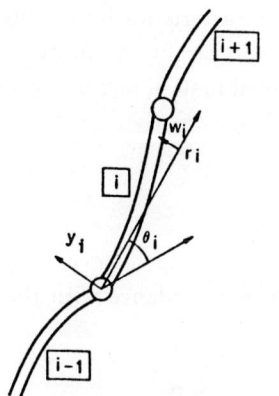
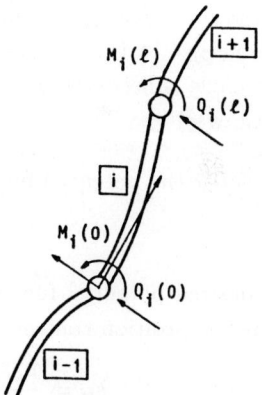

Fig.1. Coordinates for n link flexible manipulator

Fig.2. Flexible manipulator in a horizontal plane

$$\begin{cases} \dot{y}_i = \dfrac{\partial H_i}{\partial p_i} \\ \dot{p}_i = -\dfrac{\partial H_i}{\partial y_i} - F(H_i) + \dfrac{\partial}{\partial r_i}\{M_i(0)\delta(r_i)\} \end{cases} \quad 0 \le r_i \le l_i \tag{10}$$

where

$$\begin{cases} F(H_i) = Q_i(0)\delta(r_i) + Q_i(l)\delta(r_i - l_i) - \dfrac{\partial}{\partial r_i}\{M_i(l)\delta(r_i - l_i)\} + \dfrac{\partial^2}{\partial r_i^2}\left(\dfrac{\partial H_i}{\partial y_i''}\right) \\ \displaystyle\sum_{i=1}^{n}\int_0^{l_i} H_i = K + V \end{cases}$$

and δ denotes Dirac's delta function.

3. Flexible Manipulator Control

Let us consider the control law from the standpoint of energy control in the system where the energy is both kinetic and potential. If the potential energy accumulates in the link, vibration naturally occurs. Such vibration is caused by the exchange of potential energy to kinetic energy. Accordingly, the authors considered that the vibration restraint would be accomplished by modifying this energy flow, so as to minimize both kinetic and potential energy at the target point.

When controlling the global motion for general rigid manipulators, such as those for most industrial robots, each joint is independently controlled by simple linear feedback. With this algorithm, Takegaki [4] showed that the system potential energy had a very large effect on both dynamic and static mechanical properties and that it was natural to attempt to improve the system characteristics by modifying the potential energy. In the case of flexible manipulators, a careful, accurate control of each joint angle is necessary for implementing the global motion and is accomplished in the same way.

Let us consider the potential function as follows:

$$v_i' = v_{1i} + v_{2i}. \tag{11}$$

This is a desired potential function, which is chosen in accordance with the control goal. v_{1i} is the position control function:

$$v_{1i} = \dfrac{1}{2}\dfrac{c_i a_i}{l_i}(\theta_i - \theta_i^*)^2 \quad a_i > 0, c_i > 0 \tag{12}$$

where θ_i^* is a target point. v_{2i} is the vibration control function:

$$v_{2i} = c_i v_i. \tag{13}$$

An additional damping D_i is considered as follows:

$$r_i D_i = -c_i b_i \dot{\theta}_i \delta(r_i) - d_i \frac{d}{dt}\{r_i F(H_i)\} \qquad b_i > 0, d_i \geq 0. \tag{14}$$

Let us denote $\bar{H}_i$ as follows:

$$\bar{H}_i = k_i + v_i + v_{1i} + v_{2i}. \tag{15}$$

From Eq. (10), we obtain

$$\begin{cases} \dot{y}_i = \dfrac{\partial \bar{H}_i}{\partial p_i} \\ \dot{p}_i = -\dfrac{\partial \bar{H}_i}{\partial y_i} - F(\bar{H}_i) + D_i \end{cases} \tag{16}$$

setting

$$\frac{\partial}{\partial r_i}\{M_i(0)\delta(r_i)\} = \frac{\partial H_i}{\partial y_i} + F(H_i) - \frac{\partial \bar{H}_i}{\partial y_i} - F(\bar{H}_i) + D_i. \tag{17}$$

Eq. (16) denotes the system dynamics after v_{1i} and v_{2i} are added. Now, denoting u_i as the actuator torque,

$$M_i(0) = u_i - J_i \ddot{\theta}_i \tag{18}$$

is satisfied at $r_i = 0$ (J_i : inertia of output axis). Then, the integral of Eq. (17) with r_i gives

$$u_i = J_i \ddot{\theta}_i + c_i[\{a_i(\theta_i^* - \theta_i) - b_i \dot{\theta}_i\} - T_i] - d_i \frac{d}{dt} T_i \tag{19}$$

where

$$T_i = E_i I_i w_i''(0, t).$$

Eq. (19) shows that the controller consists of the local position and torque feedback loop, as indicated in Fig. 3. The authors named this law the LTIP method (Local Torque feedback In the Position loop). Let us consider $\bar{H}$ as a Lyapunov function for the system.

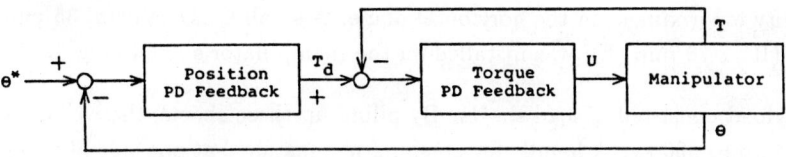

Fig.3. Control blockdiagram

$$\bar{H} = \sum_{i=1}^{n} \int_{0}^{l_i} \bar{H}_i dr_i \tag{20}$$

Differentiating $\bar{H}$ along the solution trajectory in Eq. (16), we obtain

$$\begin{aligned}\dot{\bar{H}} &= \sum_{i=1}^{n} \int_{0}^{l_i} \{(\frac{\partial \bar{H}_i}{\partial y_i})\dot{y}_i + (\frac{\partial \bar{H}_i}{\partial p_i})\dot{p}_i + (\frac{\partial \bar{H}_i}{\partial y_i''})\dot{y}_i''\}dr_i \\ &= \sum_{i=1}^{n}(-b_i c_i \dot{\theta}_i^2 - d_i E_i I_i \dot{\theta}_i \dot{w}_i''(0,t) - d_i \int_{0}^{l_i} E_i I_i \dot{w}_i''^2 dr_i) \\ &\leq \sum_{i=1}^{n}[-(b_i c_i - \frac{1}{2}d_i^2)\dot{\theta}_i^2 - \frac{1}{2}E_i I_i \{2d_i \int_{0}^{l_i} \dot{w}_i''^2 dr_i - E_i I_i \dot{w}_i''^2(0,t)\}].\end{aligned} \tag{21}$$

Therefore, if $d_i = 0$ or

$$2d_i \int_{0}^{l_i} \dot{w}_i''^2 dr_i > E_i I_i \dot{w}_i''^2(0,t), \qquad b_i c_i > \frac{1}{2}d_i^2 \tag{22}$$

is satisfied, asymptotic stability is proved by Lyapunov's Second Method [5].

4. Experimental Setup

An experimental equipment was built to investigate the validity of this method. The authors named this equipment TESRA-1 (Teleoperated Elastic Space Robot Arm). This equipment consisted of a two dimensional air suspended flexible manipulator, payload, and controller. Fig. 4 shows this equipment.

The flexible manipulator was about 1.5 m long. It had two flexible links and three d.o.f. (shoulder, elbow, wrist). An actuator was installed at each joint. It consisted of a DC motor and a planetary gear reducer (1:100 reduction ratio). The sensor system consisted of a potentiometer for sensing the joint angle, a tachogenerator for sensing the motor velocity, and the strain gages at the base of each link for sensing the joint torque. Flexible links for this manipulator were made from stainless steel. The link diameter was 6 mm. The total weight for each joint and hand were 4 kg and 1 kg. This arm floated on an acrylic plate base, using four air bearings so as to simulate a no gravity environment in the horizontal plane. A small CCD camera, 35 mm (W) x 43 mm (H) x 70 mm (L), was installed on the manipulator's hand (Fig. 5).

The payload consisted of lead sheets. By piling up these sheets, the weight could be changed up to 300 kg. A handle for grasping was installed at one side of this payload and a target marker was attached on it. This marker consisted of a rectangle formed by 4 LED points, 40 mm (W) x 30 mm (H).

This manipulator was controlled by a MOTOROLA digital computer VME-10 system as the main computer. Its MPU was the 16-bit 68010, and the VERSAdos multi-tasking system was used as the operating system. Sensor outputs were sampled at 15 msec intervals through a 32 ch A/D board. Commands were fed to the servo drivers through the 4 ch D/A board. Fig. 6 shows this system composition.

5. Experimental Result

The LTIP method was applied to a typical space environment robot task; automatic handling of a very large mass payload. Fig. 7 shows this task sequence. First, the manipulator was located about 40 cm from the payload. When the start command was actuated, the manipulator searched for the target marker with the CCD camera. After detecting this marker, the manipulator started to approach the payload, using position and attitude data obtained from the relative positions of the 4 LED points. When there were no vibration control for the manipulator, the image of the marker from CCD camera vibrated due to link vibration, therefore it was very difficult to detect the correct position and attitude. For example, Fig. 8 shows the manipulator motion without the LTIP. Although each joint moved along the planned path, vibration occurred. Fig. 9 shows the motion with the LTIP, effective vibration restraint is obvious. The high frequency vibration in the angle record results from the resolution of the data sensing system.

The manipulator grasps the handle of the payload within 2 mm positioning accuracy. Finally, the manipulator transports the 40 kg payload about 80 cm and positions it using the LTIP method. If this experiment were carried out without vibration restraint control, the link would begin to vibrate at about 0.1 Hz. This vibration continues for a minute and a half. However, using the LTIP method, no structural vibration occurs such that smooth and quick positioning can be realized. Fig. 10 shows the results of the LTIP measured from the time when the manipulator approached until grasping the handle. The manipulator arrived at the transient target point after 16 seconds with minimum vibration. This vibration results from the wrist actuator movement which pivots the CCD camera so that it is always pointing toward the target. Here, control gains were chosen by trial and error.

6. Conclusions

In this study, the authors proposed the LTIP method for dynamic control of flexible manipulators from the energy flow view-point. In the LTIP, desired torque is calculated from a comparison between desired angle and actual output. Then the actuator is controlled so that joint torque becomes equal to the desired torque. In the experiment demonstrated here, the authors used a SCARA manipulator to disregard the torsional vibration. However, this method is also adaptable for general manipulators, and the experiment on 3D arm was implemented at Miura laboratory, The Univ. of Tokyo [6].

References

1. Y. Sakawa, F. Matsuno and F. Fukushima : Modeling and Feedback Control of a Flexible Arm, J. of Robotic Systems, 2(4), 453-472, (1985)
2. R. H. Cannon, Jr. and E. Schmits : Initial Experiments on the End-Point Control of a Flexible One-Link Robot, The International Journal of Robotic Research, 3(3), 62-75, (1984)
3. C. L. Dym and I. H. Shames : Solid Mechanics, McGraw-Hill, New York, (1973)
4. M. Takegaki and S. Arimoto : A New Feedback Method for Dynamic Control of Manipulators, ASME Journal of Dynamic Systems, Measurement, and Control, Vol.102, 119-125, (1981)
5. J. LaSalle and S. Lefschez : Stability by Lyapunov's Direct Method with Applications, Academic Press, (1961)
6. J. L. Wu, I. Shimoyama, H. Miura, S. Iikura, T. Komatsu and M. Uenohara : Development of Space Manipulation System, Proc. of the SAIRAS 3-6, (1987) (in Japanese)

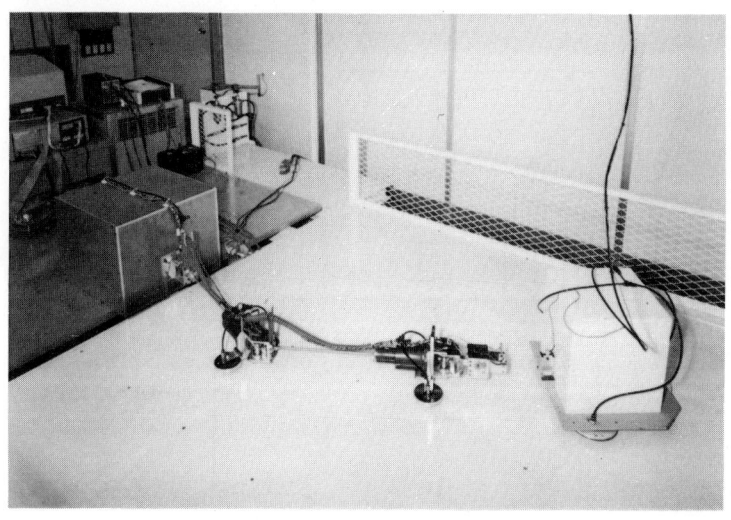

Fig.4. TESRA-I configuration

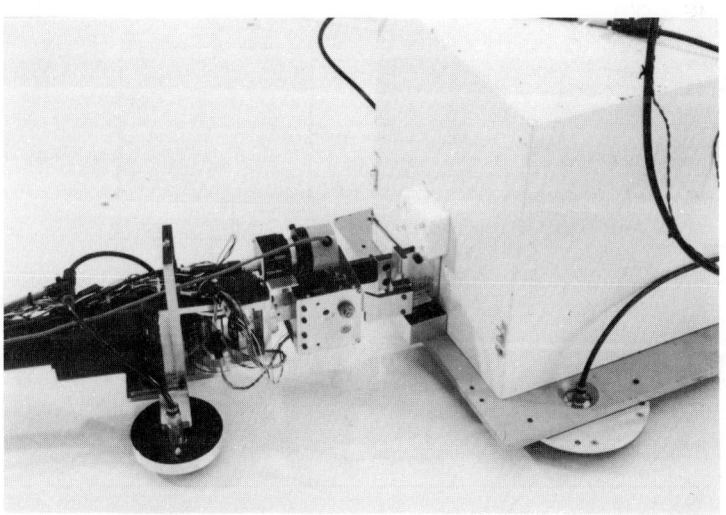

Fig.5. CCD camera and target marker

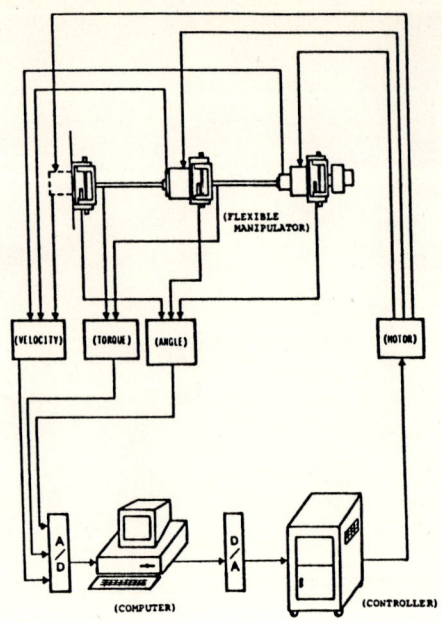

Fig.6. TESRA-I composition

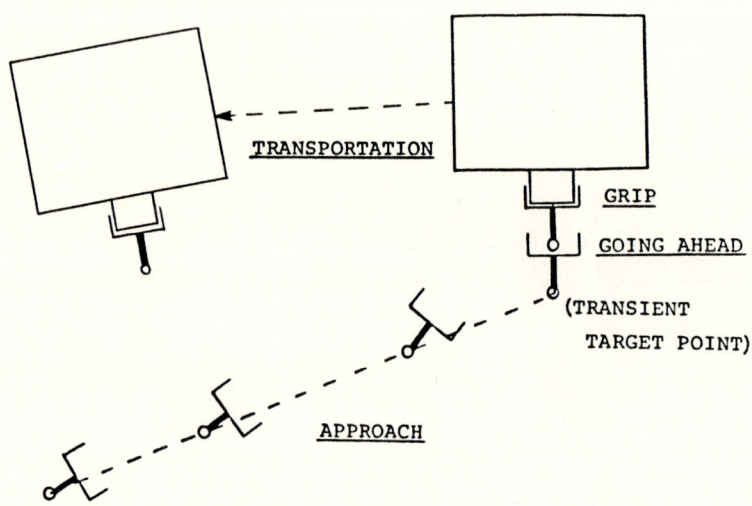

Fig.7. Task sequence

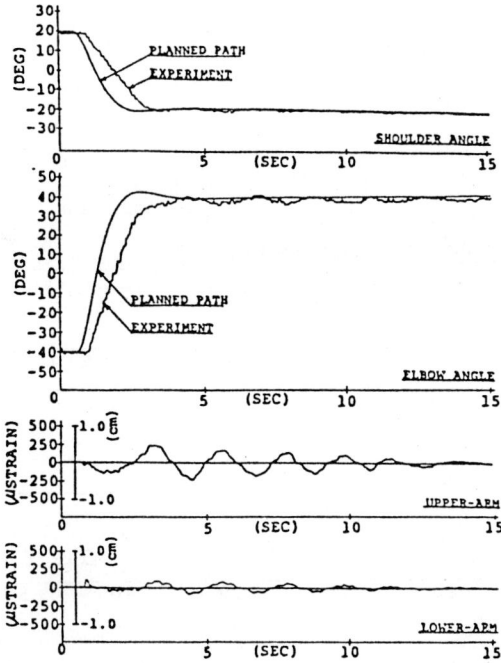

Fig.8. Experimental result (without LTIP)

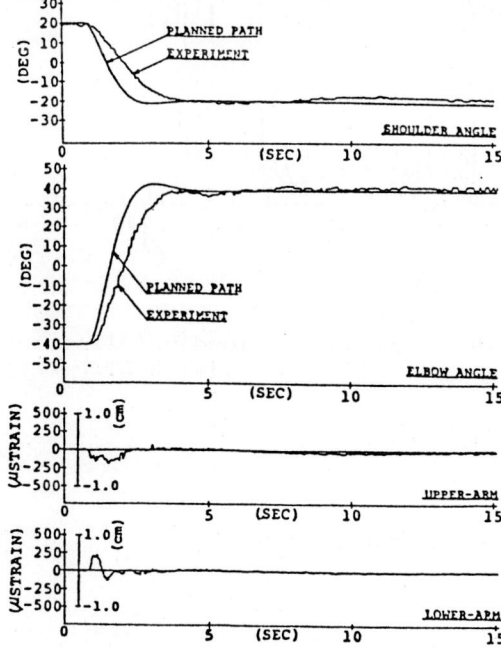

Fig.9. Experimental result (with LTIP)

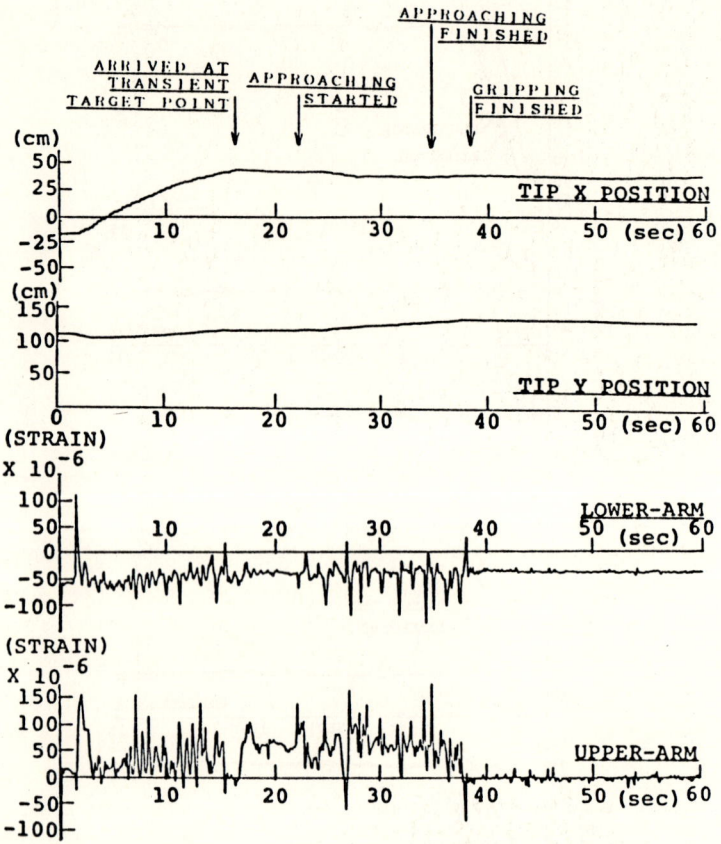

Fig.10. Automatic payload handling result
(with LTIP method)

Orientation of Large Orbital Stations

V.A. Sarychev, M.Yu. Belyaev, V.V. Sazonov, T.N. Tyan
Keldysh Institute of Applied Mathematics,
USSR Academy of Sciences

Abstract

The single-axis gravitational orientation mode is considered for the Salyut 6 and 7 orbital stations. An integral statistical technique is described for determining the real rotational motion of the stations in this mode by the solar and magnetic sensor indications. The technique is illustrated by computations of residual microaccelerations aboard the station; their knowledge is important for an analysis of some technological experiments.

Introduction

The Soviet orbital stations Salyut 6 and 7 represent elongated structures with large lateral moments of inertia. This fact allowed an extensive use of the single-axis gravitational orientation mode of the station [1, 2]. In this mode the station performs the oscillatory or rotational motion around the longitudinal axis directed approximately along the local vertical. To carry out some scientific experiments it is necessary to know the station orientation more exactly. Below, an integral statistical technique is described for determining the attitude motion of the Salyut 6 and 7 stations in the gravitational orientation mode by using the solar and magnetic sensor indications [3]. At given times the sensors allow the onboard measurements of the Earth's magnetic field strength $\vec{H}$ and the unit vector $\vec{S}$ indicating the direction to the Sun (measurements of $\vec{S}$ are possible only on the illuminated part of orbit). A set of measurements performed on some time interval is processed by using the least square method and integrating the motion equations of the station with respect to the center of mass. The developed technique allowed solving a number of scientific problems that required knowledge of the station's attitude motion. To illustrate this technique we estimate microaccelerations aboard the station, which is necessary to analyze some technological experiments [4].

Equations of attitude motion of the station

The time interval, on which the processing is performed, is approximately equal to a period of the station revolution along the orbit. Also it is known that during this interval the station is in the state of gravitational orientation. These circumstances allow using rather simple motion equations for statistical processing. The equations are derived under the following assumptions. The station is assumed to be a rigid body whose configuration is a cylinder with three attached inertia-free plates - the solar batteries. Each battery has a single degree of freedom - it can rotate about its axis that crosses the center of battery and an axis of the cylinder. On the part of orbit illuminated by the Sun the batteries are turned so that angles of incidence of solar rays on their surfaces are minimal. In the Earth's shadow the batteries are fixed with respect to the cylinder and take the positions they occupied at the time when the station entered into the shadow.

The orbit of the station's center of mass is circular and invariable in the absolute space. The gravitational and restoring aerodynamic torque effects are taken into account. It is assumed that atmosphere is fixed in the absolute space, its density along orbit is constant and the air molecules suffer an absolutely nonelastic collision at the interaction with the station. Moreover, it is assumed that the longitudinal axis of the station slightly deviates from the local vertical.

In order to write the motion equations we introduce two right-hand Cartesian systems of coordinates: the $Ox_1x_2x_3$ system formed by the principal central axes of inertia of the station, and the orbital system $OX_1X_2X_3$. The axis x_1 coincides with the longitudinal axis of the station and is directed to the service module, the axis X_3 is parallel to the geocentric radius vector of point O, the axis X_1

is directed along tangent to the orbit towards the station motion. Orientation of the system $Ox_1x_2x_3$ with respect to $OX_1X_2X_3$ is given by angles γ, δ and β [1-3]. The angles are determined in the following way. In order to transform the system $OX_1X_2X_3$ into $Ox_1x_2x_3$ it should be turned first by angle $\delta + \pi/2$ around the second axis, then by angle β around the third axis and finally by angle γ around the first axis. Angles δ and β give the direction of the x_1 axis in the orbital system of coordinates, while angle γ gives the station turning around this axis. When $\delta = \beta = 0$, the x_1 axis is directed to the Earth's center.

Let $\vec{a}$ be an arbitrary vector; a_1, a_2 and a_3 are its components in one of the introduced coordinate systems. If these components are related to the $Ox_1x_2x_3$ system we may write $\vec{a} = (a_1, a_2, a_3)_x$; if they are related to the $OX_1X_2X_3$ system, we write $\vec{a} = (a_1, a_2, a_3)_X$. For the arbitrary vector $\vec{a} = (a_1, a_2, a_3)_x = (A_1, A_2, A_3)_X$ the relations

$$A_i = \sum_{k=1}^{3} a_{ik} a_k \quad (i = 1, 2, 3)$$

are valid, where $\|a_{ik}\|_{i,k=1}^{3}$ is the matrix of transformation of the system $Ox_1x_2x_3$ into $OX_1X_2X_3$. Elements of this matrix are expressed through angles γ, δ and β. The first and third rows of the matrix have the form

$a_{11} = -\sin\delta \cos\beta$, $\qquad a_{31} = -\cos\delta \cos\beta$,

$a_{12} = \cos\delta \sin\gamma + \sin\delta \sin\beta \cos\gamma$, $\quad a_{32} = -\sin\delta \sin\gamma + \cos\delta \sin\beta \cos\gamma$,

$a_{13} = \cos\delta \cos\gamma - \sin\delta \sin\beta \sin\gamma$, $\quad a_{33} = -\sin\delta \cos\gamma - \cos\delta \sin\beta \sin\gamma$.

We introduce the designations: t is the time; t_1 is an initial point of the interval on which the measurements are processed; ω_0 is the angular velocity of orbital motion; $\tau = \omega_0(t-t_1)$ is the dimensionless time; $\vec{\omega} = \omega_0(\Omega_1, \Omega_2, \Omega_3)_x$ is the absolute angular velocity of the station; $w_2 = \Omega_2 \cos\gamma - \Omega_3 \sin\gamma$, $w_3 = \Omega_2 \sin\gamma + \Omega_3 \cos\gamma$; A, B and C are the moments of inertia of the station with respect to the axes x_1, x_2 and x_3 ; $\lambda = A/C$, $\mu = (B-C)/A$. The motion equations have the form [3]:

$$\dot{\gamma} = \Omega_1 - w_2 \operatorname{tg}\beta , \quad \dot{\delta} = \frac{w_2}{\cos\beta} - 1, \quad \dot{\beta} = w_3 ,$$

$$\dot{\Omega}_1 = \mu(\Omega_2 \Omega_3 - 3 a_{32} a_{33}) - a_{13} P_2 ,$$

$$\dot{w}_2 = w_2 w_3 \operatorname{tg}\beta - 3\sin\delta\cos\delta\cos\beta + \lambda Q_\delta , \quad (1)$$

$$\dot{w}_3 = -w_2^2 \operatorname{tg}\beta - 3\cos^2\delta \sin\beta \cos\beta + \lambda Q_\beta ,$$

$$Q_\delta = Q\cos\gamma - Q'\sin\gamma, \quad Q_\beta = Q\sin\gamma + Q'\cos\gamma ,$$

$$Q = -[(1+\mu)(\Omega_1 \Omega_3 - 3 a_{31} a_{33}) - a_{13} P_1]/(1+\lambda\mu) ,$$

$$Q' = (1-\mu)(\Omega_1 \Omega_2 - 3 a_{31} a_{32}) + a_{11} P_2 - a_{12} P_1 ,$$

$$P_1 = \mu_2 + \mu_3 [u_0 \max(|a_{12}|,|a_{13}|) + u_1 |a_{13}| + (u_0 + 2u_2)|a_{12}|] ,$$

$$P_2 = \mu_1 |a_{13}|^{-1} [u_0 \max(a_{13}^2 - a_{12}^2, 0) + u_1 a_{13}^2] ,$$

$$u_0 = \min(v_1, v_2), \quad u_1 = \max(v_1 - v_2, 0),$$

$$u_2 = \max(v_2 - v_1, 0),$$

$$v_1 = |S_3|(S_1^2 + S_3^2)^{-1/2}, \quad v_2 = |S_2|(S_1^2 + S_2^2)^{-1/2},$$

$$S_j = \sum_{i=1}^{3} S_i a_{ij} \quad (j=1,2,3).$$

Here the point denotes differentiation in τ ; Ω_2 and Ω_3 should be expressed through w_2 and w_3 ; μ_1, μ_2 and μ_3 are the dimensionless aerodynamic parameters; $(S_1, S_2, S_3)_x = \vec{S}$ is the unit vector directed from point O to the Sun. If the station is illuminated by the Sun we calculate v_1 and v_2 by the above formulae. If the station is in the Earth's shadow, the v_1 and v_2 preserve the values they had when the station entered into the shadow. In order the values of v_1 and v_2 were determined at the beginning of motion, the time t_1 should be always chosen on the illuminated part of the orbit. The parameters μ_i have the form: $\mu_i = q\sigma_i d_i / A\omega_0^2$ $(i=1,2,3)$, where σ_i and d_i are characteristic areas and coordinates in the $Ox_1 x_2 x_3$ system of some elements of the station surface, $q = const$ is an absolut value of the aerodynamic drag force acting on the unit area of the station surface perpendicular to the free air stream. To determine equations (1) finally it is necessary to give the functions $S_i = S_i(\tau)$ $(i=1,2,3)$, and the criterion for the station stay in the Earth's shadow. According to [3], we shall assume that the Sun is fixed with respect to the orbit. Then

$$S_1 = S_{10} \cos(\tau-\tau_0) - S_{30} \sin(\tau-\tau_0), \quad S_2 = S_{20},$$
$$S_3 = S_{10} \sin(\tau-\tau_0) + S_{30} \cos(\tau-\tau_0), \quad (2)$$

where τ_0 and S_{i0} $(i=1,2,3)$ are constants, $S_{10}^2 + S_{20}^2 + S_{30}^2 = 1$. The station will be in the Earth's shadow if $S_3 \leq -\sqrt{1-r_0^2 \omega_4^{4/3} \mu_E^{-2/3}}$, where $r_0 = 6378$ km is the Earth's radius, $\mu_E = 398603$ km^{3}s^{-2} is its gravitational parameter. Equations (1) contain five parameters: λ, μ and μ_i $(i=1,2,3)$. The values of λ and μ are known rather accurately, while the values of μ_i just approximately. Therefore, at the statistical processing of sensor indications the parameters λ and μ are supposed to be known, and the parameters μ_i $(i=1,2,3)$ are considered as unknown and determined by the processing together with the unknown initial conditions of the motion.

The single-axis gravitational orientation mode.

The single-axis gravitational orientation mode of the station is called its motion when $|\beta| + |\sin\delta| \ll 1$, i.e. the angle between the x_1 and x_3 axes is near zero or 180°. The orbital stations Salyut 6 and 7 together with the docked Soyuz and Progress spacecraft have the form of an elongated structure which is characterized by a small parameter λ.

As a rule, $\lambda \lesssim 0.05$, $0 < \mu \lesssim 0.1$, $0 < \mu_1 \lesssim 1$, $|\mu_2| \lesssim 1$, $|\mu_3| \lesssim 1$. At $\lambda = 0$, i.e. for the station in the form of a rod with longitudinal axis x_1, the system (1) admits two families of particular solutions of a special form. In these families

$$\sin \delta = 0, \beta = 0, w_2 = 1, w_3 = 0 \qquad (3)$$

and the variables γ and Ω_1 are defined by the equations

$$\dot{\gamma} = \Omega_1, \quad \dot{\Omega}_1 = -\mu \sin\gamma \cos\gamma - \mu_1 \cos\delta [u_1 \cos\gamma |\cos\gamma| + \\ + u_o \max(\cos 2\gamma, 0) \operatorname{sign}(\cos\gamma)]. \qquad (4)$$

One family of solutions is obtained from (3), (4) at $\delta = 0$ the other at $\delta = \pi$. In the both families the axis x_1 coincides with x_3. In the family where $\delta = \pi$ these axes have the same direction; in the family with $\delta = 0$ the opposite directions.

At $\lambda \neq 0$ the system (1) do not have solutions of the form of (3), (4); however, if $\lambda \ll 1$ its solutions with initial conditions satisfying the relation $\sin^2 \delta(0) + \beta^2(0) + [w_2(0)-1]^2 + w_3^2(0) \ll 1$ will differ slightly from the ones of (3), (4) on a fixed time interval. This circumstance is due to the fact that the solutions of the system (1) continuously depend on the parameters and initial conditions. The station motion in the single-axis gravitational orientation mode is described by just the above solutions. Continuous dependence of the solutions of system (1) on the parameters and initial conditions ensures the existence of this orientation mode on the finite, generally short, time interval. Special theoretical and experimental studies are needed to ascertain whether the gravitational orientation can exist on long time intervals [1, 2, 5]. Below the solutions of system (1) ar used for approximation of the station motion on time intervals not more than 2 hours.

Determining the motion parametres.
The station is provided with the sensors that allow at given times measuring the Earth's magnetic field strength $\vec{H} = (h_1, h_2, h_3)_x$ and the Sun position vector $\vec{S} = (s_1, s_2, s_3)$. The time interval on which the processing is made approximately equals to the orbital period and contains several tens of

points for which the measurements are available. The processing consists of a few stages.

At the first stage, for the actual orbit of the station and the times, at which the vectors $\vec{S}$ and $\vec{H}$ were measured, the components of these vectors are calculated in the orbital system of coordinates: $\vec{S} = (S_1, S_2, S_3)_X$ and $\vec{H} = (H_1, H_2, H_3)_X$. As a result, we obtain a set of numbers:

$$t_n, s_i^{(n)}, h_i^{(n)}, S_i^{(n)}, H_i^{(n)} \quad (n=1,\ldots,N; \; i=1,2,3). \qquad (5)$$

They indicate the results of measuring the values of s_i, h_i and the calculated values of S_i, H_i at the time t_n. In (5) $t_1 < t_2 < \ldots < t_N$, and for the times t_n when the station is in the Earth's shadow $s_i^{(n)} = S_i^{(n)} = 0$ ($i=1,2,3$). At the second stage, we determine ω_0 and the constants τ_0, S_{i0} in (2) by using the values t_n and $S_i^{(n)}$. For this we consider the function

$$\Psi(\Omega, k) = \sum_{n=1}^{N} \left\{ [S_1^{(n)} - S_1^{(k)} \cos\Omega(t_n - t_k) + S_3^{(k)} \sin\Omega(t_n - t_k)]^2 + \right.$$
$$\left. + [S_2^{(n)} - S_2^{(k)}]^2 + [S_3^{(n)} - S_1^{(k)} \sin\Omega(t_n - t_k) - S_3^{(k)} \cos\Omega(t_n - t_k)]^2 \right\},$$

where $\Omega > 0$, $k=1,2,\ldots,N$, $|S_1^{(k)}| + |S_3^{(k)}| > 0$. Then we determine $(\Omega^*, m) = \arg\min \Psi(\Omega, k)$ and accept $\omega_0 = \Omega^*$, $\tau_0 = \omega_0 (t_m - t_1)$, $S_{i0} = S_i^{(m)}$ ($i=1,2,3$).

As soon as ω_0 is found, the values t_n in (5) are replaced by $\tau_n = \omega_0(t_n - t_1)$, and the measured and calculated values of $\vec{H}$ are normed to unity. Such a transformation provides the reduction of processed data to the dimensionless form and introduces the same scale for components of $\vec{S}$ and $\vec{H}$.

The final stage of the data processing consists of obtaining the solution of system (1) that would bring into agreement the measured and calculated values of $\vec{S}$ and $\vec{H}/|\vec{H}|$. At the given λ and μ the solution of this system is determined by the vector $\alpha \in R^9$, whose first six components are initial conditions at point $\tau_1 = 0$ and the last three components are μ_1, $\lambda\mu_2$, and $\lambda\mu_3$ (the parameters μ_2, μ_3 enter into the system (1) as the products $\lambda\mu_2$, $\lambda\mu_3$). On the solutions of (1) we define the function

$$\Phi(\alpha) = w_S \sum_{n=1}^{N} \sum_{i=1}^{3} \left[s_i^{(n)} - \sum_{k=1}^{3} S_k^{(n)} a_{ki}(\tau_n) \right]^2 +$$

$$+ w_H \sum_{n=1}^{N} \sum_{i=1}^{3} \left[\tilde{h}_i^{(n)} - \sum_{k=1}^{3} \tilde{H}_k^{(n)} a_{ki}(\tau_n) \right]^2,$$

where w_S and w_H are positive constants, $\tilde{h}_i^{(n)}$ and $\tilde{H}_i^{(n)}$ are normed measured and calculated components of $\vec{H}$ in the coordinate systems $Ox_1x_2x_3$ and $OX_1X_2X_3$ respectively. According to the least square method we take $\hat{\alpha} = \text{argmin}\, \Phi(\alpha)$ as an estimate of the vector α. Minimization of $\Phi(\alpha)$ is carried out by using first a random search and then the Marquardt method. The standard errors of the motion parameter estimates obtained by the least square method with measurements on the time interval $\sim$ 90 min are $\sim 0.5°$ in angles and $\sim 0.0015°/s$ in angular velocities [6]. By processing the star photometer indications we could obtain an independent estimate of accuracy in determining the motion parameters [7]. If a star gets into the photometer field of vision, we may estimate an error in the knowledge of the station orientation by comparing the calculated and actual positions of the star on the celestial sphere. The maximal error thus obtained is 3.3°. In most cases it did not exeed 2°.

As an example we present the results of processing of information (5) obtained on board of the Salyut 7 station and referring to revolution 1595 (29.07.1982). The station motion determined as a result of processing is shown in Fig. 1, where the plots of the functions $\gamma(t)$, $\delta(t)$, $\beta(t)$ and $\omega_i = \omega_0 \Omega_i(t)$ $(i = 1, 2, 3)$ are presented. In this case $t_1 = 8^h 26^m 26^s$ in the decret Moscow time, $N = 43$, the number of measurements S is 26, $w_S = 1$, $w_H = 0.4$, $\Phi(\hat{\alpha}) = 0.0867$. As it is seen from the figure the station, in fact, regularly rotates about its axis x_1 (in angle γ) performing small oscillations with respect to the axis X_3. The period of rotation is about 23 min; the oscillation periods of axis x_1 in angles δ (in the orbital plan) and β (in the direction perpendicular to the orbital

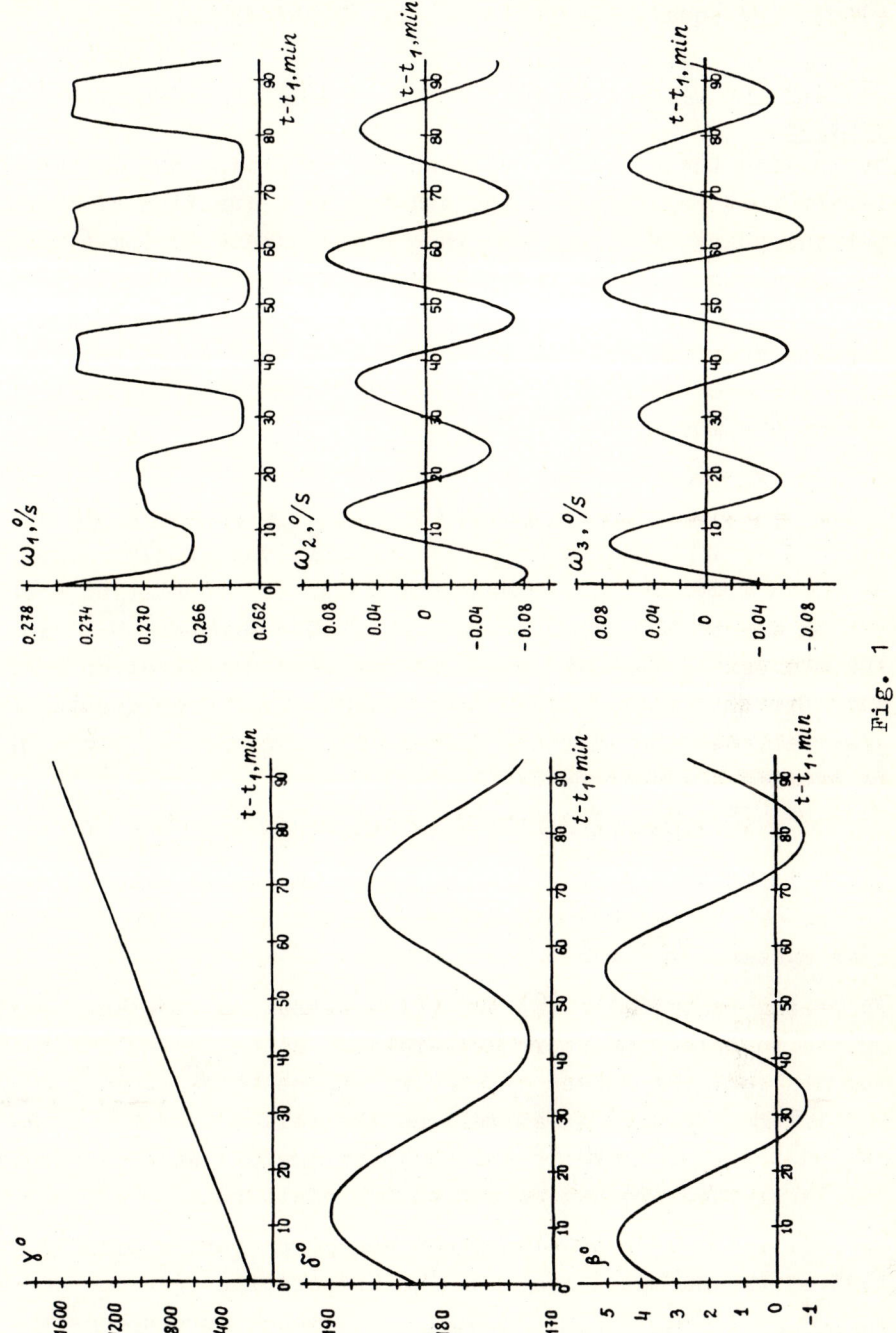

Fig. 1

plane) are equal, respectively, to 55 and 48 min.

Calculation of microaccelerations on board of the orbital station.

By knowing the station motion we may determine the microacceleration at any point of the station as a function of time. Let the point P be fixed in the coordinate system $Ox_1x_2x_3$ and given in it by the radius vector $\vec{\rho}$. The microacceleration at the point P is called a difference between the Earth's gravitational field strength at this point and the absolute acceleration of it. The microacceleration is calculated by the formulae [4]

$$\vec{n} = \vec{n}^o + \vec{n}^a, \quad \vec{n}^a = d\vec{E}_1, \qquad (6)$$
$$\vec{n}^o = \vec{\rho} \times (d\vec{\omega}/dt) + (\vec{\omega} \times \vec{\rho}) \times \vec{\omega} + \omega_o^2 [3\vec{E}_3(\vec{E}_3 \cdot \vec{\rho}) - \vec{\rho}]$$

where $\vec{E}_i = (a_{i1}, a_{i2}, a_{i3})_x$ is the unit vector along the axis X_i ($i=1,2,3$), d is an absolute value of the station acceleration due to the aerodynamic drag. The component $\vec{n}^a$ in the expression for $\vec{n}$ is equal to this acceleration taken with opposite sign, while the component $\vec{n}^o$ is due to the gravitational and inertial forces. For convenience, from $\vec{n}^o$ we extract the component

$$\vec{n}^z = \vec{\rho} \times (d\vec{\omega}/dt) + (\vec{\omega} \times \vec{\rho}) \times \vec{\omega} - \omega_o^2 (\vec{E}_2 \times \vec{\rho}) \times \vec{E}_2 \qquad (7)$$

which appears as a result of the station motion with respect to the orbital coordinate system. If the station is fixed in this system, $\vec{n}^z = 0$.

By basing on formulae (6) and (7) a computing code was constructed to determine microaccelerations at any given point of the station. The values of vector components $\vec{n}^z$, $\vec{n}^o$, $\vec{n}^a$ and $\vec{n} = (n_1, n_2, n_3)_x$ as well as the values of $|\vec{n}^z|$, $|\vec{n}^o|$ and $|\vec{n}|$ ($|\vec{n}^a| = d$ is the input parameter) were printed in the tabulated form and output on the plotter.

In Fig. 2 the microaccelerations are given for revolution 1595 of the Salyut 7 station. The computations were performed for $\vec{\rho} = (2.5m, 1.45m, -0.8m)_x$, which corresponds to the technological device "Splav", and $d = 7.257 \cdot 10^{-6}$ ms^{-2}. As it is seen from Fig. 2 the microaccelerations undergo ra-

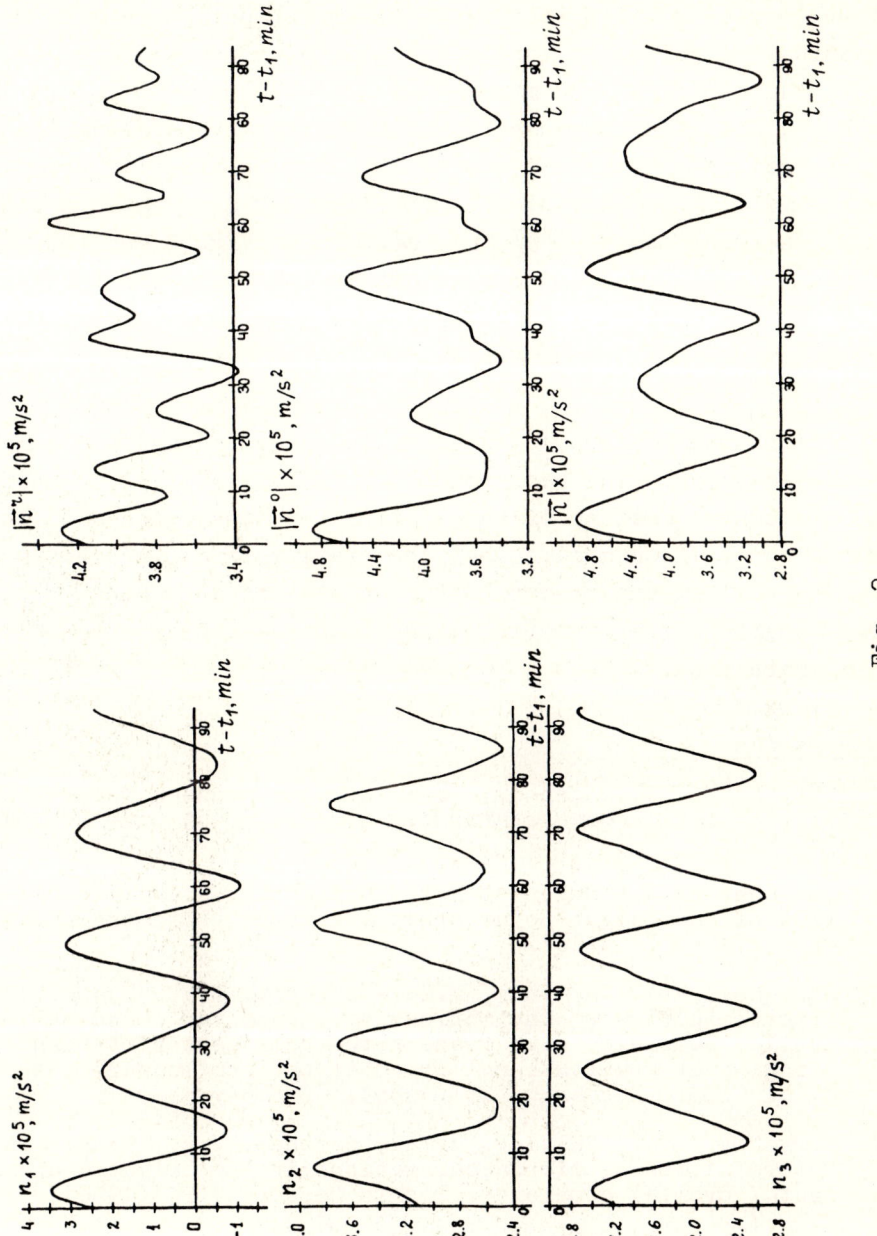

Fig. 2

ther regular oscillations. A component of period 23 min is distinct on all plots, and, in addition, a component of period 12 min $\approx$ 0.5·23 min can be also observed in the curve $|\vec{n}^{\tau}|$. Analysing formulae (6) and (7) we may conclude that such oscillations appear due to the station rotation about axis x_1. This conclusion is confirmed by a comparison between the functions $n_i(t)$ ($i=1,2,3$) in Fig. 2 and the functions $\omega_2(t)$, $\omega_3(t)$ in Fig. 1. While analysing the two last functions we should take into account that in the gravitational orientation mode $\omega_2(t) \approx \omega_o \cos\delta(t)$, $\omega_3(t) \approx -\omega_o \sin\delta(t)$ (compare relations (3)).

The microacceleration computed by formulae (6) and (7) is an averaged value - the background. In fact, oscillations caused by various vibrations of the station body are unposed on this background. If we assume that the station is a rigid body these oscillations cannot be taken into account. However, during technological experiments special measures are taken to eliminate them.

References

1. Sarychev, V.A.; Sazonov, V.V.: Gravity gradient stabilization of large space stations. Acta Astronautica 8 (1981) 549-573.

2. Sarychev, V.A.; Sazonov, V.V.: Gravity gradient stabilization of the Salyut-Soyuz orbital complex. Acta Astronautica 11 (1984) 435-447.

3. Sarychev, V.A.; Belyaev, M.Yu.; Sazonov, V.V.; Tyan, T.N.: A determination of the motions of the Salyut 6 and Salyut 7 orbital complexes with respect to their center of mass in a regime of gravitational orientation from measurements data. Cosmic Research, 23 (1985), 672-654.

4. Sarychev, V.A.; Belyaev, M.Yu.; Sazonov, V.V.; Tyan, T.N.: Determination of microaccelerations in the Salyut 6 and Salyut 7 orbital complexes. Cosmic Research, 24 (1986), 267-273.

5. Sarychev, V.A.; Belyaev, M.Yu.; Kuz'min, S.P.; Sazonov,V.V. Tyan, T.N.: Investigation of attitude motion of the Salyut 7 orbital station for long time intervals. Acta Astronautica 16 (1987) 165-192.

6. Sazonov, V.V.; Belyaev, M.Yu.; Kuz'min, S.P.; Tyan, T.N.: Investigation of an accuracy in determining the rotational motion of Salyut 6 and Salyut 7 orbital stations in the gravitational orientation mode by using measurement data. Preprint, Keldysh Inst. Appl. Math., USSR Ac.Sci. (1986), N 192.

7. Sarychev, V.A.; Sazonov, V.V.; Belyaev, M.Yu.; Efimov, N.I.; Tyan, T.N.; Sheffer, E.K.; Sklyankin, V.A.: Specifying the rotational motion of the Salyut 7 orbital station by using the star photometer indications. Preprint, Keldysh Inst. Appl. Math., USSR Ac.Sci. (1987), N 233.

Attitude Stability of a Flexible Asymmetric Dual Spin Spacecraft

Kazuo TSUCHIYA*, Katsuhiko YAMADA*
Brij N. AGRAWAL**

* : Central Research Lab., Mitsubishi Electric Corp., Amagasaki, Japan
** : Spacecraft R & D, INTELSAT, Washington, D.C., U.S.A.

Summary

Stability of an attitude motion of a large dual spin spacecraft is studied; the effects of various kinds of asymmetries of the spacecraft and energy dissipations in the spacecraft are examined. Attitude instabilities due to interactions between the asymmetries and the interactions between the asymmetries and the energy dissipations are examined in detail. The analysis is based on the method of multiple time scales. The results are verified by numerical solutions based on the Floquet's theorem.

Introduction

A dual spin spacecraft consists of two bodies, a rotor and a stator. The stator is despun to keep its mission equipments fixed in an inertia space. The rotor is spun at high spin rate to exert a gyroscopic stiffness to the system. The spacecraft, on which an angular momentum is exerted, has a lot of energy as a kinetic energy of rotation. When this kinetic energy of rotation is transferred through various processes into other degrees of freedom, the attitude motion becomes unstable. One means of energy transfer is a parametric resonance due to asymmetries in the spacecraft, i.e., unequal moments of inertia of the rotor and the stator about the axes perpendicular to the spin axis and unequal bending stiffness of the shaft, which connects the rotor and the stator, in the transverse directions etc.[1],[2]. Besides these asymmetries, there is another asymmetry to be considered; an unequal reaction torque of a moving part in the spacecraft to the main bodies also causes an attitude motion unstable. The other means of energy transfer is a frictional force between a moving part and the main bodies of the spacecraft[3].

The attitude motion may also become unstable through the interactions between the asymmetries and the interactions between the asymmetries and the energy dissipations. This paper deals with various kinds of unstable attitude motion due to the asymmetries of the spacecraft and the energy dissipations in the spacecraft. The analysis is based on the method of multiple time scales. The results are verified by numerical solutions based on the Floquet's theorem.

Equations of Motion

A spacecraft model is shown in Fig.1. The spacecraft consists of a stator and a rotor which are connected by a shaft. The shaft which is fixed on the rotor has a joint with two degrees of freedom of rotation. The rotor is spun at a constant speed ω_r, while the stator is fixed in an inertia space. The moments of inertia of the rotor and the stator about the axes perpendicular to the spin axis are supposed to be unequal. Unequal bending stiffness of the shaft at the joint is also supposed.

Reference axes $O-XYZ$ are set in such a way that the origin O coincides with the mass center of the spacecraft, the axis Z coincides with the nominal spin axis of the spacecraft and the axes $X\ Y\ Z$ rotate about the Z axis with the angular velocity ω_r. An attitude of the rotor is denoted by a rotation ψ_X

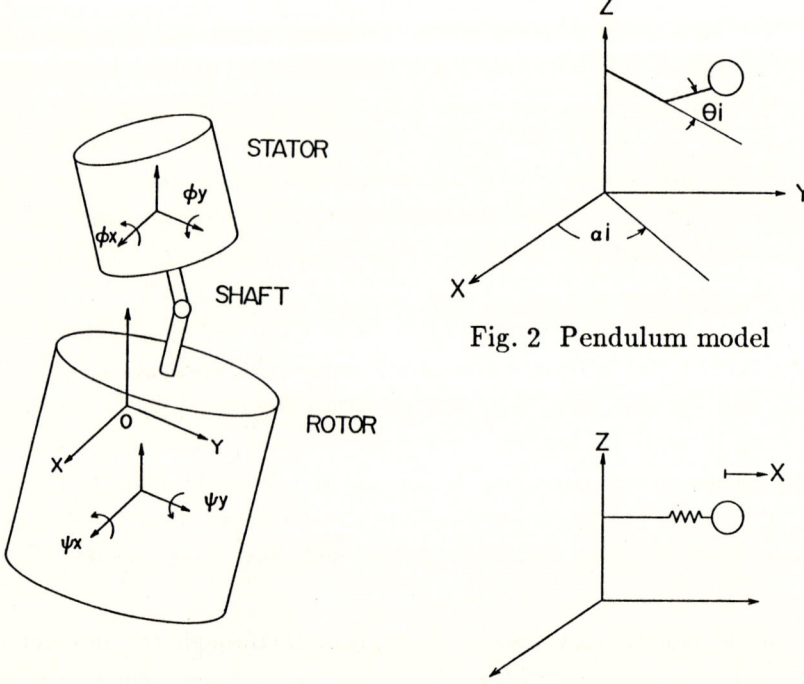

Fig. 1 Spacecraft model

Fig. 2 Pendulum model

Fig. 3 Mass-Spring model

about the X axis and a rotation ψ_Y about the Y axis. An attitude of the stator is also denoted by a rotation ϕ_X about the X axis and a rotation ϕ_Y about the Y axis. A moving part in the rotor, e.g., fuel in a tank and a flexible appendage is modeled as a pendulum; n pendulums are set in the rotor which rotate along an axis perpendicular to the spin axis (Fig.2). An angle of rotation of pendulum i is denoted by θ_i. On the other hand, a moving part in the stator is modeled as a mass and a spring (Fig.3). The mass moves along a line perpendicular to the spin axis. A displacement of the mass is denoted by x. The moving parts exert frictional forces on the main bodies. Usually, a dual spin spacecraft is composed of a heavy rotor and a light stator, and so, the asymmetry of the stator is assumed to be small. The frictional forces which act between the moving parts and main bodies are assumed to be small.

Equations of motion of the system are derived by adopting the variables ψ_X, ψ_Y, ϕ_X, ϕ_Y, θ_i and x as generalized coordinates. By neglecting higher order terms, the equations of motion are given as follows:

$$(M^{(0)} + \varepsilon M^{(1)})\ddot{U} + (G^{(0)} + \varepsilon G^{(1)})\dot{U} + (K^{(0)} + \varepsilon K^{(1)})U = \varepsilon F(x, U)$$
$$\ddot{x} + 2\zeta_s\omega_s\dot{x} + \omega_s^2 x = W(U) \quad (1)$$

where

$$U^T = (\Psi^{(+)}, \Phi^{(+)}, \Theta^{(+)}, \Psi^{(-)}, \Phi^{(-)}, \Theta^{(-)})$$

$$\Psi^{(+)} = \psi_X + i\psi_Y, \quad \Psi^{(-)} = \psi_X - i\psi_Y, \quad \Phi^{(+)} = \phi_X + i\phi_Y,$$

$$\Phi^{(-)} = \phi_X - i\phi_Y, \quad \Theta^{(+)} = \sum_i \theta_i e^{i\alpha_i}, \quad \Theta^{(-)} = -\sum_i \theta_i e^{-i\alpha_i}$$

Parameter ε is a small parameter expressing a magnitude of the asymmetry of the stator and frictional forces; matrices $M^{(1)}$, $G^{(1)}$ and $K^{(1)}$ are mass, gyro and stiffness matrices relating to the asymmetry of the stator and vector F expresses a reaction torque of the moving part in the stator and frictional forces of moving parts in the rotor.

Analysis

Stability of a solution of Eq.(1) is examined by using the method of multiple time scales[4]. Consider the following eigenvalue problem

$$H(\lambda_i)\hat{U}_i = 0 \quad (2)$$
$$\hat{U}_i^T \hat{U}_j = \delta_{ij}$$

where $H(\lambda_i) = M^{(0)}\lambda_i^2 + G^{(0)}\lambda_i + K^{(0)}$, λ_i is an eigenvalue, and $\hat{U}_i^T = (\hat{\Psi}_i^{(+)}, \hat{\Phi}_i^{(+)}, \hat{\Theta}_i^{(+)}, \hat{\Psi}_i^{(-)}, \hat{\Phi}_i^{(-)}, \hat{\Theta}_i^{(-)})$ is a corresponding eigenvector.

Introduce the following transformation based on the solutions (2).

$$t_0 = \varepsilon t$$
$$t_i = \lambda_i t$$
$$U = \sum_{n=0} \sum_{i=1} \varepsilon^n U_i^{(n)}(t_0) \exp(t_i) \qquad (3)$$
$$x = \sum_{n=0} \varepsilon^n x^{(n)}(t_0, t_i)$$

Substituting Eq.(3) into Eq.(1) and equating the coefficients of like powers of ε in each coefficient of $\exp(t_i)$ to zero, we obtain the following equations to order ε^0.

$$H(\lambda_i) U_i^{(0)}(t_0) = 0 \qquad (4)$$
$$\ddot{x}^{(0)} + 2\zeta_s \omega_s \dot{x}^{(0)} + \omega_s^2 x^{(0)} = W(U_j^{(0)}(t_0) \exp(t_j))$$

A zeroth order approximation solution to U is given by

$$U = \sum_{i=1} \hat{U}_i g_i^{(0)}(t_0) \exp(t_i) \qquad (5)$$

where $g_i^{(0)}(t_0)$ is a scalar function of t_0, which is determined to the next approximation.

Solution (5) is expressed by superposition of the modes of attitude motion. The modes of the attitude motion have the following characteristics. When the asymmetries in the rotor do not exist, the eigenvalues of the modes are pure imaginaries and the modes are classified into two groups, $(+)$ mode and $(-)$ mode, where the modes whose components corresponding to $\Psi^{(-)}$, $\Phi^{(-)}$ and $\Theta^{(-)}$ are zero are called $(+)$ mode and the modes whose components corresponding to $\Psi^{(+)}$, $\Phi^{(+)}$ and $\Theta^{(+)}$ are zero are called $(-)$ mode. When the asymmetries in the rotor exist, the separation of the components of the modes become incomplete. In the case where the natural frequencies of two modes are close, these two modes may become unstable due to the asymmetries in the rotor, i.e., the asymmetry of the rotor, the asymmetry of the shaft and the asymmetry of the reaction torques of the moving parts in the rotor. This instability can be examined by the zeroth order solution (5).

On the other hand, instabilities of attitude motion due to the asymmetry of the stator and energy dissipations in the rotor and stator are examined to the first order approximation. From Eqs.(1) and (3), equations for $U^{(1)}$ and $x^{(1)}$ are given as follows:

$$H(\lambda_i) U_i^{(1)} = L(\lambda_i) \hat{U}_i \frac{dg^{(0)}(t_0)}{dt_0} + F(x^{(0)}, U^{(0)})$$
$$\ddot{x}^{(1)} + 2\zeta_s\omega_s\dot{x}^{(1)} + \omega_s^2 x^{(1)} = W(U_j^{(1)}(t_0)\exp(t_j)) \quad (6)$$

where

$$L(\lambda_i) = -\frac{\partial H(\lambda_i)}{\partial \lambda_i}$$

From the condition that $U^{(1)}$ must be bounded, the following equation is derived.

$$\hat{U}_i^T L(\lambda_i) \hat{U}_i \frac{dg^{(0)}(t_0)}{dt_0} + \hat{U}_i^T F(x^{(0)}, U^{(0)}) = 0 \quad (7)$$

When any two modes of attitude motion are not in resonance, $\lambda_i - \lambda_j \neq 2i\omega_r$, a change in the attitude motion is determined by the energy dissipations in the rotor and the stator. In this case, Eq.(7) is reduced to

$$a_i \frac{dg_i^{(0)}}{dt} + b_i g_i^{(0)} = 0 \quad (8)$$

where

$$a_i = \hat{U}_i^T L(\lambda_i) \hat{U}_i, \quad b_i = b_i^{(+)} + b_i^{(-)}$$

$$b_i^{(+)} = -c_r\lambda_i\hat{\Theta}_i^{(+)2} - \frac{i\, m_{ds}^2}{2}\left\{(a_r\hat{\Phi}_i^{(+)} + b_r\hat{\Psi}_i^{(+)})^2 X_i^{(+)}\right\}$$

$$b_i^{(-)} = -c_r\lambda_i\hat{\Theta}_i^{(-)2} - \frac{i\, m_{ds}^2}{2}\left\{(a_r\hat{\Phi}_i^{(-)} + b_r\hat{\Psi}_i^{(-)})^2 X_i^{(-)}\right\}$$

$$X_i^{(\pm)} = (\lambda_i \pm i\omega_r)^4\left\{(\lambda_i \pm i\omega_r)^2 + 2\zeta_s\omega_s(\lambda_i \pm i\omega_r) + \omega_s^2\right\}^{-1}$$

The eigenvalue λ_{ii} of Eq.(8) is given by

$$\lambda_{ii} = \lambda_{ii}^{(+)} + \lambda_{ii}^{(-)} = -b_i^{(+)}/a_i - b_i^{(-)}/a_i \quad (9)$$

From Eqs.(5),(9), the damping factor δ_i of the i-th mode is given by

$$\delta_i = -\text{Re}(\lambda_{ii}^{(+)} + \lambda_{ii}^{(-)}) \quad (10)$$

Stability of the i-th mode is determined by the sign of the damping factor δ_i.

When any two modes of attitude motion are in resonance, $\lambda_i - \lambda_j \approx i2\omega_r$, a change in attitude motion is determined by the asymmetry of the stator as well as energy dissipations in the rotor and the stator. In this case, Eq.(7) is reduced to

$$a_i \frac{d\hat{g}_i^{(0)}}{dt} + (b_i - ia_i\Delta_{ij})\hat{g}_i^{(0)} + c_{ij}^{(+)}\hat{g}_j^{(0)} = 0$$
$$a_j \frac{d\hat{g}_j^{(0)}}{dt} + (b_j + ia_j\Delta_{ij})\hat{g}_j^{(0)} + c_{ji}^{(-)}\hat{g}_i^{(0)} = 0 \quad (11)$$

where

$$\lambda_i - \lambda_j = 2i(\omega_r + \Delta_{ij})$$
$$g_i^{(0)} = \hat{g}_i^{(0)}\exp(-i\Delta_{ij}t)$$
$$c_{ij}^{(+)} = -i_{ms}(\lambda_j - i\omega_r)^2 \hat{\Phi}_i^{(+)}\hat{\Phi}_j^{(-)}$$
$$c_{ji}^{(-)} = -i_{ms}(\lambda_j + i\omega_r)^2 \hat{\Phi}_j^{(-)}\hat{\Phi}_i^{(+)}$$

The eigenvalue determined by Eq.(11) is denoted by λ_{ij}, and then, the damping factor δ_i of the i-th mode is given by

$$\delta_i = -\mathrm{Re}(\lambda_{ij})$$

Stability of the i-th mode is determined by the sign of δ_i.

Numerical Examples and Discussions

The results obtained are applied to a spacecraft model. The spacecraft has two pendulums in the rotor which rotate about an axis parallel to the Y axis. Parameters of the spacecraft are listed in Table 1. Figure 4 shows the

Table 1 Parameters of the spacecraft

$\hat{i}_{pr}$	397 [kg m^2]	k_p	2×10^4 [Nm / rad]
$\hat{i}_{ps}$	235 [kg m^2]	k_r	$15\omega_r^2$ [Nm / rad]
i_r	4.2 [kg m^2]	m_{ds}	2 [kg]
$\hat{m}$	73.4 [kg m^2]	b_r	0.34 [m]
$\hat{m}_{dr}$	15 [kg m^2]	a_r	1.53 [m]
$\hat{i}_{zr}$	300 [kg m^2]		

natural frequencies of the modes of attitude motion in the case where all the asymmetries in the spacecraft are reduced to zero, where (+) (or (−)) sign indicates that the mode belongs to (+) (or (−)) mode. The precession mode, which corresponds to the drift of the angular momentum vector and is an integral of motion, is omitted.

From Fig.4, it is possible that the 2nd and the 7th modes become unstable due to the asymmetries in the rotor in the vicinity of $\omega_r = 22$ rad/s. Figure 5 shows the damping factor δ_2 of the 2nd mode obtained by the zeroth order solution (5). In the case I, where the asymmetries of the rotor and the shaft

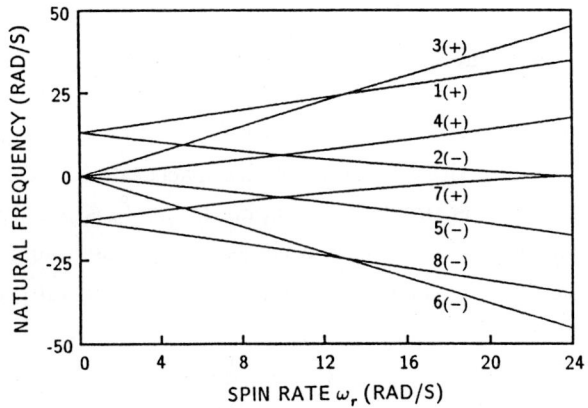

Fig. 4 Natural frequencies of the modes of attitude motion

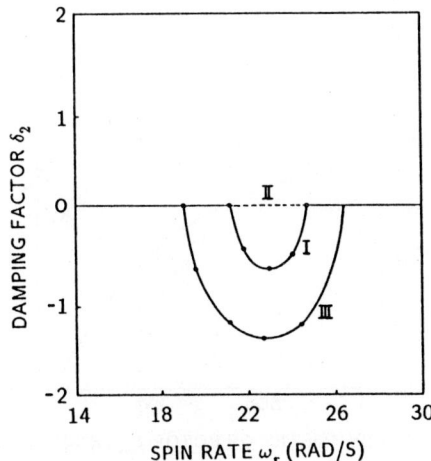

Fig. 5 Damping factor δ_2 of the 2nd mode

are reduced to zero, the asymmetry of the reaction torque of the moving parts in the rotor, which exerts about only the Y axis, causes the 2nd mode unstable. In the case II, where the asymmetry of the rotor is set up, i.e., the moment of inertia of the rotor about the Y axis increases $\hat{i}_{mr} = -15$ kgm^2, the asymmetries of the rotor and the reaction torque balance out and the instability disappears. In the case III, where the moment of inertia of the rotor about the Y axis decreases $\hat{i}_{mr} = 15$ kgm^2, the asymmetries of the rotor and the reaction torque are superimposed and the instability region extends. The numerical results of the damping factor calculated by the Floquet's theorem are shown with solid circles. They are in good agreement with analytical results.

Figure 6 shows the damping factor δ_2 of the 2nd mode as a function of the asymmetry of the rotor, where the energy dissipation is assumed to occur only on the stator. In this case, the stability rule derived by the energy sink method says that the attitude motion of the spacecraft is stable. When the asymmetry of the rotor is small, the damping factor δ_2 is positive. However, when the asymmetry of the rotor becomes large, the damping factor δ_2 becomes negative. The reason is as follows: from Eqs.(9),(10), damping factor δ_i is given by

$$\delta_i = \delta_i^{(+)} + \delta_i^{(-)}$$

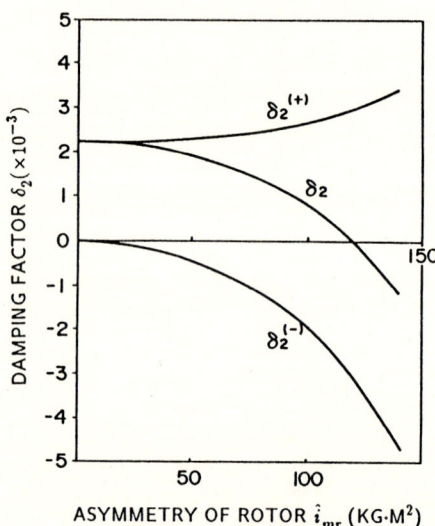

Fig. 6 Damping factor δ_2 of the 2nd mode

where

$$\delta_i^{(+)} = \text{Re}(b_i^{(+)}/a_i), \quad \delta_i^{(-)} = \text{Re}(b_i^{(-)}/a_i)$$

Factors $\delta_i^{(+)}$ and $\delta_i^{(-)}$ are the contributions of $(+)$ and $(-)$ mode components of the i-th mode to the damping factor δ_i, respectively. From Eq.(10), it is found that, if the i-th mode belongs to $(+)$ (or $(-)$) mode and the natural frequency of the mode is greater than the spin rate of the body on which a frictional force exerts, $\delta_i^{(+)}$ (or $\delta_i^{(-)}$) is positive and $\delta_i^{(-)}$ (or $\delta_i^{(+)}$) is negative. Figure 6 shows factors $\delta_2^{(+)}$ and $\delta_2^{(-)}$ as a function of the asymmetry of the rotor. For the value of the asymmetry $\hat{i}_{mr}$ greater than 130 kg m^2, $\delta_2^{(-)}$ becomes dominant and the 2nd mode turns out to be unstable. The stability rule derived by the energy sink method no longer holds for a dual spin spacecraft with a large asymmetry.

From Fig.4, it is possible that the 3rd and the 7th modes become resonant and unstable due to the asymmetries of the stator near $\omega_r = 13.7$ rad/s. Figure 7 shows the damping factor δ_3 of the 3rd mode based on the first order approximation solution (12). In the case I, where a large amount of asymmetry of the rotor is set up, an unstable region appears near $\omega_r = 13.7$ rad/s. On the contrary, in the case II, where the asymmetries of the rotor and the reaction torque balance out, the unstable region disappears. The

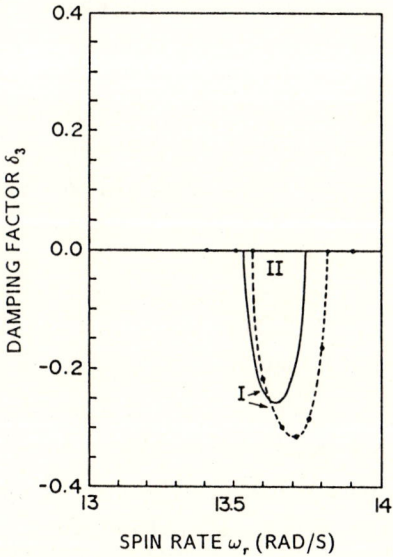

Fig. 7 Damping factor δ_3 of the 3rd mode

reason is as follows: it is found, from Eq.(11), that this type of instability depends on mainly parameter $c_{ij}^{(+)}$ and $c_{ji}^{(-)}$. Since parameters $c_{ij}^{(+)}$ and $c_{ji}^{(-)}$ are the functions of $\hat{\Phi}_i^{(+)}$, $\hat{\Phi}_j^{(-)}$, if the modes i and j belong to the same mode ((+) mode or (−) mode), parameters $c_{ij}^{(+)}$ and $c_{ji}^{(-)}$ become very small when the asymmetries in the rotor are reduced. If any two modes belong to the same mode ((+) mode or (−) mode), the instability due to asymmetry needs the interactions between the asymmetries in the rotor and the stator.

Conclusion

Stability of an attitude motion of a large dual spin spacecraft is studied. Main conclusions are as follows:

(1) An attitude motion of the spacecraft may become unstable due to an unequal reaction torque of moving parts in the spacecraft perpendicular to the spin axis.

(2) The effect of the energy dissipation in the spacecraft on the attitude stability depends on the asymmetries in the spacecraft; stability rule obtained by the energy sink method, in some cases, may no longer give a correct result.

(3) An interaction between asymmetries in the rotor and the stator may cause the attitude motion unstable.

The analysis is based on the method of multiple time scales. The results are verified by numerical solutions based on the Floquet's theorem.

Acknowledgement

This paper is based on work performed, in part, under the sponsorship and technical direction of the International Telecommunications Satellite Organization (INTELSAT). Any views expressed are not necessarily those of INTELSAT.

References

1. Tsuchiya, K. : Attitude Behavior of a Dual Spin Spacecraft Composed of Asymmetric Bodies. J. Guidance & Control, Vol.2, No.4 (1979) 328-333.
2. Agrawal, B.N. : Attitude Stability of Flexible Asymmetric Dual Spin Spacecraft. AIAA paper, 83-2177 (1983).
3. Likins, P.W. : Attitude Stability Criteria for Dual Spin Spacecraft. J. Spacecraft & Rocket, Vol.4 (1967) 1638-1643.
4. Nayfeh, A.H.; Mook, D.H. : Nonlinear Oscillations. John Wiley & Sons, 1979.

Appendix -- The elements of Eq.(1)

$$M^{(0)} = \begin{bmatrix} M^{(0)}_{(++)} & M^{(0)*}_{(-+)} \\ M^{(0)}_{(-+)} & M^{(0)}_{(++)} \end{bmatrix}, \quad M^{(0)}_{(++)} = \begin{bmatrix} \hat{i}_{pr} & \hat{m} & \hat{m}_{dr} \\ \hat{m} & \hat{i}_{ps} & 0 \\ \frac{n}{2}\hat{m}_{dr} & 0 & \hat{i}_r \end{bmatrix}, \quad M^{(0)}_{(-+)} = \begin{bmatrix} \hat{i}_{mr} - i\,\hat{i}_{cr} & 0 & 0 \\ 0 & 0 & 0 \\ -\frac{1}{2}\hat{m}_{dr}\sum_j e^{-i2\alpha_j} & 0 & 0 \end{bmatrix}$$

$$G^{(0)} = \begin{bmatrix} G^{(0)}_{(++)} & 0 \\ 0 & G^{(0)*}_{(++)} \end{bmatrix}, \quad G^{(0)}_{(++)} = \begin{bmatrix} -i(2\hat{i}_{pr}-\hat{i}_{zr})\omega_r & -2i\hat{m}\omega_r & 0 \\ -2i\hat{m}\omega_r & -2i\hat{i}_{ps}\omega_r & 0 \\ 0 & 0 & 0 \end{bmatrix}$$

$$K^{(0)} = \begin{bmatrix} K^{(0)}_{(++)} & K^{(0)*}_{(-+)} \\ K^{(0)}_{(-+)} & K^{(0)}_{(++)} \end{bmatrix}$$

$$K^{(0)}_{(++)} = \begin{bmatrix} k_p - (\hat{i}_{pr}-\hat{i}_{zr})\omega_r^2 & -(k_p + \hat{m}\,\omega_r^2) & \hat{m}_{dr}\omega_r^2 \\ -(k_p + \hat{m}\,\omega_r^2) & k_p - \hat{i}_{ps}\omega_r^2 & 0 \\ \frac{n}{2}\hat{m}_{dr}\omega_r^2 & 0 & k_r \end{bmatrix}$$

$$K^{(0)}_{(-+)} = \begin{bmatrix} k_m + (\hat{i}_{mr} - i\,\hat{i}_{cr})\omega_r^2 & -k_m & 0 \\ -k_m & 0 & 0 \\ -\frac{1}{2}\hat{m}_{dr}\sum_j e^{-2i\alpha_j} & 0 & 0 \end{bmatrix}$$

$$(M^{(1)})_{5,2} = (M^{(1)})^*_{2,5} = i_{ms}e^{2i\omega_r t}$$

$$(G^{(1)})_{5,2} = (G^{(1)})^*_{2,5} = 2i\,i_{ms}e^{2i\omega_r t}$$

$$(K^{(1)})_{5,2} = (K^{(1)})^*_{2,5} = -i_{ms}\omega_r^2 e^{2i\omega_r t}$$

$$F^T = (-i\,m_{ds}b_r\ddot{x}\exp(-i\omega_r t), -i\,m_{ds}a_r\ddot{x}\exp(-i\omega_r t), -c_r\dot{\Theta}^{(+)},$$

$$+i\,m_{ds}b_r\ddot{x}\exp(i\omega_r t), +i\,m_{ds}a_r\ddot{x}\exp(i\omega_r t), -c_r\dot{\Theta}^{(-)})$$

$$W = \sum_j [\frac{i}{2}(a_r\hat{\phi}_j^{(+)} + b_r\hat{\psi}_j^{(+)})(\lambda_j^2 + 2i\,\omega_r\lambda_j - \omega_r^2)\exp(t_j + i\,\omega_r t)$$

$$-\frac{i}{2}(a_r\hat{\phi}_j^{(-)} + b_r\hat{\psi}_j^{(-)})(\lambda_j^2 - 2i\,\omega_r\lambda_j - \omega_r^2)\exp(t_j - i\,\omega_r t)]$$

M^* means that all the element of M are replaced by its complex conjugate.

Robotics

Robot Control in Cartesian Space with Adaptive Nonlinear Dynamics Compensation

Hanspeter Faessler

Institut für Mechanik, ETH Zürich, Switzerland

Abstract
A motion control scheme is presented that allows for controller design directly in terms of task specific cartesian variables, rather than in joint variables. The overall control structure is separated into a discrete-time position controller plus an underlying dynamics compensator which assures the extended cartesian plant - the manipulator and the dynamics compensator - to behave like a set of decoupled unit masses in cartesian space. The cartesian compensator matrices are computed based on a model reference adaptive control scheme rather than using the explicit dynamical equations. The control structure was tested successfully in simulations and experiments for a three degrees of freedom high speed robot.

1. Introduction

Mechanical manipulators belong to a class of multibody systems which exhibit a dynamic behaviour that must be described by strongly nonlinear differential equations. These nonlinearities are associated both with positional variables and with velocity variables. Nevertheless, the control strategies for practically all industrial manipulators currently in use are based on classical linear control theory. Although a few more advanced industrial robot controllers compensate for some of the position-dependent nonlinear terms, such as gravity terms, they all neglect the velocity dependent nonlinear terms in the controller design, which restricts the manipulators to slow motions.

In several publications more advanced control schemes have been suggested which actively compensate for all nonlinear dynamic forces [1,8]. These schemes require the online evaluation of complex expressions in the dynamical equations, a task that gives rise to a considerable amount of computational effort. Furthermore, the control is generally done in joint space, rather than in a task-specific coordinate-space, such as a cartesian space. This requires the transformation of the task description (the desired motion) and of the controller specifications (maximum allowable deviations etc.) from task-specific space into joint space.

Recently, a scheme was proposed by Khatib [6], where the dynamical equations are first cast in terms of joint variables, but are then transformed into task-specific cartesian variables, whereupon they are used for the design of an underlying nonlinear dynamics compensator, the result being a linear closed loop system. This procedure offers three significant advantages over those used conventionally:

1) It allows for a simple controller design, since for the motion controller the total system to be controlled (manipulator plus compensator) should behave like a set of decoupled unit masses.
2) It allows for a controller design directly in terms of task-specific variables, rather than in joint variables. Among other things, this leads to a control structure

that permits a natural partition into complementary position-controlled and force-controlled subspaces.
3) It eliminates the need for any inverse kinematic transformation, i.e. a transformation from the task-specific cartesian variables into joint variables.

However, there are at least two obstacles which, so far, have limited the practical application of the above approach:
1) The unit-mass behaviour is guaranteed only so long as there is a perfect match between the real system (the manipulator) and the dynamic model on which the compensator is based. This implies that all geometric and inertia parameters, e.g., the load mass, must be known at all times.
2) Since the compensator design is based on a continuous time, zero delay model, the expressions underlying the dynamic model must be evaluated so quickly that discretization effects do not degrade performance. However, even with a computationally efficient formulation of the dynamical equations, the number of operations for a complete 6 d.o.f. manipulator model is still very large.

In this paper, a solution is suggested for these two problems. Instead of designing a cartesian dynamics compensator on the basis of the explicit dynamical equations, the compensation is formulated as a nonlinear model reference adaptive controller (MRAC) in cartesian space. The reference model consists of a set of n unit masses in cartesian space, where n is the number of degrees of freedom of the manipulator. The adaptive algorithm forces the dynamic response of the total system - the manipulator plus its adaptive compensator - to converge to that of the reference model. The algorithm can be applied to any multibody system; it is not confined to manipulators, and no restrictions are placed on the choice of generalized coordinates or generalized speeds. Also, it does not require any knowledge about the inertia properties of the system.

The approach taken in this paper was inspired by [4]. Their scheme, however, applies to joint space control only and is both more involved and less general than the one presented here. For other approaches to manipulator control based on adaptive control theory the reader may refer to [2,5].

The remainder of the paper is organized as follows: First, the dynamical equations in cartesian variables and the nonlinear dynamics compensator for a manipulator are presented in a general form. The discrete time controller for motion control in cartesian variables is described next. The controller is designed to take into account time delays for computations, AD/DA-conversions, etc. Thereafter, the MRAC scheme for cartesian dynamics compensation is outlined. A proof for the asymptotic stability of the adaptive compensator is given in the Appendix. In sections 6 and 7, both the adaptive compensation and the deterministic compensation based on the complete dynamical equations are applied to a three-degree-of-freedom high speed robot. The performance of both compensator types are demonstrated and discussed on the basis of simulations and experimental results. Finally, the extension of the presented control concept to simultaneous control of position and contact forces, often called hybrid control [1], will be discussed.

2. Dynamical equations in cartesian space

To obtain the dynamical equations of a general manipulator in cartesian variables, one may start with the dynamical equations in terms of n independent joint coordinates q_i and n independent joint speeds $\dot{q}_i$, where n is the number of degrees of freedom of the manipulator. The dynamical equations in joint variables can be written in the general form

$$\mathbf{M(q)\ddot{q} + V(q,\dot{q}) = F} \tag{1}$$

where $\mathbf{M}(\mathbf{q})$ denotes the $n \times n$ inertia matrix of the manipulator, $\mathbf{V}(\mathbf{q}, \dot{\mathbf{q}})$ contains all velocity dependent terms arising from inertia forces and inertia torques, as well as gravity force terms, and the n joint forces and joint torques are included in $\mathbf{F}$. In joint space control, the controller determines directly the driving joint forces and joint torques in $\mathbf{F}$, doing so on the basis of error signals generated by differences between actual and desired values for joint coordinates q_i and joint speeds $\dot{q}_i$ ($i = 1, \ldots, n$). In cartesian space control, the controller outputs $\mathbf{F_C}$ are imaginary cartesian force- and torque-components acting directly on the end effector, and the controller inputs are differences between actual and desired values for end effector (TCP) position ${}^c\mathbf{x}$ and velocity ${}^c\dot{\mathbf{x}}$ in a cartesian reference frame C fixed in space (Fig. 1). To derive the dynamical equations of the manipulator in cartesian space, i.e., of the cartesian plant shown in Fig. 1, one uses the following well known relationships [1]:

$$ {}^c\dot{\mathbf{x}} = {}^C\mathbf{J}(\mathbf{q})\dot{\mathbf{q}} \tag{2} $$

$$ \mathbf{F_C}(\mathbf{q}) = {}^C\mathbf{J}^{-T}(\mathbf{q})\mathbf{F} \tag{3} $$

where ${}^C\mathbf{J}$ denotes the Jacobian of the end effector TCP for the cartesian frame C. By taking the derivative with respect to time of Eq. (2) to obtain $\ddot{\mathbf{q}}(\mathbf{q}, \dot{\mathbf{q}}, {}^c\ddot{\mathbf{x}})$, and using Eqs. (1...3), the dynamical equations in cartesian space can be written as

$$ \mathbf{M_C}(\mathbf{q}){}^c\ddot{\mathbf{x}} + \mathbf{V_C}(\mathbf{q}, \dot{\mathbf{q}}) = \mathbf{F_C} \tag{4} $$

where *

(1...4)
$$ \mathbf{M_C}(\mathbf{q}) = {}^C\mathbf{J}^{-T}(\mathbf{q})\mathbf{M}(\mathbf{q}){}^C\mathbf{J}^{-1}(\mathbf{q}) \tag{5} $$

(1...5)
$$ \mathbf{V_C}(\mathbf{q}, \dot{\mathbf{q}}) = -\mathbf{M_C}(\mathbf{q}){}^C\dot{\mathbf{J}}(\mathbf{q})\dot{\mathbf{q}} + {}^C\mathbf{J}^{-T}(\mathbf{q})\mathbf{V}(\mathbf{q}) \tag{6} $$

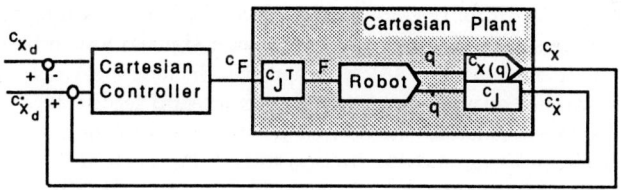

Figure 1: *Cartesian control system*

3. Compensation of nonlinear dynamics

Equation (4), describing the cartesian plant shown in Fig. 1, is inherently nonlinear, both in position and in velocity variables. This makes it impossible to design a cartesian controller based on linear control theory. However, since Eq. (4) should represent the nonlinear plant dynamics correctly, we use this model to actively compensate for all nonlinearities. To this end, we imagine that the cartesian controller output, which shall now be called ${}^c\mathbf{u}$, is input to an extended cartesian plant (Fig. 2) incorporating the previous cartesian plant plus a nonlinear dynamics compensator. The compensator is chosen such that the input $\mathbf{F_C}$ to the cartesian plant is determined by

$$ \mathbf{F_C} = \mathbf{M_C}(\mathbf{q}){}^c\mathbf{u} + \mathbf{V_C}(\mathbf{q}, \dot{\mathbf{q}}) \tag{7} $$

---

* Numbers in parentheses to the left of an equal sign refer to previous equations used to form the equation under consideration.

Assuming that $\mathbf{M}_C(\mathbf{q})$ and $\mathbf{V}_C(\mathbf{q},\dot{\mathbf{q}})$ model the dynamics of the cartesian plant perfectly, the extended cartesian plant dynamics is given by

$$^c\ddot{\mathbf{x}}(t) = {}^c\mathbf{u}(t) \qquad (8)$$

As a result, the cartesian controller can be designed to control a set of n completely decoupled unit masses.

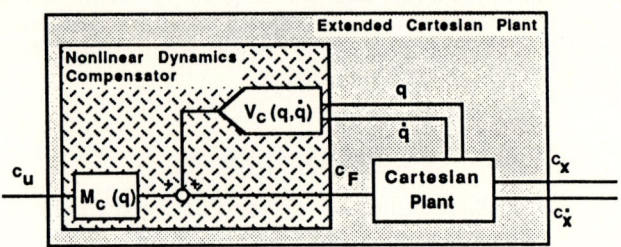

Figure 2: *Extended cartesian plant*

4. Discrete time cartesian controller

The design of the cartesian controller is based on the continuous-time extended cartesian plant model given in Eq. (8). However, in a practical implementation on a microprocessor system the controller output $^c\mathbf{u}(t)$ will not be updated continuously, but only at discrete times. The rate at which $^c\mathbf{u}$ is updated is determined by the time delays introduced through finite computation times for the control algorithm, for AD-conversions if analog sensor signals (e.g., tachometer signals) are to be converted, etc. Therefore, the plant model of Eq. (8) is extended to include a zero order hold element acting on the controller output. Additionally, an integral feedback on position errors is used to ensure convergence to the desired position even in the presence of constant disturbing forces, such as unmodeled friction forces. For each of the n unit masses to be controlled in cartesian space the following discrete-time state space representation then can be derived as

$$\begin{aligned}(x_1)_{k+1} &= (x_1)_k + T(x_2)_k + \frac{T^2}{2}(x_3)_k \\ (x_2)_{k+1} &= (x_2)_k + T(x_3)_k \\ (x_3)_{k+1} &= {}^c u_k \\ (x_4)_{k+1} &= (x_4)_k - (x_1)_k \end{aligned} \qquad (9)$$

The four state variables x_1, x_2, x_3, x_4 correspond to cartesian position, cartesian velocity, delayed controller output, and position error integral, respectively. Since all four states are directly accessible, a complete state controller can be used:

$$^c u_k = -(k_1[(x_1)_k - {}^c x_d] + k_2[(x_2)_k - {}^c \dot{x}_d] + k_3(x_3)_k + k_4(x_4)_k) \qquad (10)$$

$^c x_d$ and $^c \dot{x}_d$ represent the desired position and velocity, respectively. The choice of numerical values for the feedback gains $k_1, \ldots, k_4$ can be based on optimal control theory or on a pole placement scheme for discrete-time linear systems. For the results presented in this paper, pole placement was used.

The discrete-time state controller of Eq. (10) is designed to control a system that can be represented by Eq. (9). Equation (9) is an exact discrete-time representation of a continuous-time unit mass as described by Eq. (8), with a zero order hold on the driving force. However, Eq. (8) is an idealized representation of the extended cartesian plant, i.e., the manipulator plus the cartesian dynamics compensation. The main assumptions underlying Eq. (8) are:
1) Equation (4) is an exact model of the manipulator dynamics.
2) The matrices $\mathbf{M_C(q)}$ and $\mathbf{V_C(q,\dot{q})}$ can be updated continuously.

Of course, neither assumption can be totally valid in a practical implementation. As regards the model, the main difficulties are posed by friction forces and by the determination of inertia parameters, such as masses and moments of inertia, especially when an unknown load mass is carried by the manipulator. On the other hand, geometrical parameters such as constant angles and lengths of links can be obtained relatively easyly and very accurately. The second assumption means that the performance of the overall control system will be strongly dependent on how quickly the two matrices can be updated. Considering the constantly increasing efficiency of computer hardware, one may anticipate that this will become less of a problem in the future. At present, the online evaluation for a six degree of freedom manipulator cannot be carried out sufficiently quickly by reasonably priced hardware.

5. Adaptive cartesian dynamics compensation

The use of a model reference adaptive controller (MRAC) for the cartesian dynamics compensation could solve most of the above mentioned problems. The MRAC described in this paper is an extension of an MRAC scheme presented in [4]. The extension includes the application to cartesian space compensation as opposed to joint space compensation. Also, no assumptions regarding the structure of $\mathbf{V_C}$ are made, which at the same time simplifies the adaptive scheme and makes it applicable to a bigger class of multibody systems. This simplification largely reduces the number of operations and the number of adjustable parameters.

The reference model in the MRAC should incorporate the desired behaviour of the extended cartesian plant as described by Eq. (8). Therefore,

(8)
$$^c\ddot{\hat{\mathbf{x}}}(t) = {}^c\mathbf{u}(t) \tag{11}$$

is used as a reference model, where $^c\hat{\mathbf{x}}(t)$ measures the position of the n decoupled reference model unit masses in cartesian space. The design goal of the MRAC scheme is to adjust adaptively the elements of matrices $\hat{\mathbf{M}}_C$ and $\hat{\mathbf{V}}_C$ shown in Fig. 3 so that the errors $^c\mathbf{e}(t)$ and $^c\dot{\mathbf{e}}(t)$ defined as

$$^c\mathbf{e}(t) \doteq {}^c\hat{\mathbf{x}}(t) - {}^c\mathbf{x}(t) \quad \text{and} \quad {}^c\dot{\mathbf{e}}(t) \doteq {}^c\dot{\hat{\mathbf{x}}}(t) - {}^c\dot{\mathbf{x}}(t) \tag{12}$$

converge to zero. This again means that the dynamic behaviour of the extended cartesian plant converges to that of the reference model. To this end, the cartesian driving forces are determined by the following control algorithm (Fig. 3):

$$\mathbf{F}_C(t) = \hat{\mathbf{M}}_C(t){}^c\mathbf{u}(t) + \hat{\mathbf{V}}_C(t) + h_p {}^c\mathbf{e} + h_v {}^c\dot{\mathbf{e}} \tag{13}$$

For the adaptation of $\hat{\mathbf{M}}_C(t)$ and $\hat{\mathbf{V}}_C(t)$ an additional error feeback,

$$^c\mathbf{y}(t) \doteq c_p{}^c\mathbf{e}(t) + c_v{}^c\dot{\mathbf{e}}(t) \tag{14}$$

is required. Then, with $\hat{m}_{ij}, {}^cy_i, {}^cu_i$ denoting the elements of $\hat{\mathbf{M}}_C, {}^c\mathbf{y}, {}^c\mathbf{u}$, respectively, the adaptation algorithm for $\hat{\mathbf{M}}_C$ is given by

$$\hat{m}_{ii}(t) = \hat{m}_{ii}(0) + k_{mi}\int_0^t {}^cy_i(\tau){}^cu_i(\tau)d\tau; \qquad i = 1,2,3$$

$$\hat{m}_{ij}(t) = \hat{m}_{ij}(0) + k_{mij}\int_0^t [{}^cy_i(\tau){}^cu_j(\tau) + {}^cy_j(\tau){}^cu_i(\tau)]d\tau; \quad i,j = 1,2,3, i \neq j \tag{15}$$

and with $\hat{v}_i$ denoting the elements of $\hat{\mathbf{V}}_C$

$$\hat{v}_i(t) = \hat{v}_i(0) + k_{vi}\int_0^t {}^cy_i(\tau)d\tau; \qquad i = 1,2,3 \tag{16}$$

Under the assumption that $\mathbf{M}_C$ and $\mathbf{V}_C$ remain constant during the adaptation, the adaptive dynamics compensation can be proven to be asymptotically stable if the following conditions on the choice of the error feedback coefficients are satisfied:

$$c_p, c_v, h_p, h_v \geq 0 \quad , \quad c_v > c_p,$$
$$c_v h_v \mathbf{I} - c_p \mathbf{M}_C > 0 \quad , \quad (c_v h_p + c_p h_v)\mathbf{I} - c_p \mathbf{M}_C > 0 \tag{17}$$

An outline of the stability proof is given in the Appendix.

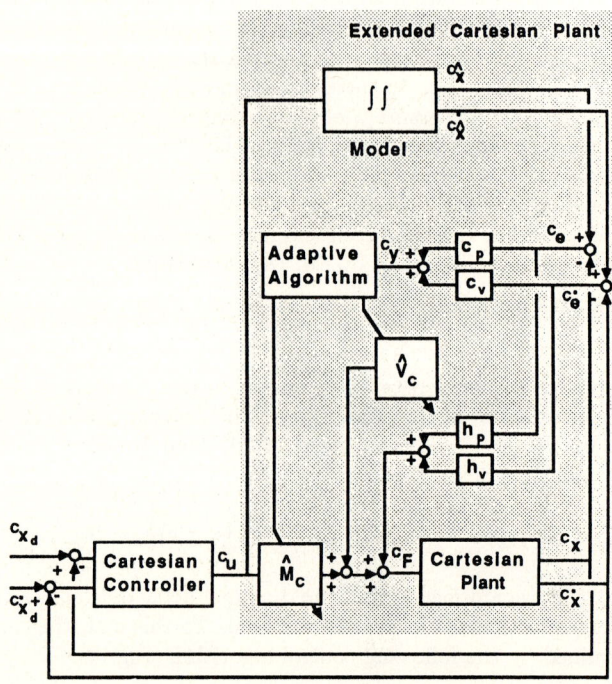

Figure 3: *Cartesian Control System with MRAC dynamics compensation*

6. Three degrees of freedom high speed robot

Position controller and dynamics compensation were tested on the cylindrical three degrees of freedom high speed robot [3] depicted in Fig. 4. A schematic representation of the robot model is shown in Fig. 5, where B_1 denotes a fixed point on the vertical axis of rotation of body 1, B_i^* denotes the center of mass of body i ($i=2,3$), and D stands for the TCP of the end effector. Three dextral sets of unit vectors are introduced: $\underline{x}_0, \underline{y}_0, \underline{z}_0$ are inertially fixed with $\underline{z}_0$ parallel to the vertical axis of rotation of body 1, $\underline{x}_1, \underline{y}_1, \underline{z}_1$ are fixed on body 3 with $\underline{z}_1$ parallel to $\underline{z}_0$ and $\underline{x}_1$ parallel to the horizontal axis of relative translation from body 3 to body 2, $\underline{x}_C, \underline{y}_C, \underline{z}_C$ are fixed on the task-dependent cartesian compliance frame C in which the robot control should be formulated. To characterize the orientation of compliance frame C relative to the inertially fixed frame 0, three orientation angles $\alpha_1, \alpha_2, \alpha_3$ are introduced. A general orientation of frame C relative to frame 0 is obtained by starting with originally coincident orientations and first rotating C about a line parallel to $\underline{x}_C$ by an amount α_1, then rotating C about a line parallel to the newly oriented unit vector $\underline{y}_C$ by an amount α_2, and finally about a line parallel to $\underline{z}_C$ by an amount α_3. The corresponding direction cosine matrix is given by

$$^0_C T = \begin{pmatrix} c\alpha_2 c\alpha_3 & -c\alpha_2 s\alpha_3 & s\alpha_2 \\ s\alpha_1 s\alpha_2 c\alpha_3 + s\alpha_3 c\alpha_1 & -s\alpha_1 s\alpha_2 s\alpha_3 + c\alpha_3 c\alpha_1 & -s\alpha_1 c\alpha_2 \\ -c\alpha_1 s\alpha_2 c\alpha_3 + s\alpha_3 s\alpha_1 & c\alpha_1 s\alpha_2 s\alpha_3 + c\alpha_3 s\alpha_1 & c\alpha_1 c\alpha_2 \end{pmatrix} \quad (18)$$

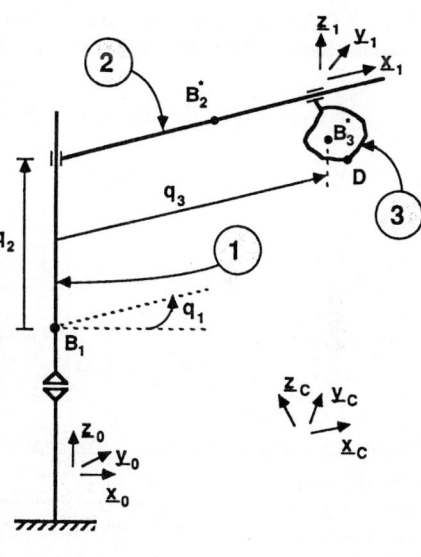

Figure 4
Three degree of freedom high speed robot

Figure 5
Robot model

The following position vectors describe the relative position of the points in Fig. 5:

$$\overrightarrow{B_1 B_2^*} = L_1 \underline{x}_1 + q_2 \underline{z}_1$$
$$\overrightarrow{B_1 B_3^*} = q_3 \underline{x}_1 + L_2 \underline{y}_1 + (q_2 + L_3)\underline{z}_1$$
$$\overrightarrow{B_3^* D} = d_1 \underline{x}_1 + d_2 \underline{y}_1 + d_3 \underline{z}_1 \tag{19}$$

The inertia properties of the three bodies are modelled such that all central principal axes are parallel to one of the unit vectors $\underline{x}_1, \underline{y}_1, \underline{z}_1$, and the center of mass of body 1 lies on the vertical axis of rotation.

Only a few intermediate results of the derivation of the dynamical equations will be given in the sequel. First, by writing the absolute velocity of tool center point D in components of frame 1, the Jacobian $^1\mathbf{J_D}$ of point D in frame 1 components can be obtained as

$$^1\mathbf{J_D} = \begin{pmatrix} -(L_2 + d_2) & 0 & 1 \\ q_3 + d_1 & 0 & 0 \\ 0 & 1 & 0 \end{pmatrix} \tag{20}$$

The Jacobian of point D in frame C components can be calculated from Eqs. (18,20) according to

$$^C\mathbf{J_D} = {}^C_0\mathbf{T}\, {}^0_1\mathbf{T}\, {}^1\mathbf{J_D} \tag{21}$$

where $^0_1\mathbf{T}$ denotes the direction cosine matrix of frame 1 relative to frame 0. The matrices $\mathbf{M(q)}$, $\mathbf{V(q,\dot{q})}$, and $\mathbf{F}$ then can be found as

$$\mathbf{M(q)} = \begin{pmatrix} I_{1tot} + I_2 + L_1^2 m_2 + I_3 + (q_3^2 + L_2^2)m_3 & 0 & -m_3 L_2 \\ 0 & m_2 + m_{2eq} + m_3 & 0 \\ -m_3 L_2 & 0 & m_{3eq} + m_3 \end{pmatrix} \tag{22}$$

$$\mathbf{V(q,\dot{q})} = \begin{pmatrix} 2 m_3 q_3 \dot{q}_1 \dot{q}_3 \\ (m_2 + m_3)g \\ -m_3 q_3 \dot{q}_1^2 \end{pmatrix}, \quad \mathbf{F} = \begin{pmatrix} F_1 \\ F_2 \\ F_3 \end{pmatrix} \tag{23}$$

where m_i is the mass and I_i the central principal moment of inertia about a vertical line of body i ($i = 2,3$), m_{ieq} is a translationally accelerated mass equivalent to the rotational inertias of the rotor of the driving motor, gears and other rotating force transmission elements, I_{1tot} is the central principal moment of inertia about a vertical line of body 1 plus equivalent inertias for driving motor, gear box etc., $F_1 \underline{z}_1$ is the driving torque vector acting on body 1 and $F_2 \underline{z}_1$, $F_3 \underline{x}_1$ are the driving force vectors acting on bodies 2 and 3, respectively.

From Eqs. (21, 22, 23) the dynamical equations (4) in cartesian space C can be obtained with the help of Eqs. (5, 6).

7. Simulations and experimental results

The performance of the control schemes under consideration was first investigated by means of a very detailed simulation model for the manipulator of Fig. 4. It included not only a realistic model of the mechanical structure, including friction forces etc., but also a model of the electric drive units and of the time delays caused by finite computation times for the different parts of the controller. The controller was programmed in MODULA-2 and implemented on two Motorola 68000 microprocessors, with the discrete-time position control of Eqs. (9,10) running on the first, and the dynamics compensation on the second. For more details on the actual implementation, the reader may refer to [2].

First, a few simulation results will be shown, where the deterministic cartesian dynamics compensation of section 3 is used. Figure 6 shows three repositioning manoeuvres, each over a distance of 30 cm in 0.3 seconds. In Figure 6a a vertical motion is simulated, in Figure 6b a horizontal motion, and in Figure 6c a motion parallel to x_C for $\alpha_1 = 45$ Deg., $\alpha_2 = 0$ Deg., $\alpha_3 = 45$ Deg. The dashed line represents the commanded trajectory and the solid line the actual trajectory. The three resulting trajectories are basically identical, which shows that the cartesian dynamics compensation works well. Here, it should again be pointed out, that the position controllers for every direction in the cartesian frame C are identical. However, the dynamics of the cartesian plant (the manipulator) depends very much on the direction in which the motion is performed. For a vertical motion the driving forces must overcome gravity and accelerate a total mass $m_2 + m_{2eq} + m_3$ of 12 kg, for a horizontal motion only a mass $m_3 + m_{3eq}$ of 5 kg is accelerated, and in Figure 6c all three axes are moving simultaneously with a nonlinear dynamic coupling between axes one and three. The largest errors occur during acceleration and deceleration phases. The main reason for this is that the command of the actual driving forces $\mathbf{F}$ acting on the manipulator is delayed by approximately 10 ms, which is the time needed for the computation of the matrices $\mathbf{M_C}$, $\mathbf{V_C}$ and ${}^C\mathbf{J}^T$ and the requisite matrix multiplications and additions shown in Figures 1 and 2. Significantly smaller errors are obtained for reduced delay times and/or for slower movements.

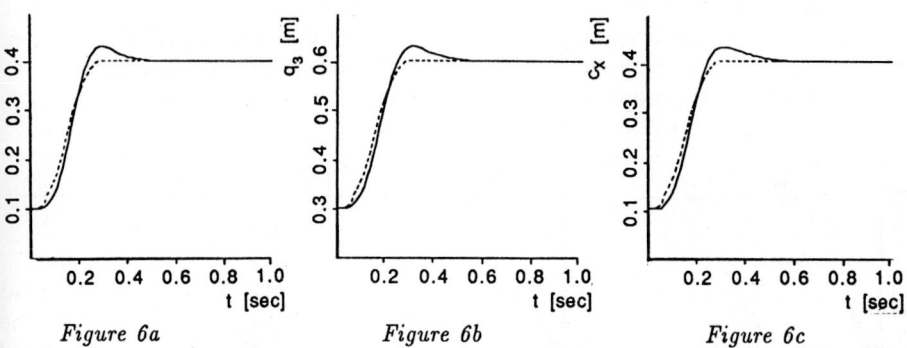

Figure 6a   Figure 6b   Figure 6c

Simulation results with deterministic dynamics compensation: Commanded and actual position vs. time

Figure 7 shows measurements on the actual robot corresponding to the simulations in Fig. 6a and 6b. The reason for the steps in the actual position measurements is that new positions are stored only at the rate at which the position controller is running, i.e., every 13 ms. Since the rotational degree of freedom was not yet in operation at the time the measurements were taken, no measurement corresponding to Fig. 6c is shown. Instead, a measurement of a repositioning maneuver over 0.3 m in vertical direction with a repositioning time of 0.6 seconds is displayed in Fig. 7c. The close coincidence between commanded and actual trajectory indicates that for more reasonable but still quite short positioning times the controller quality is very acceptable.

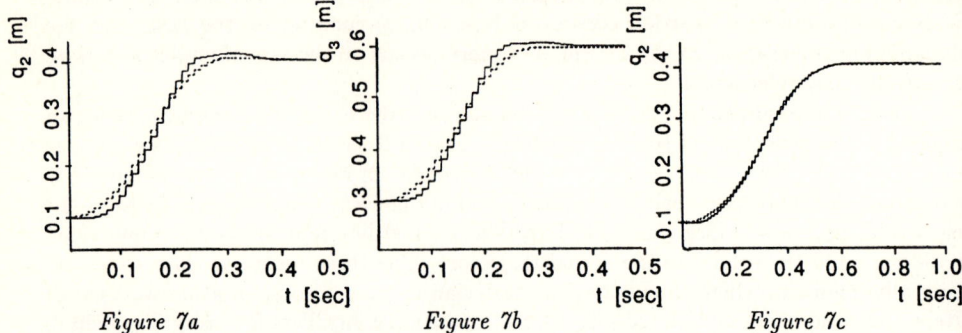

Measured results with deterministic dynamics compensation: Commanded and actual position vs. time

Figure 8 shows simulation results of the same repositioning maneouvers as in Fig. 6. For these simulations, however, the adaptive dynamics compensation of section 5 was used. The matrices $\hat{M}_C$ and $\hat{V}_C$ were initialized with zeroes, which represents a worst case assumption. The adaptive compensation ran at a rate of 1 KHz, whereas the sampling time of the position controller was kept at 13 ms. Again, the results for the three movements in different directions are almost identical, which indicates the proper functioning of the adaptive compensator. The undesireable overshoot can be reduced or even completely eliminated by taking one or several of the following measures: A more realistic initialization of $\hat{M}_C$ and $\hat{V}_C$; an increase in repositioning time; a different choice of the commanded trajectory in the end phase of the motion; probably a better choice of the feedback coefficients c_p, c_v, h_p and h_v, and an increase in sampling rate for the adaptive compensator.

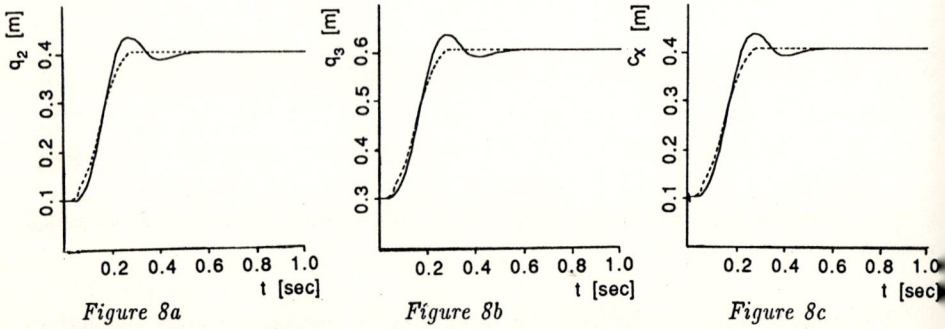

Simulation results with adaptive dynamics compensation : Commanded and actual position vs. time

Because the stability proof for the adaptive compensation is based on the assumption that $\mathbf{M_C}$ and $\mathbf{V_C}$ remain constant during the adaptation, a few cases were simulated where the load mass was instantaneously changed, but no unacceptable responses were obtained.

Here, it should be made clear again that, although the cartesian adaptive compensation does not require a model of the manipulator dynamics, it does require a correct model for the manipulator geometry underlying the transformations $^C\mathbf{J(q)}$ and $^c\mathbf{x(q)}$ between joint space and cartesian space. The overall behavior of the cartesian plant is made to converge towards the desired unit mass behavior, but errors in the transformations within the cartesian plant (Fig. 1) will lead to errors in the end effector position.

8. Extension to simultaneous control of position and contact forces

The concept of cartesian control can be applied very naturally to the problem of simultaneous control of position and contact forces [1,6]. In [2], the control concepts presented in this paper were extended to simultaneously control position along the unconstrained directions of the cartesian frame C, e.g., the ones in a cartesian subspace spanned by $\underline{x}_C$ and $\underline{y}_C$, and contact forces along the constrained directions of a complementary cartesian subspace, e.g., the one spanned by $\underline{z}_C$. As for the type of contact, it is modelled to be stiff, which means that kinematical constraints are introduced. Consequently, the matrices in the cartesian dynamical equations (4) must be reformulated to describe motion in the position controlled subspace only. However, motion in the position controlled subspace may lead to dynamic forces in the force controlled subspace. Therefore, a model equivalent to Eq. (4) for the force controlled subspace must be derived, which is then used to design a dynamics compensator analogous to Eq.(7). The resulting extended cartesian plant for the force controlled subspace can be represented by a simple zero order hold element for each force controlled direction in frame C. A discrete time force controller is then derived to control a zero order hold plant. Finally, the cartesian dynamics compensator for the force controlled subspace was also formulated as an MRAC, and the whole control scheme was successfully tested on the three degrees of freedom high speed robot described in section 6 [2].

9. Conclusions

A control scheme was presented that allows for controller design directly in terms of task specific cartesian variables, rather than in joint variables. The overall control structure was separated into a discrete-time position controller and an underlying dynamics compensator which assures the extended cartesian plant - the manipulator plus the dynamics compensator - to behave like a set of decoupled unit masses in cartesian space. To overcome the problems of unknown inertia parameters as well as to shorten the cycle time of the dynamics compensator, a model reference adaptive control scheme was presented for updating the cartesian compensator matrices rather than computing these matrices based on the explicit dynamical equations. The control structure was applied to and tested on a three degrees of freedom high speed robot. Simulations and experimental results showed that even for very fast positioning maneouvres the resulting responses are well behaved.

Appendix

Stability proof for adaptive nonlinear dynamics compensation

First, the definitions and basic results of hyperstability theory that will be used in the stability proof are summarized [7,9]:

The standard multivariable feedback system depicted in Fig. 9 is formed by a linear time-invariant feedforward block and a nonlinear time-varying feedback block. A nonlinear time-varying feedback block $\mathbf{w}(^c\mathbf{y}, t)$ is denoted as belonging to the class $\{P\}$ if it satisfies the Popov integral inequality

$$\int_0^t \mathbf{w}^T(\tau)^c\mathbf{y}(\tau)d\tau \geq -\gamma_0^2 \quad \text{for all} \quad t \geq 0 \tag{A.1}$$

with γ_0 being a positive constant depending only on initial conditions.

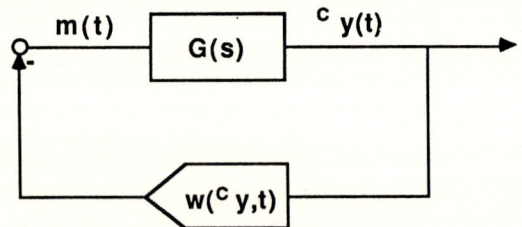

Figure 9: *Standard multivariable feedback system*

A standard nonlinear feedback system is said to be asymptotically hyperstable if it is globally asymptotically stable for all feedback blocks $\mathbf{w}(^c\mathbf{y},t) \in \{P\}$. Once the feedback block satisfies inequality (A.1), the hyperstability properties of the total feedback system will depend only on the characteristics of the feedforward block. The necessary and sufficient condition for a standard nonlinear feedback system with class $\{P\}$ nonlinearity to be asymptotically hyperstable is that the transfer matrix of the linear feedforward block be strictly positive real.

The first step in the stability proof is to show that the MRAC dynamics compensator as shown in Fig. 3 can be represented in the form of a standard nonlinear feedback system. To this end, the equations describing the robot dynamics, the reference model dynamics, and the adaptive compensator structure are combined to

$$\begin{aligned}
(4, 11) \quad \mathbf{M}_C(t)^c\ddot{\mathbf{x}}(t) + \mathbf{V}_C(t) &= \mathbf{F}_C(t) + \mathbf{M}_C(t)^c\ddot{\mathbf{x}}(t) - \mathbf{M}_C(t)^c\mathbf{u}(t) \\
(13) \quad &= \hat{\mathbf{M}}_C(t)^c\mathbf{u}(t) + \hat{\mathbf{V}}_C(t) + h_p{}^c\mathbf{e}(t) + h_v{}^c\dot{\mathbf{e}}(t) \\
&\quad + \mathbf{M}_C(t)^c\ddot{\mathbf{x}}(t) - \mathbf{M}_C(t)^c\mathbf{u}(t)
\end{aligned} \tag{A.2}$$

$$(A.2, 12) \quad \mathbf{M}_C(t)^c\ddot{\mathbf{e}}(t) + h_v{}^c\dot{\mathbf{e}}(t) + h_p{}^c\mathbf{e}(t) = [\mathbf{M}_C(t) - \hat{\mathbf{M}}_C(t)]^c\mathbf{u}(t) + \mathbf{V}_C(t) - \hat{\mathbf{V}}_C(t) \tag{A.3}$$

After $\mathbf{m}(t)$ has been defined as

$$\mathbf{m}(t) \doteq [\mathbf{M_C}(t) - \hat{\mathbf{M}}_\mathbf{C}(t)]^c\mathbf{u}(t) + \mathbf{V_C}(t) - \hat{\mathbf{V}}_\mathbf{C}(t) \quad (A.4)$$

the transfer matrix $\mathbf{G}(s)$ from input $\mathbf{m}$ to output $^c\mathbf{y}$ as defined in Eq. (14) is given by

$$\mathbf{G}(s) \doteq {}^c\mathbf{y}(s)\mathbf{m}^{-1}(s)$$

(14, A.3, A.4)
$$= [c_p\mathbf{I} + c_v\mathbf{I}s][\mathbf{M_C}s^2 + h_v\mathbf{I}s + h_p\mathbf{I}]^{-1} \quad (A.5)$$

with $\mathbf{I}$ being the unit matrix. Equation (A.5) represents the linear block in the standard nonlinear feedback system. Since the linear block must be time invariant, the robot inertia matrix $\mathbf{M_C}$ must be assumed to remain constant during the adaptation process. The nonlinear block is then defined as

(A.4)
$$\mathbf{w}(t) \doteq -\mathbf{m}(t)$$
$$= [\hat{\mathbf{M}}_\mathbf{C}(t) - \mathbf{M_C}(t)]^c\mathbf{u}(t) - \mathbf{V_C}(t) + \hat{\mathbf{V}}_\mathbf{C}(t) \quad (A.6)$$

Next, $\mathbf{w}(t)$ must be shown to satisfy inequality (A.1). Writing the inner product of input and output as a summation from 1 to the number of degrees of freedom n, one obtains

(A.6)
$$\int_0^t \mathbf{w}^T(\tau)^c\mathbf{y}(\tau)d\tau = \sum_{j=1}^n \sum_{i=1}^n \int_0^t [\hat{m}_{ji}(\tau) - m_{ji}]^c u_i(\tau)^c y_j(\tau)d\tau$$
$$+ \sum_{i=1}^n \int_0^t [\hat{v}_i(\tau) - v_i(\tau)]^c y_i(\tau)d\tau \quad (A.7)$$

The validity of inequality (A.1) can be shown separately for each value of the summation indices i and j. First, the second term in the right-hand side of Eq. (A.7) is integrated using Eq. (16):

(16)
$$\int_0^t [\hat{v}_i(\tau) - v_i(\tau)]^c y_i(\tau)d\tau = \int_0^t [\hat{v}_i(0) - v_i + k_{vi}\int_0^\tau {}^c y_i(\sigma)d\sigma]^c y_i(\tau)d\tau$$
$$\geq -[\hat{v}_i(0) - v_i]^2/2k_{vi} \quad (A.8)$$

For the right-hand side of inequality (A.8) to be constant, $\mathbf{V_C}$ must be regarded as constant during the adaptation process. Inequality (A.8) can be proven as follows: Define auxiliary quantities

$$a \doteq \hat{v}_i(0) - v_i \quad , \quad k \doteq k_{vi} > 0 \quad , \quad g(t) \doteq {}^c y_i(t) \quad , \quad h(t) \doteq g(t) - g(0) \quad (A.9)$$

and rewrite inequality (A.8) in terms of these quantities:

$$\int_0^t [a + k\int_0^\tau \dot{g}(\sigma)d\sigma]\dot{g}(\tau)d\tau = a[g(t) - g(0)] - kg(0)[g(t) - g(0)] + \frac{k}{2}[g^2(t) - g^2(0)]$$

(A.9)
$$= ah(t) + \frac{k}{2}h^2(t)$$
$$= [\frac{a}{\sqrt{2k}} + \sqrt{\frac{k}{2}}h(t)]^2 - \frac{a^2}{2k}$$
$$\geq -\frac{a^2}{2k} \quad (A.10)$$

The same reasoning can be used to prove analogous inequalities obtained for every value of i, j in the double summation in Eq. (A.7).

The last step in the stability proof is to show that the transfer matrix $G(s)$ as defined in Eq. (A.5) is strictly positive real. The conditions on the gains c_p, c_v, h_p, and h_v as stated in Eq. (17) are chosen such that this requirement is fulfilled for all symmetric positive definite matrices M_C. For more details, the reader may refer to [2].

References

[1] Craig, J.J.: *Introduction to Robotics: mechanics and control.*
Addison-Wesley, Reading (Mass.),1986
[2] Faessler, Hp.: Concurrent Position- and Force-Control of Robot Manipulators.
Ph.D. Thesis, Swiss Federal Institute of Technology (ETH), Zurich, 1988
[3] Faessler,Hp.; Buffinton, K.; Nielsen, E.: Design of a High Speed Robot Skilled in The Play of Ping Pong. *18th International Symposium on Industrial Robots.*
Lausanne (Switzerland), April 1988
[4] Horowitz, R.; Tomizuka, M.: An Adaptive Control Scheme for Mechanical Manipulators - Compensation of Nonlinearity and Decoupling Control.
Journal of Dynamic Systems, Measurement, and Control, Vol. 108, June 1986, pp. 127-135
[5] Hsia, T.C.; Adaptive Control of Robot Manipulators - A Review.
IEEE Int. Conference on Robotics and Automation. San Francisco, April 1986
[6] Khatib, O.: A Unified Approach for Motion and Force Control of Robot Manipulators: The Operational Space Formulation.
IEEE Journal of Robotics and Automation Vol. RA-3, No. 1, 1987, pp. 43-53
[7] Landau, Y.D.: *Adaptive Control: The Model Reference Approach.*
Marcel Dekker, New York, 1979
[8] Patzelt, W.: Zur Lageregelung von Industrierobotern bei Entkopplung durch das inverse System. *Regelungstechnik* No. 12, 1981, pp. 411-422
[9] Popov, V.M.: *Hyperstability of Control Systems.* Springer Verlag, Berlin, 1973

Modeling and Control of Elastic Robot Arm with Prismatic Joint

M. GÜRGÖZE [*], P. C. MÜLLER

Safety Control Engineering
University of Wuppertal
D-5600 Wuppertal 1, FRG

Summary

A reasonable modeling and a suitable design of a control system for the translational motion of an elastic robot arm with a prismatic joint is a still open problem. In this paper the dynamic behaviour of such an elastic beam is described with respect to control requirements. A complex control system is obtained represented approximately by a set of ordinary linear time-variant differential equations of variable order. Certain approaches of designing a feedback control are discussed.

Introduction

The application of industrial robots to advanced manufacturing tasks requires highly accurate position and/or force control. Actual limitations to these requirements are mainly caused by elasticity, Coulomb friction and backlash in the system. A basic problem is to develop a control feedback for damping out the elastic vibrations such that the end-effector can perform its tasks without delay. In almost all the investigations so far, elastic robots with revolute joints have been considered only, i. e. the flexible members of the robot has been assumed to have fixed lengths. Surveys and recent results on the fast control of elastic robots with rotational degrees of freedom are given by Henrichfreise [1] and Ackermann [2].

Just recently first results on translational moving flexible robot arms with prismatic joints were published. Lilov and Wittenburg [3] presented a general formalism to model the dynamics of chains of rigid bodies and elastic rods with revolute

---
[*] On leave of the Technical University of Istanbul, Turkey, by a fellowship of the Alexander von Humboldt-Foundation.

and prismatic joints. By Riemer and Wauer [4] the equations of motions were derived for a planar two-body system with beam-shaped substructures and with a revolute-prismatic joint. Wang and Wei considered the vibrations of a moving flexible robot arm with prismatic joint [5] and its feedback control [6]. Here, the vibrations of the robot arm were composed of two bending motions. The torsional motion as well as gravitational effects were neglected. The driving motion of the prismatic joint was assumed to consist only of a translation along and a rotation about the vertical axis.

In this paper a more general problem is discussed. A robot with a flexible arm is considered as shown in Fig. 1. The prismatic

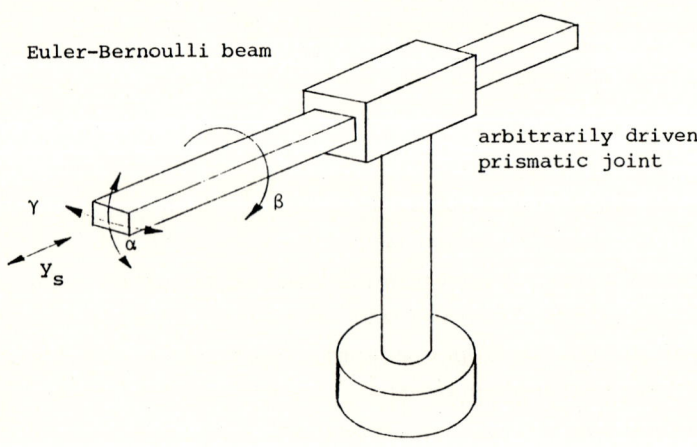

Fig. 1. Sketch of elastic robot arm

joint connecting the elastic beam and - in general - the preceeding robot link is built such that the beam axis and the link axis are made to coincide at two or more points by bearings which allow only relative translational motion $y_S(t)$. The orientation and the motion of the preceeding link and with that the orientation and the motion of the joint may be arbitarily given; it characterizes orientation and motion of a reference coordinate system, cf. $\{x_B, y_B, z_B\}$ and $v_B(t)$, $\omega_B(t)$ in Fig. 2.

The elastic vibrations of the beam are composed of bending motions in the x- and z-directions perpendicular to the y-axis of the beam and of a torsional motion about the y-direction.

Based on the rigid body model of the non-elastic robot arm in the following two approaches of modeling the motion of the elastic robot arm will be considered: continuum model and multibody model. Subsequently certain remarks on the problem of reduction of model order will be presented to obtain a suitable model for control design. Finally the design of a control feedback for an active damping of the elastic vibrations will be discussed.

Rigid Body Model

Firstly we consider the problem of a rigid robot arm with prismatic joint. Here, only the equation of translational motion of the arm relatively to the prismatic joint has to be derived. This results in

$$m \ddot{y}_S - m(\omega_{Bx}^2 + \omega_{By}^2) y_S =$$
$$- m(\dot{v}_{By} + v_{Bx} \omega_{Bz} - v_{Bz} \omega_{Bx})$$
$$- m g(\sin \bar{\alpha} \cos \bar{\gamma} + \cos \bar{\alpha} \sin \bar{\beta} \sin \bar{\gamma})$$
$$+ Q_y + F(t) \tag{1}$$

where m is the mass of robot arm; v_{Bx}, v_{By}, v_{Bz} and ω_{Bx}, ω_{By}, ω_{Bz} are the components of velocity v_B and angular velocity ω_B of the joint represented in the joint-fixed coordinate system; the angles $\bar{\alpha}$, $\bar{\beta}$, $\bar{\gamma}$ characterize the actual orientation of the joint and are needed to represent gravitational effects; the normal force Q_y arises due to a dynamic end load, and finally $F(t)$ is the axial control force.

Usually the task of the robot will define certain time functions $\bar{\alpha}(t)$, $\bar{\beta}(t)$, $\bar{\gamma}(t)$ for the orientation and $v_B(t)$, $\omega_B(t)$ for the motion of the joint as well as $y_S(t)$ for the relative position of the robot arm. Then equation (1) defines the required control force F(t).

Continuum Model

The vibrations of the elastic robot arm are composed of bending motions w_x, w_z and torsional motion β, cf. Fig. 2. To derive

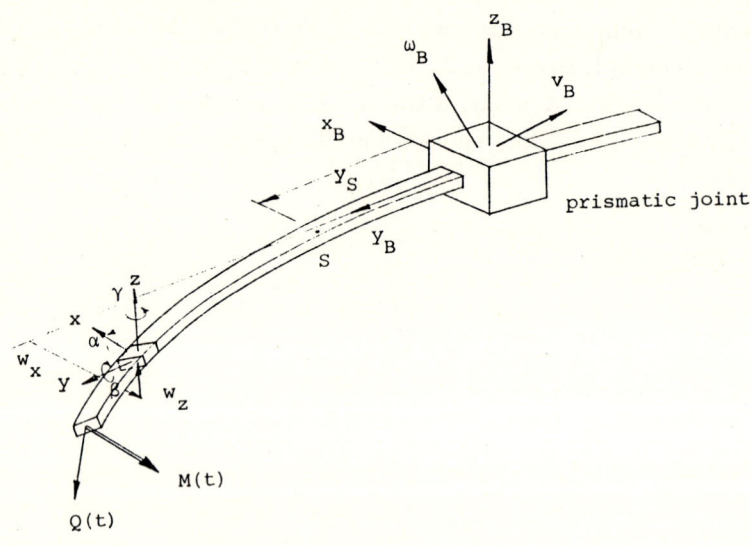

Fig. 2. Elastic robot arm

the equations of motion the generalized Hamilton's principle is applied:

$$\int_{t_1}^{t_2} [\delta(T - V) + \delta'A] \, dt = 0 \, . \tag{2}$$

The kinetic energy T is due to translation and rotation of each element of the beam. The potential energy V has to be regarded with respect to bending and torsion as well as to gravitational effects and axial forces. The virtual work $\delta'A$ has to be calculated due to the axial control force F(t) at the joint and the end load consisting of force Q(t) and torque M(t).

According (2) a coupled set of one ordinary differential equation for the driven translational motion and three partial differential equations for the two bending and the torsional vibrations is derived. Additionally, boundary conditions and time-variant intermittency conditions depending on $y_S(t)$ will appear. For example, the bending $w_x(y, t)$ is gouverned by the partial differential equation

$$\rho A \{\ddot{w}_x - \omega_{BZ}\dot{y}_S - \dot{\omega}_{BZ}(y+y_S) + \gamma_1$$

$$- [\ddot{y}_S + \gamma_2 - s_{g2} + S_{22} y_S] \cdot [w_x''(y-\frac{L}{2}) + w_x']$$

$$- \frac{1}{2} S_{22}[w_x'' (y^2-\frac{L^2}{4}) + 2yw_x'] \}$$

$$+ E I_z w_x''''$$

$$+ \rho [b_{11}w_x'' + b_{21}w_z'' + b_{31}\dot{\beta}' + b_{41}w_x'' + b_{61}\ddot{\beta}$$

$$- (b_{14}w_x')^{\cdot\cdot} - (b_{24}w_z')^{\cdot\cdot} - (b_{34}\beta)^{\cdot\cdot}$$

$$- (b_{44}\dot{w}_x')^{\cdot\cdot} - (b_{54}\dot{w}_z')^{\cdot\cdot} - (b_{64}\dot{\beta})^{\cdot\cdot}] = 0. \quad (3)$$

Here, γ_i are accelerations according to the body velocity v_B, s_{g2} has regard to gravitational effects, S_{22} contains squares of components of ω_B and charcterizes centrifugal effects, and b_{ij} are abbreviations of transformed moments of inertia of cross-sectional areas. As usual, $E I_z$ means flexural rigidity, ρ is mass density, and A denotes cross-sectional area.

The intermittency conditions related to w_x are

$$w_x(y, t)\Big|_{y= -y_S(t)} = 0, \quad w_x'(y, t)\Big|_{y= -y_S(t)} = 0. \quad (4)$$

The partial differential equation of the bending $w_z(y, t)$ looks similarly to (3). The equation of torsional vibration is simpler than (3) and is not represented. The ordinary differential equation of the translational motion of the robot arm is a modification of (1):

$$\ddot{y}_s + s_{22} y_s = -\gamma_2 + s_{g2} + \dot{w}_z \omega_{Bx} - \dot{w}_x \omega_{Bz}$$

$$+ \frac{1}{\rho AL} [Q_y + F(t)] . \tag{5}$$

This set of coupled differential equations describes a unconventional and troublesome control problem. Considerung the components of v_B and ω_B as kinematical control inputs and $F(t)$ as force control input then we have a nonlinear control problem also including time derivatives of the control inputs. In the section after next some remarks on reduction of model order and of an approximate simplification will be presented.

Multibody Model

Another approach modeling the dynamic behaviour of an elastic robot arm is based on the theory of multibody systems. For this, the beam will be physically discretized and it will be considered as a chain of small beam-like rigid subbodies coupled by fictitious Cardan joints and ficitious springs and dampers representing elasticity and material damping of the beam, cf. Fig. 3.

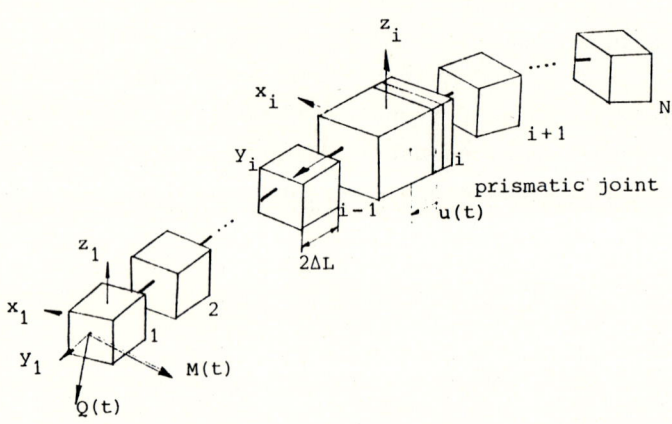

Fig. 3. Multibody model of elastic robot arm

The coefficients of the elastic springs are determined according to the correspondence of static deformation under static end load. This results in equal coefficients

$$k_\alpha = \frac{3\,E\,I_x}{L} \frac{N}{i^3} \sum_{m=1}^{i-1} m^2, \qquad (6)$$

$$k_\beta = (i-1)\,G\,I_T, \qquad (7)$$

$$k_\gamma = \frac{3\,E\,I_z}{L} \frac{N}{i^3} \sum_{m=1}^{i-1} m^2 \qquad (8)$$

of the bending springs (k_α, k_γ) and the torsional spring (k_β) between the subbodies j+1 and j for j = i-1,......,1 if subbody no. i contacts the prismatic joint.

Assuming small relative angels α_j, β_j, γ_j between the subbodies the theory of multibody systems can be applied, cf. e. g. Schiehlen [7]. For example, the kinematic of each subbody has to be determined. With regard to small angles a typical result is

$$\omega_j = \omega_i + \dot{z}_{ji} + \tilde{\omega}_i\, z_{ji} \qquad (9)$$

where $\omega_i = \omega_B$ is the angular velocity of subbody no. i, ω_j is the angular velocity of subbody no. j, and

$$z_{ji}^T = \sum_{m=j}^{i-1} z_m^T = \sum_{m=j}^{i-1} [\alpha_m\ \beta_m\ \gamma_m]. \qquad (10)$$

More complicated is the expression of acceleration a_j of the center of mass of subbody no. j, which is not written down here.

Applying Newton's and Euler's equations of motion to each subbody, the following equations are obtained including the constraint forces R_j:

$$m_j\,a_j = G_j + R_j - S_{j-1,j}\,R_{j-1}, \qquad (11)$$

$$I_j\,\dot{\omega}_j + \tilde{\omega}_j\,I_j\,\omega_j = -K(z_j - z_{j-1}) +$$
$$\Delta\tilde{L}(R_j + S_{j-1,j}R_{j-1}) \qquad (12)$$

where $S_{j-1,j}$ is linearized direction cosine matrix relating the two body-fixed bases of subbodies no. j-1 and j, G_j characterizes gravitational effects, and K denotes a diagonal matrix of the spring coefficients (6 - 8). Eliminating the constraint forces, and introducing state space notation by a state vector

$$x_i = [z_1^T \ldots z_{i-1}^T \ y_s \ \dot{z}_1^T \ldots \dot{z}_{i-1}^T \ \dot{y}_s]^T, \tag{13}$$

finally a set of differential equations of first order is obtained:

$$\dot{x}_i = A_i(v_B, \omega_B, \dot{v}_B, \dot{\omega}_B, F) \ x_i + b_i \ F. \tag{14}$$

This mathematical model applies as long as subbody no. i contacts the prismatic joint. The system matrix A_i depends on the kinematic control inputs v_B and ω_B and its time derivaties as well as on the axial control force F. Again a untypical control problem has been encountered. A change of the description no. i to that of no. i-1 or no. i+1 will appear if y_s cross the values

$$y_s = 2i-2-N \quad \text{or} \quad y_s = 2i-N. \tag{15}$$

It has to be noted that the dimension of the state vector x_i depends on i:

$$\dim x_i = 6(i-1) + 2. \tag{16}$$

Model Reduction

Neither the continuum model (3 - 5) nor the multibody model (14) are suitable for a feedback control design. Therefore, adequate simplifications are needed. Firstly looking on the continuum model, a Ritz-Galerkin approach may be applied. Assuming certain known shape functions $w_{xi}(y)$, $w_{zi}(y)$, $w_{\beta i}(y)$, the approximations

$$w_x(y,t) = \sum_{m=1}^{n_x} a_{xm}(t) \ w_{xm}(y),$$

$$w_z(y,t) = \sum_{m=1}^{n_z} a_{zm}(t) w_{zm}(y), \quad (17)$$

$$\beta(y,t) = \sum_{m=1}^{n_\beta} a_{\beta m}(t) w_{\beta m}(y)$$

lead to

$$\dot{x} = A(v_B, \omega_B, \dot{v}_B, \dot{\omega}_B, F; y_S, \dot{y}_S) x + b F \quad (18)$$

with a state vector

$$x = [a_{xm}\ a_{zm}\ a_{\beta m}\ y_S\ \dot{a}_{xm}\ \dot{a}_{zm}\ \dot{a}_{\beta m}\ \dot{y}_S]. \quad (19)$$

Comparing (18) with (14) an additional dependence of A on y_S and $\dot{y}_S$ can be noted. But this is nothing else than the substitution of the index i of A_i.

The suitable choice of shape functions is very difficult because of the intermittency condition (4). Therefore, these functions usually depend on $y_S(t)$. A possible selection are the instanteneous natural modes for the motionless joint, i. e. for $v_B = 0$, $\omega_B = 0$.

To get a model (14) or (18) of low order, the number of subbodies or the number of shape functions has to be low. This is essentially the question for a very good choice of shape functions which is still unsolved for our problem.

An additional simplification may be a piecewise approximation of the system matrices in (14) or (18) considering typical time histories of the kinematical control inputs. Very often the point-to-point control is realized by trapezoidal time functions. During a start interval $0 \leq t \leq t_1$ the system matrix A and analogously A_i of (14) may be replaced by

$$A \approx A_1 = A(\varepsilon v_{Bo}, \varepsilon \omega_{Bo}, 0, 0, \varepsilon F_o; y_{So}, 0) \quad (20)$$

where v_{Bo}, ω_{Bo}, F_o represent maximum constant values of trapezoidal functions and $\varepsilon \approx 0.1 ./. 0.2$. Afterwards, during $t_1 \leq t \leq t_2$ the approximation is

$$A \approx A_2 = A(v_{Bo}, \omega_{Bo}, 0, 0, F_o; y_{sm}, 0) \qquad (21)$$

where $2y_{sm} = y_{yo} + y_{s1}$. In the last period $t_2 \leq t \leq t_3$ of braking the robot motion, the system matrix is approximated by

$$A \approx A_3 = A(\varepsilon\, v_{Bo}, \varepsilon\, \omega_{Bo}, 0, 0, \varepsilon\, F_o; y_{s1}, 0). \qquad (22)$$

Here y_{so} and y_{s1} denote the start and the end position of the translational motion of the robot arm. Summarizing, the state equations (14) or (18) can be represented by a family of conventional time-invariant systems

$$\dot{x}_{(j)}(t) = A_j\, x_{(j)}(t) + b_{(j)}\, F(t), \quad t_{j-1} \leq t < t_j, \qquad (23)$$

where $F(t)$ is the usual control input.

Control Concept

Looking for a suitable method for the control design it has to be noticed again that the gouverning equations of motion are very unconvenient for usual design methods. For example, the robust decentralized control algorithms [1, 2] successfully developed and implemented for elastic robots with rotational degrees of freedom cannot be applied. There is a different structure of control inputs and additionally there is a certain loss of controllability of the elastic vibrations in the neighbourhood of rest positions.

Therefore the authors shall apply two different control design methods. On the one hand the design of a robust control with respect to the multi-model-problem (23) will be considered according to Ackermann [8]. On the other side a suitable determination of the input functions will be considered regarding the method of Meckel and Seering [9] to avoid the excitation of the elastic motions. The authors are hopefully looking forward to reducing the elastic vibrations of a robot arm with prismatic joint by one of these design methods.

References

1. Henrichfreise, H.: Aktive Schwingungsdämpfung an einem elastischen Knickarmroboter. Dissertation, Universität - GH Paderborn 1988.

2. Ackermann, J.: Positionsregelung reibungsbehafteter, elastischer Industrieroboter. Dissertation, Bergische Universität - GH Wuppertal 1988.

3. Lilov, L.; Wittenburg, J.: Dynamics of chains of rigid bodies and elastic rods with revolute and prismatic joints. In: Bianchi, G.; Schiehlen, W. (eds.): Dynamics of Multibody Systems, Berlin, Heidelberg, New York, Tokyo: Springer-Verlag 1986, pp. 141 - 152.

4. Riemer, M.; Wauer, J.: Equations of motion for hybrid industrial robot models with revolute and prismatic joints. Poster presentation, 1st Int. Conf. on Industrial and Applied Mathematics (ICIAM), Paris, June 29 - July 3, 1987, and private communication.

5. Wang, P. K. C.; Wei, J.-D.: Vibrations in a moving flexible robot arm. J. Sound Vibr. 116 (1987) 149 - 160.

6. Wang, P. K. C.; Wei, J.-D.: Feedback control of vibrations in a moving flexible robot arm with rotary and prismatic joints. Proc. 1987 IEEE Int. Conf. on Robotics and Automation, Rayleigh, NC. pp. 1683 - 1689, Mar.-Apr. 1987.

7. Schiehlen, W.: Technische Dynamik. Stuttgart: B. G. Teubner 1986.

8. Ackermann, J.: Robuste Regelung: Beispiele - Parameterraum-Verfahren. In: Robuste Regelung, GMA-Bericht 11, Düsseldorf: VDI/VDE-GMA 1986.

9. Meckl, P. H.; Seering, W. P.: Reducing residual vibration in systems with time-varying resonances. Proc. 1987 IEEE Int. Conf. on Robotics and Automation, Rayleigh, NC, pp. 1690 - 1695, Mar. - Apr. 1987.

A Decentralized and Robust Controller for Robots

Dr. L. Guzzella, Dr. A.H. Glattfelder

Corporate R&D, Sulzer bros., CH-8401 Winterthur, Switzerland

1. Introduction

There seems to be a rather broad agreement in the robotics researcher community that a good solution of the robot motion-control problem would be given by some sort of compensation of all nonlinear effects [1]. After that the well known and powerful linear systems theory could be used to control robots. Unfortunately this appealing solution has some still unresolved problems the main two of which are the parameter sensitivity of the compensators and the large amount of on-line computations. This paper presents a new controller structure which is able to cope with both problems.

The first idea is to separate the robot in two subsystems, viz. the arm-system (large workspace and inertia) and the hand-system. The arm-system is modeled assuming a fixed hand, i.e. the hand is modeled as a passive payload. The hand-system is modeled assuming a constant arm-position. For both systems a decentralized nonlinear compensator is proposed. This decentralization produces simpler compensators of smaller order and is well suited to a parallel controller-structure (thus the real-time implementation becomes feasible).

The second idea is to eliminate the parameter sensitivity of the compensators by using a reference model and a variable structure controller (VSC) which guarantees a zero error between plant and model states, i.e. the VSC eliminates all unwanted couplings between the arm and the hand system. In the literature some algorithms for robots using VSC's have already been presented [2], [3]. All of them require a rather cumbersome stability analysis. This paper will show that with some reasonable assumptions the stability analysis can be performed in a simpler way.

The following section will give the formal problem statement and will introduce some definitions. Section 3 will introduce the design procedure giving the structure of the controller. In Section 4 the ideas introduced in Section 3 will be used studying the system in the "sliding mode". In Section 5 the stability analysis is done. The last Section 6 gives an example of the complete design and analysis procedure and shows some digital simulations of a 3 degree of freedom robot.

2. PROBLEM DESCRIPTION AND DEFINITIONS

A dynamic system is here defined to be a robot if it can be described by the following differential equation:

$$M(y)\ddot{y}(t) = f(y(t),\dot{y}(t)) + u(t); \quad y(t), u(t) \in \mathbb{R}^p \quad (1)$$

The vectors y(t) and u(t) represent generalized coordinates respectively forces of the robot (1). The matrix M(y) is the mass-matrix and therefore symmetric and positive definite for all possible y(t). The vectorfunction $f(y(t),\dot{y}(t))$ represents nonlinear couplings due to centrifugal and Coriolis forces and also the gravitational effects. The actuators are assumed to be very fast (neglected actuator dynamics).

Using the following definitions:

$$x_j(t) = y_i(t), \quad j = 1, 3, \ldots n-1, \quad i = 1, 2, \ldots p, \quad n = 2p$$
$$x_j(t) = \dot{y}_i(t), \quad j = 2, 4, \ldots n, \quad i = 1, 2, \ldots p$$

and introducing the structural matrices $Q \in \mathbb{R}^{n \times n}$ and $B \in \mathbb{R}^{n \times p}$:

$$Q = \begin{pmatrix} Q_o & \\ & \ddots \\ & & Q_o \end{pmatrix} \quad Q_o = \begin{pmatrix} 0 & 1 \\ 0 & 0 \end{pmatrix} \quad B = \begin{pmatrix} b_o & \\ & \ddots \\ & & b_o \end{pmatrix} \quad b_o = \begin{pmatrix} 0 \\ 1 \end{pmatrix}$$

the second degree equation (1) can be transformed in a first order one :

$$\dot{x}(t) = Q\,x(t) + B\,M(x)^{-1}[f(x) + u(t)] \quad x(t) \in \mathbb{R}^n \quad (2)$$

In the sequel it is assumed that the robot has m degrees of freedom (dof) in the arm system and q dof in the hand system (m+q=p). The vectors x(t) and u(t) can now be partitioned into the following two parts:

$$x(t) = [x^A(t), x^H(t)]^T \quad x^A \in \mathbb{R}^{2m}; x^H \in \mathbb{R}^{2q} \quad \text{and} \quad u(t) = [u^A(t), u^H(t)]^T \quad u^A \in \mathbb{R}^m; u^H \in \mathbb{R}^q$$

This separation introduces four sub-matrices in the matrix M(x) and two coupling functions :

$$M(x) = \begin{pmatrix} M_{11}(x) & M_{12}(x) \\ M_{21}(x) & M_{22}(x) \end{pmatrix} \quad f(x) = \begin{pmatrix} f^A(x) \\ f^H(x) \end{pmatrix}$$

With this partitions equation (2) can be rewritten as follows:

$$\dot{x}^A(t) = Q^A x^A(t) + B^A [M_{11}(x) + \delta M^A(x)]^{-1} [f^A(x) + \delta f^A(x,u^H) + u^A(t)] \quad (3a)$$

$$\dot{x}^H(t) = Q^H x^H(t) + B^H [M_{22}(x) + \delta M^H(x)]^{-1} [f^H(x) + \delta f^H(x,u^A) + u^H(t)] \quad (3b)$$

with:
$$\delta M^A(x) = -M_{12}(x) M_{22}(x)^{-1} M_{21}(x) \qquad \delta M^A(x) \in \mathbb{R}^{m \times m}$$
$$\delta f^A(x, u^H) = -M_{12}(x) M_{22}(x)^{-1} \{f^H(x) + u^H(t)\} \qquad \delta f^A(x) \in \mathbb{R}^m$$
$$\delta M^H(x) = -M_{21}(x) M_{11}(x)^{-1} M_{12}(x) \qquad \delta M^H(x) \in \mathbb{R}^{q \times q}$$
$$\delta f^H(x, u^A) = -M_{21}(x) M_{11}(x)^{-1} \{f^A(x) + u^A(t)\} \qquad \delta f^H(x) \in \mathbb{R}^q$$

The matrices $Q^A \in \mathbb{R}^{2m \times 2m}$, $Q^H \in \mathbb{R}^{2q \times 2q}$, $B^A \in \mathbb{R}^{2m \times m}$ and $B^H \in \mathbb{R}^{2q \times q}$ have the same structure as the matrices Q and B defined above. Due to the symmetry and positive definiteness of M the matrices M_{11} and M_{22} are symmetric and positive definite, too (i.e. their inverses exist). Also the matrices $M_{11} + \delta M^A$ and $M_{22} + \delta M^H$ have to be symmetric. In addition it can be shown that this matrices are positive definite such that the existence of the corresponding inverses is guaranteed.

The aim of this work is to find controllers $u^A(t)$ and $u^H(t)$ which force the states $x^A(t)$ and $x^H(t)$ to follow some desired independent motions $x_d^A(t)$ and $x_d^H(t)$. Moreover the designer should be able to prescribe in a natural way the dynamics of this motion, i. e. the overall system should be linear with arbitrarily placeable poles.

3. CONTROLLER DESIGN

The first step in the design procedure is the formulation of a decoupled reference model of the system (3). Of course this reference model should be as close to the real system as possible since this will reduce the controller effort for matching both systems. The reference models are given by the following nonlinear equations (4a/b):

$$\dot{z}^A(t) = Q^A z^A(t) + B^A N^A(z^A)^{-1} [g^A(z^A) + w^A(t)]; \quad z^A \in \mathbb{R}^{2m} \tag{4a}$$

$$\dot{z}^H(t) = Q^H z^H(t) + B^H N^H(z^H)^{-1} [g^H(z^H) + w^H(t)]; \quad z^H \in \mathbb{R}^{2q} \tag{4b}$$

The models (4) are required to have the same dimensions as the robot (1), no other assumptions are necessary. If the inputs $w^A \in \mathbb{R}^m$ and $w^H \in \mathbb{R}^m$, which will be defined later, are fed to both the robot (3) and the reference models (4) the system depicted in the next figure is formed (onlyone half of the system is shown the other half being analogous). Due to the neglected couplings and to the imperfect reference models the state-error will not vanish.

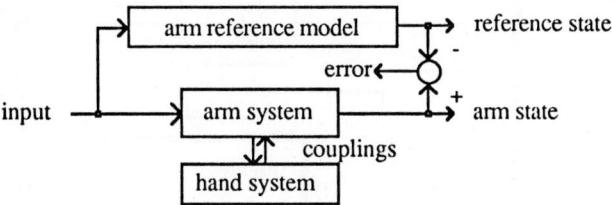

Figure 1, plant and reference model for the arm sub-system

At this point the "variable structure controller" (VSC) [4] can be introduced. Its purpose is to suppress the state-errors between robot and reference models thus producing a perfectly known and decoupled input-output behaviour of the subsystem shown in Figure 2 (the hand sub-system is not shown since it is completely analogous).

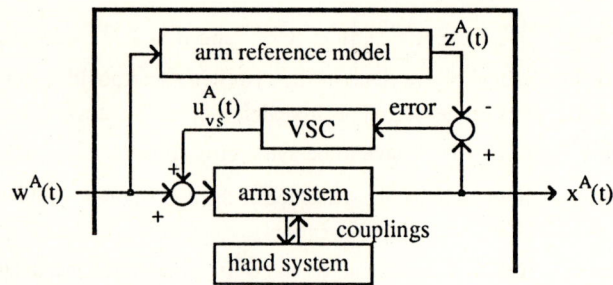

Figure 2, first controller-shell producing a perfect plant-model matching

The VSC have the following form:

$$s^A(t) = C^A (x^A(t) - z^A(t)) \quad\quad u^A_{vs}(t) = - (d^A_x + d^A_{wA} |w^A(t)| + d^A_{wH} |w^H(t)|) \, \text{sign}(s^A(t)) \quad (5a)$$

$$s^H(t) = C^H (x^H(t) - z^H(t)) \quad\quad u^H_{vs}(t) = - (d^H_x + d^H_{wA} |w^A(t)| + d^H_{wH} |w^H(t)|) \, \text{sign}(s^H(t)) \quad (5b)$$

The vectors $s^A(t) \in \mathbb{R}^m$ and $s^H(t) \in \mathbb{R}^q$, which will play an important role in the next section, are the so called switching variables. The matrices C^A and C^H can always be chosen to be orthogonal to the matrices B^A and B^H respectively [5]. Since this fact simplifies the derivations without hiding the main ideas, it is assumed in the sequel that this special choice has been adopted.

The vectors $u^A_{vs}(t) \in \mathbb{R}^m$ and $u^H_{vs}(t) \in \mathbb{R}^q$ are nonlinear functions of the state-error between the plant and the reference-model. In Section 5 conditions will be given for the gains d of the VSC which will make sure that the state-error between the reference model and the plant vanishes. Assuming this zero state-error the compensator can be defined using only the reference model characteristics (4) which of course are perfectly known (again only the arm system is shown).

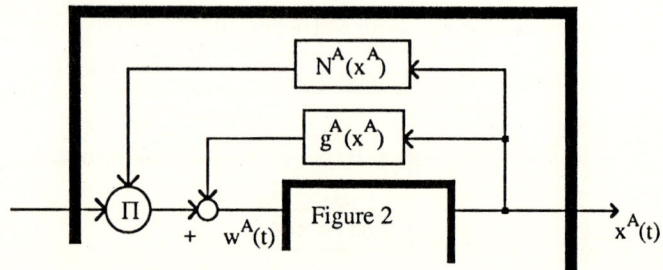

Figure 3, second controller-shell producing a linear input-output-behaviour

From the outside this second shell has a linear transfer function, i.e. it consists of m decoupled integrator pairs. The last step represents a classical linear system design, e.g. a pole placement by state feedback. Since no integral part is assumed in the linear controller an additional gain-matrix R^A can be introduced to produce a zero steady-state error $x^A(\infty) - x_d^A(\infty)$.

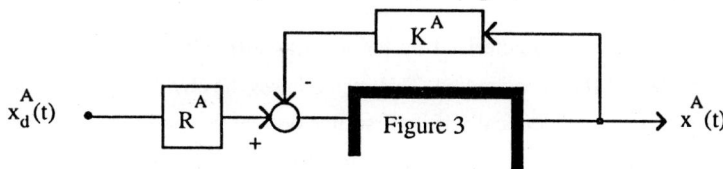

Figure 4, third controller-shell producing an overall behaviour with arbitrary poles

In order to use consistent representations the original system equations (3) are reformulated using the matrices N^A and N^H respectively the vectors g^A and g^H introduced in equation (4a/b):

$$\dot{x}^A(t) = Q^A x^A(t) + B^A [N^A(x^A) + \delta N^A(x)]^{-1} [g^A(x^A) + \delta g^A(x,u^H) + u^A(t)] \qquad (6a)$$

$$\dot{x}^H(t) = Q^H x^H(t) + B^H [N^H(x^H) + \delta N^H(x)]^{-1} [g^H(x^H) + \delta g^H(x,u^A) + u^H(t)] \qquad (6b)$$

with: $\quad \delta N^A(x) = M_{11}(x) - N^A(x^A) + \delta M^A(x) ; \quad \delta g^A(x,u^H) = f^A(x) - g^A(x^A) + \delta f^A(x,u^H) \quad$ (7a)

$\quad \delta N^H(x) = M_{22}(x) - N^H(x^H) + \delta M^H(x) ; \quad \delta g^H(x,u^A) = f^H(x) - g^H(x^H) + \delta f^H(x,u^A) \quad$ (7b)

The structure of the controller is now completely defined and can be summarized in the Figure 5.

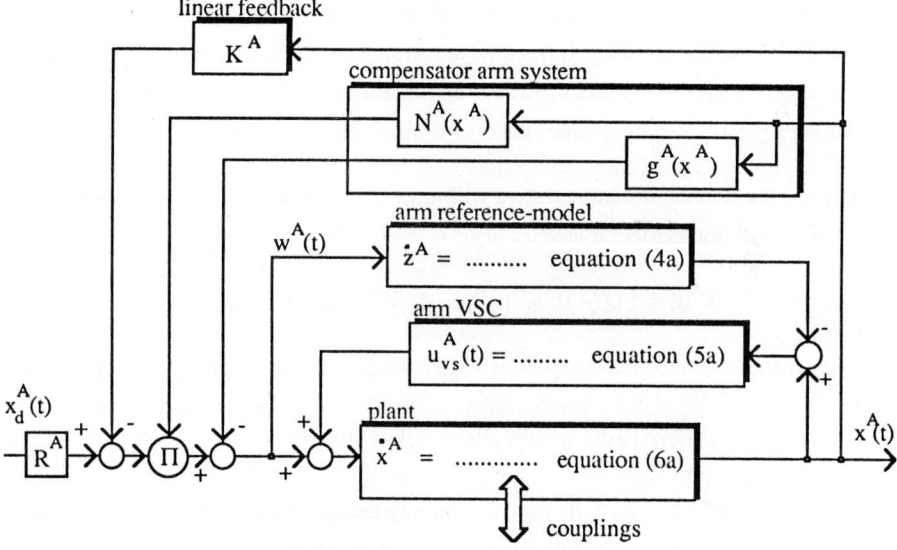

Figure 5, complete controller-structure for the arm sub-system (hand sub-system analogous)

4. SYSTEM BEHAVIOUR IN THE SLIDING MODE

The derivations in this chapter are shown for the arm-subsystem only. The equations of the hand-subsystem are almost identical, in fact the only difference consists in the super- or subscripts which have to be changed. Therefore only the final result is given for both parts.

The sliding mode is characterized by the identity $s^A(t) \equiv 0$. Using this relation the "equivalent control method" [4] can be applied. This method uses the obvious fact that the switching variable $s^A(t)$ can only vanish identically if its first time derivative vanishes, too. From this fact a control $u^A_{eq}(t)$ can be calculated which is equivalent to $u^A_{vs}(t)$ (C^A is chosen to be orthogonal to B^A):

$$u^A_{eq}(t) = -[N^A + \delta N^A]C^A\{Q^A x^A - Q^A z^A - B^A N^{A-1}[g^A + w^A]\} - [g^A + \delta g^A + w^A] \tag{8}$$

If this equivalent control vector (8) is applied on the arm system (4a) and (6a) the following very simple arm error-dynamics are obtained ($e^A(t) = x^A(t) - z^A(t)$):

$$\dot{e}^A(t) = [I - B^A C^A] Q^A e^A(t) = \Pi^A Q^A e^A(t) \tag{9a}$$

Therefore the error-dynamics of the arm is governed by a simple linear differential equation. Moreover the poles of the matrix $\Pi^A Q^A$ are determined by the entries of the matrix C^A only [5] (note that those poles of (9a) lying in the origin do not affect the behaviour of the sliding system [6]).

Thus, for the moment supposing that the sliding mode is stable, the error between both hand and arm reference model vanishes with an exponential decay rate which is arbitrarily chosen by the designer. For some sufficiently large times $t > t^{**}$ the error $e^A(t)$ can be assumed to be virtually zero. An other approach would be to introduce a start-up procedure which guarantees zero errors. In this case, assuming a persistent sliding mode, the error remains zero for all times.

For times $t > t^{**}$ the nonlinear compensators will have the desired effect on the real robot. The compensated and controlled robot-dynamics in the sliding mode are given by:

$$\dot{x}^A(t) = [Q^A - B^A K^A] x^A(t) + B^A R^A x^A_d(t) \tag{10a}$$

Using the same argumentation the compensated and controlled dynamics of the hand system can be found:

$$\dot{x}^H(t) = [Q^H - B^H K^H] x^H(t) + B^H R^H x^H_d(t) \tag{10b}$$

Since the pairs $\{Q^A, B^A\}$ and $\{Q^H, B^H\}$ are structurally completely controllable the designer can specify an arbitrary pole-placement by choosing appropriate matrices K^A respectively K^H [8] (of course any other synthesis method can be used now, e.g. LQG or frequency-domain approaches).

5. STABILITY ANALYSIS

Before starting the stability analysis a lemma used below is introduced.

Lemma: If $M = M^T > 0 \Rightarrow M_{11} + \delta M^A > 0$ and $M_{22} + \delta M^H > 0$

Proof: $M > 0 \Rightarrow x^{A^T} M_{11} x^A + x^{A^T} M_{12} x^H + x^{H^T} M_{21} x^A + x^{H^T} M_{22} x^H > 0$

since x^A and x^H are arbitrary one can choose $x^H = -M_{22}^{-1} M_{21} x^A$

$\Rightarrow x^{A^T} \{M_{11} - M_{12} M_{22}^{-1} M_{21} - M_{12} M_{22}^{-1} M_{21} + M_{12} M_{22}^{-1} M_{22} M_{22}^{-1} M_{21}\} x^A$

$= x^{A^T} \{M_{11} - M_{12} M_{22}^{-1} M_{21}\} x^A = x^{A^T} \{M_{11} + \delta M^A\} x^A > 0 \quad \Omega$

The second assertion is proved by choosing $x^A = -M_{11}^{-1} M_{12} x^H \quad \Omega$

The general stability analysis is rather cumbersome and the resulting stability conditions are quite conservative. Here a reasonable assumption is adopted in order to simplify things. In fact **zero initial errors** $e^A(t_o) = x^A(t_o) - z^A(t_o)$ and $e^H(t_o) = x^H(t_o) - z^H(t_o)$ are assumed. This is not a restrictive assumption, since at start-up most robots perform some kind of reference-mark localization which can be used to initialize the reference model.

In the sequel only the stability analysis of the arm-system is shown. The hand-system can be analyzed in the same way such that only the results will be given here. Since the error $e^A(t)$ is assumed to be zero the dynamics of the switching variable $s^A(t)$ are described by the following equation (the vector $z^A(t)$ is substituted by $x^A(t)$):

$$\dot{s}^A(t) = [N^A + \delta N^A]^{-1}[g^A + \delta g^A + w^A + u_{vs}^A] - N^{A-1}[g^A + w^A] \tag{11}$$

In this equation the input $u^H(t)$ is still involved (δg^A is a function of $x(t)$ and $u^H(t)$). In order to be able to calculate the derivative of the switching function the vector $u^H(t)$ is replaced by its equivalent value $u_{eq}^H(t)$. This corresponds to the "hierarchical control"-principle introduced in [4]. After some algebraical manipulations the following $u_{eq}^H(t)$ is found ($\Xi = I - M_{21} M_{11}^{-1} M_{12} M_{22}^{-1}$):

$$u_{eq}^H(t) = -f^H + M_{22} N^{H-1} g^H + M_{21} N^{A-1} g^A + (M_{22} N^{H-1} - \Xi^{-1}) w^H + (M_{21} N^{A-1} - \Xi^{-1} M_{21} M_{11}^{-1}) w^A \tag{12}$$

The stability proof uses the following Lyapunov-function:

$$v^A(t) = s^A(t)^T [M_{11}(x) + \delta M^A(x)] s^A(t) = s^A(t)^T [N^A(x^A) + \delta N^A(x)] s^A(t) = s^A(t)^T P(x) s^A(t) \tag{13}$$

Due to the lemma the matrix $P(x)$ is symmetric and positive definite, therefore $v^A(t)$ is a valid Lyapunov-function candidate. The time derivative of $v^A(t)$ is given by the equation (14):

$$\dot{v}^A(t) = 2 s^A(t)^T P(x) \dot{s}^A(t) + s^A(t)^T \dot{P}(x) s^A(t) \tag{14}$$

The second term in (14) can be neglected since it is of second order small (remember the error $e^A(t)$ is virtually zero therefore $s^A(t)$ is small, too). Using equations (7a), (11) and (12) the derivative of the Lyapunov function (13) can be calculated:

$$\dot{v}^A(t) = 2s^A(t)^T \{ \rho_1^A(x) + \rho_2^A(x)w^A(t) + \rho_3^A(x)w^H(t) - (d_x^A + d_{wA}^A |w^A(t)| + d_{wH}^A |w^H(t)|) \text{sign}(s^A(t)) \} \quad (15)$$

The functions $\rho_1^A(x) \in \mathbb{R}^m$, $\rho_2^A(x) \in \mathbb{R}^{m \times m}$ and $\rho_3^A(x) \in \mathbb{R}^{m \times q}$ used in (15) are defined by:

$$\rho_1^A(x) = f^A - M_{11} N^{A-1} g^A - M_{12} N^{H-1} g^H \quad (16a)$$

$$\rho_2^A(x) = I - M_{11} N^{A-1} + M_{12} [M_{22} + \delta M^H]^{-1} M_{21} M_{11}^{-1} \quad (16b)$$

$$\rho_3^A(x) = M_{12} [M_{22} + \delta M^H]^{-1} - M_{12} N^{H-1} \quad (16c)$$

A sufficient stability condition for the VSC-gains is given by the following bounds:

$$d_x^A > \max_x \{|\rho_1^A(x)|\} \qquad d_{wA}^A > \max_x \{\|\rho_2^A(x)\|\} \qquad d_{wH}^A > \max_x \{\|\rho_3^A(x)\|\} \quad (17)$$

The norm operators used here are the Euclidean length of ρ_1^A and the greatest singular value of ρ_2^A and ρ_3^A. Using exactly the same argumentation sufficient condition for gains of the hand VSC can be found. Definition of $\rho_1^H(x) \in \mathbb{R}^q$, $\rho_2^H(x) \in \mathbb{R}^{q \times q}$ and $\rho_3^H(x) \in \mathbb{R}^{q \times m}$:

$$\rho_1^H(x) = f^H - M_{22} N^{H-1} g^H - M_{21} N^{A-1} g^A \quad (18a)$$

$$\rho_2^H(x) = I - M_{22} N^{H-1} + M_{21} [M_{11} + \delta M^A]^{-1} M_{12} M_{22}^{-1} \quad (18b)$$

$$\rho_3^H(x) = M_{21} [M_{11} + \delta M^A]^{-1} - M_{21} N^{A-1} \quad (18c)$$

Stability conditions for the gains of the hand VSC:

$$d_x^H > \max_x \{|\rho_1^H(x)|\} \qquad d_{wH}^H > \max_x \{\|\rho_2^H(x)\|\} \qquad d_{wA}^H > \max_x \{\|\rho_3^H(x)\|\} \quad (19)$$

The maximum values of the VSC-gains have to be found over the entire set of planned trajectories. This can be done by simulating the compensated and controlled reference system and inserting z(t), which is equal to x(t), into the stability conditions (17) respectively (19).

The main result of this work is summarized in the following Theorem:

<u>Theorem:</u> If the gains of both VSC (5a) and (5b) fulfill the stability conditions (17) respectively (19) and if the system depicted in Figure 5 starts with zero plant-model state-errors then this errors will remain zero for all times $t > t_o$.

Proof: For the arm-system: If the VSC-gains are chosen greater as imposed by condition (17) and if the robot starts at t_o with zero arm-error (and thus in sliding mode) the derivative of the Lyapunov-function (13) will be smaller than zero for all times $t>t_o$. Since the Lyapunov-function (13) has a minimum for $s^A(t)=0$ the sliding mode will be stable for all $t>t_o$, too. But if this is true the error is governed by the equation (9a) and therfeore the error has to remain zero for all times $t>t_o$ (of course the designer is supposed to choose such a matrix C^A which produces a stable error system).

Of course the same argumentation is valid for the hand-system, too. Ω

6. EXAMPLE

The presented example has 2 dof in the arm system and 1 dof in the hand system. This robot is able to reach a certain point in its workplane and to produce a desired orientation of the tool. The geometry is defined in the following sketch:

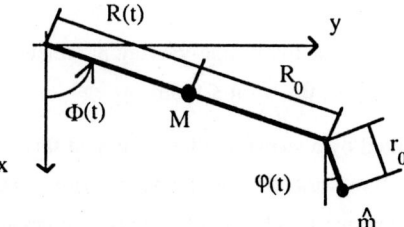

Figure 6, sketch of the analyzed robot with 3 dof

The differential equation (1) for this robot is given by the following expressions (in order to avoid overloaded equations in the sequel the time dependencies of variables are often ommitted):

$$\begin{pmatrix} MR^2+\hat{m}[R+R_0]^2 & 0 & \hat{m}[R+R_0]r_0\cos(\Phi-\varphi) \\ 0 & M+\hat{m} & \hat{m}r_0\sin(\Phi-\varphi) \\ \hat{m}[R+R_0]r_0\cos(\Phi-\varphi) & \hat{m}r_0\sin(\Phi-\varphi) & \hat{m}r_0^2 \end{pmatrix} \begin{pmatrix} \ddot{\Phi}(t) \\ \ddot{R}(t) \\ \ddot{\varphi}(t) \end{pmatrix}$$

$$= \begin{pmatrix} -2\{MR+\hat{m}[R+R_0]\}\dot{R}\dot{\Phi} - \hat{m}[R+R_0]r_0\dot{\varphi}^2\sin(\Phi-\varphi) \\ \hat{m}r_0\dot{\varphi}^2\cos(\Phi-\varphi) + \{MR+\hat{m}[R+R_0]\}\dot{\Phi}^2 \\ \hat{m}v[R+R_0]r_0\dot{\Phi}^2\sin(\Phi-\varphi) - 2\hat{m}r_0\dot{R}\dot{\Phi}\cos(\Phi-\varphi) \end{pmatrix} + \begin{pmatrix} u_\Phi(t) \\ u_R(t) \\ u_\varphi(t) \end{pmatrix}$$

For the sake of simplicity no inertia at $R(t) = 0$ is introduced; therefore $R(t)$ has always to be greater than a certain minimal value R_{min}.

The reference models (4a/b) are defined by the following two equations.

Arm system:

$$\begin{pmatrix} MR^2 + m[R+R_0+r_0]^2 & 0 \\ 0 & M+m \end{pmatrix} \begin{pmatrix} \ddot{\Phi}(t) \\ \ddot{R}(t) \end{pmatrix} = \begin{pmatrix} -2\{MR(t) + m[R+R_0+r_0]\}\dot{R}(t)\dot{\Phi}(t) \\ \{MR + m[R+R_0+r_0]\}\dot{\Phi}^2 \end{pmatrix} + \begin{pmatrix} u_\Phi(t) \\ u_R(t) \end{pmatrix}$$

Hand system: $\quad (mr_0^2)\ddot{\varphi}(t) = u_\varphi(t)$

The nominal values of the parameters are :

$$M = 10 \text{ (kg)} \quad m = 1 \text{ (kg)} \quad R_0 = 1 \text{ (metre)} \quad r_0 = 0.2 \text{ (metre)}$$

The pay-load $\hat{m}-m$ is assumed to be the main time-varying parameter (e.g. pick-and-place tasks) and its variation range is assumed to be $0 \leq \hat{m}-m \leq 0.5m$.

The desired motion is represented by a step-function starting at $\Phi(t_o)=\varphi(t_o)=0$ and $R(t_o)=0.5$ (at stand-still) with the set-points $\Phi(\infty)=\pi/4$, $R(\infty)=1.0$ and $\varphi(\infty)=\pi/2$. The dynamics of the closed-loop system are determined by the matrices K^A and K^H. This matrices are choosen in the following way:

$$K^A = \begin{pmatrix} 2 & 2 & 0 & 0 \\ 0 & 0 & 8 & 4 \end{pmatrix} \quad \text{arm-poles } -1\pm j, -2\pm 2j \qquad K^H = (\ 32 \ \ 8 \ ) \quad \text{hand-poles } -4\pm 4j$$

The following figure shows the behaviour of the system with no VSC and maximum pay-load.

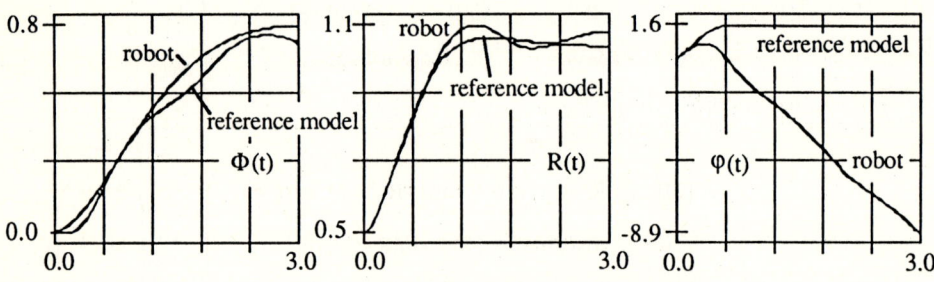

Figure 7, closed-loop behaviour without VSC

As expected the neglected nonlinear couplings and the parameter errors cause the robot to diverge from the reference model.

The VSC is determined by the choice of the matrices C^A and C^H and by the calculation of the VSC-gains. Matrices C^A and C^H determine the error-dynamics which are choosen to produce stable error poles in the sliding mode:

$$C^A = \begin{pmatrix} 5 & 1 & 0 & 0 \\ 0 & 0 & 5 & 1 \end{pmatrix} \qquad C^H = (5\ 1)$$

For the calculation of the VSC-gains the following parts of equation (3) are needed.

Arm subsystem:

$$M_{11} = \begin{pmatrix} MR^2 + \hat{m}[R+R_0]^2 & 0 \\ 0 & M+\hat{m} \end{pmatrix}$$

$$\delta M^A = \begin{pmatrix} -\hat{m}[R+R_0]^2\cos^2(\Phi-\varphi) & -\hat{m}[R+R_0]\sin(\Phi-\varphi)\cos(\Phi-\varphi) \\ -\hat{m}[R+R_0]\sin(\Phi-\varphi)\cos(\Phi-\varphi) & -\hat{m}\sin^2(\Phi-\varphi) \end{pmatrix}$$

Hand subsystem:

$$M_{22} = \left(\hat{m}\ r_0^2\right)$$

$$\delta M^H = \left(-\{\hat{m}^2[R+R_0]^2 r_0^2 \cos^2(\Phi-\varphi)\}/\{MR^2+\hat{m}[R+R_0]^2\} - \{\hat{m}^2 r_0^2 \sin^2(\Phi-\varphi)\}/\{M+\hat{m}\}\right)$$

With that the stability conditions (16) and (18) can be applied to this example. The qualitative behaviour of the resulting VSC-gains for the planned trajectory are shown in the next figure:

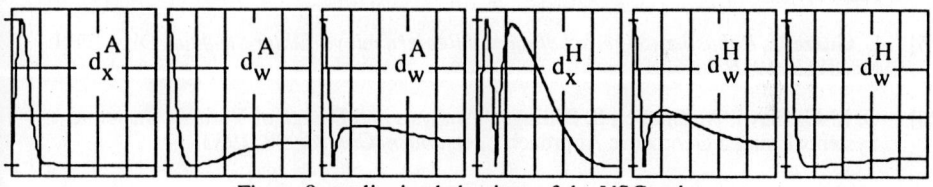

Figure 8, qualitative behaviour of the VSC-gains

The explicit numerical values are:

$d_x^A \geq 5.974..$ $d_{wA}^A \geq 1.260..$ $d_{wH}^A \geq 10.95..$ $d_x^H \geq 0.04995..$ $d_{wA}^H \geq 0.850..$ $d_{wH}^H \geq 0.09651..$

With that the VSC is completely specified and its effect on the system is shown in the last figure. As expected the VSC produces a perfect model-plant matching.

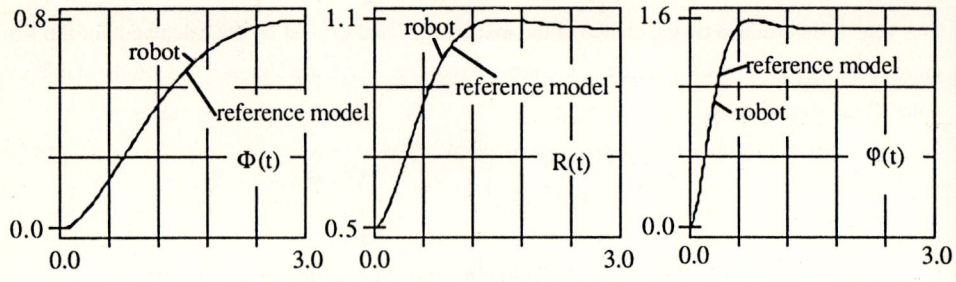

Figure 9, closed-loop behaviour with VSC

References :

[1] J. J. Craig,*Introduction to Robotics*, Addison-Wesley, 1985

[2] K. K. Young, "Controller Design for Manipulators Using Theory of Variable Structure Systems", *EEE Trans. on Systems, Man and Cybernetics*, SMC-8, 1978

[3] J. Slotine, "The Robust Control of Robot Manipulators", *Int. J. of Robotics Res.*, vol. 4, No. 2, 1985

[4] V. Utkin,"Variable Structure Systems With Sliding Modes", *IEEE Trans. Autom.Contr*, AC-23, 1977

[5] L. Guzzella, *Robustheitseigenschaften von Reglern mit variabler Struktur*, Diss. ETH Nr. 8163, 1986

[6] O. El-Gezhawi, A. Zinober, S. Billings, "Analysis and Design of Variable Structure Systems Using a Geometric Approach", *Int. Journ. Control*, 38, 1983

[7] T. Kailath, *Linear Systems*, Prentice Hall, Enlewood Cliffs, N.J., 1982

Isotropic and Uniform Inertial and Acceleration Characteristics: Issues in the Design of Redundant Manipulators

Oussama Khatib and Sunil Agrawal

Robotics Laboratory
Computer Science Department
Stanford University

Abstract

The paper investigates the dynamic characterization of redundant manipulators and formalizes the problem of dynamic optimization in manipulator design. The dynamic performance of a manipulator is described by both inertial and acceleration characteristics as perceived at the end-effector operational point. The inertial characteristics at this point are given by the operational space kinetic energy matrix (pseudo-kinetic energy matrix for a redundant manipulator) which is dependent on the kinematic and inertial parameters of the manipulator and varies with its configuration. The acceleration characteristics of the end-effector are described by a joint torque/acceleration transmission matrix. In addition to their dependency on the kinematic and inertial parameters, the acceleration characteristics depend on the velocities and actuator torque bounds. The dynamic optimization is formalized in terms of finding the design parameters under the various constraints to achieve the smallest most isotropic and most uniform end-effector inertial properties, while providing the largest, most isotropic, and most uniform bounds on the magnitude of end-effector acceleration. This approach is used in the design of ARTISAN, a ten-degree-of-freedom manipulator currently under development at Stanford University.

Introduction

Over the past two decades, an important research effort has been devoted to the development of robot systems. This effort has produced significant improvements in dexterity, workspace, and kinematic characteristics of robot mechanisms. Research in kinematics has developed means for the analysis of workspace characteristics [8,9], and the evaluation of kinematic performance [2,6,11].

Manipulators are highly nonlinear and coupled systems. During motion a manipulator is subject to inertial, centrifugal, and Coriolis forces. The magnitude of these dynamic forces cannot be ignored when large accelerations and fast motions are considered. The dynamic characterization is, therefore, an essential consideration in the analysis, design, and control of these mechanisms. One of the most significant characteristics in evaluating manipulator performance is associated with the dynamic behavior of its end-effector. The end-effector is indeed the part most closely linked to the task. These characteristics cannot be found in the manipulator joint space dynamic model, as it provides a description of joint motion dynamics. The description, analysis and control of manipulator systems with respect to the dynamic characteristics of their end-effectors has been the basic motivation in the development of *the operational space formulation* [3,5]. The end-effector dynamic model is a fundamental tool for the analysis and dynamic characterization of manipulator systems.

The inertial characteristics at some point on the end-effector or the manipulated object are given by the operational space kinetic energy matrix. The kinetic energy matrix, or the generalized inertia ellipsoid [1], establishes the relationship between end-effector forces and accelerations. However, this relationship does not relate the actual actuator torque input to the end-effector accelerations. The description of the acceleration characteristics is an essential requirement for the evaluation of the dynamic performance of manipulators. The operational space dynamic model has been used to establish [4], for different regimes, the input/output relationships between joint forces and end-effector acceleration. A similar relationship has been used to establish a measure of dynamic manipulability [12].

The joint torque/acceleration transmission matrix has been used in the design of manipulators with improved dynamic characteristics. An optimal selection of the design parameters has been shown [4] to significantly improve the end-effector dynamic characteristics by providing large, isotropic, and uniform end-effector accelerations.

In this paper, the dynamic characterization integrates both inertial and acceleration properties. The dynamic optimization is aimed at obtaining the smallest, most isotropic and most uniform end-effector inertial characteristics, while providing the largest, most isotropic, and most uniform bounds on the magnitude of end-effector acceleration. The approach is extended to redundant manipulator systems and used in the design of ARTISAN, a ten-degree-of-freedom redundant manipulator.

End-Effector Equations of Motion

The end-effector position and orientation, with respect to an inertial reference frame $\mathcal{R}_O$ is described by the relationship between $\mathcal{R}_O$ and a coordinate frame $\mathcal{R}_\odot$ of origin $\odot$ attached to this effector. $\odot$ is called the *operational point*. It is with respect to this point that motions and active forces of the effector are specified. An *operational coordinate system* associated with an m-degree-of-freedom effector and a point $\odot$, is a set $\mathbf{x}$ of m *independent* parameters describing the effector position and orientation in $\mathcal{R}_O$. For a non-redundant n-degree-of-freedom manipulator, i.e. $n = m$, these parameters form a set of of *generalized operational coordinates*. The effector equations of motion in operational space [3,5] are given by

$$\Lambda(\mathbf{x})\ddot{\mathbf{x}} + \mu(\mathbf{x}, \dot{\mathbf{x}}) + \mathbf{p}(\mathbf{x}) = \mathbf{F}; \tag{1}$$

where $\Lambda(\mathbf{x})$ designates the kinetic energy matrix, and $\mathbf{p}(\mathbf{x})$ and $\mathbf{F}$ are respectively the gravity and the generalized operational force vectors. $\mu(\mathbf{x}, \dot{\mathbf{x}})$ represents the vector of centrifugal and Coriolis forces. The dynamic decoupling and motion control of the manipulator in operational space is achieved by selecting the control structure

$$\mathbf{F} = \Lambda(\mathbf{x})\mathbf{F}^* + \mu(\mathbf{x}, \dot{\mathbf{x}}) + \mathbf{p}(\mathbf{x}); \tag{2}$$

and the end-effector becomes equivalent to a *single unit mass*, I_m, moving in the m-dimensional space,

$$I_m \ddot{\mathbf{x}} = \mathbf{F}^*. \tag{3}$$

$\mathbf{F}^*$ is the input of the decoupled end-effector. This provides a general framework for the implementation of various control structures at the level of decoupled end-effector. The generalized joint forces Γ needed to produce the operational forces $\mathbf{F}$ of (eq. 2) are given, using the Jacobian matrix $J(\mathbf{q})$, by

$$\Gamma = J^T(\mathbf{q})\mathbf{F}; \tag{4}$$

where $\mathbf{q}$ represents the vector of generalized joint coordinates.

Redundant Manipulators

A set of operational coordinates, which describes the end-effector position and orientation, is not sufficient to completely specify the configuration of a redundant manipulator. Therefore, the dynamic behavior of the entire system cannot be described by a dynamic model in operational coordinates. With respect to a system of generalized joint coordinates, the equations of motion of a manipulator can be written in the form

$$A(\mathbf{q})\ddot{\mathbf{q}} + \mathbf{b}(\mathbf{q}, \dot{\mathbf{q}}) + \mathbf{g}(\mathbf{q}) = \Gamma; \tag{5}$$

where $\mathbf{b}(\mathbf{q}, \dot{\mathbf{q}})$, $\mathbf{g}(\mathbf{q})$, and Γ, represent the Coriolis and centrifugal, gravity, and generalized forces in joint space; and $A(\mathbf{q})$ is the $n \times n$ joint space kinetic energy matrix.

While the dynamics of the entire system cannot be described in operational coordinates, the dynamic behavior of the end-effector itself, can still be described, and its equations of motion in operational space can still be established. In fact, the structure of the effector dynamic model is identical to that obtained in the case of non-redundant manipulators (eq. 1). In the redundant case, however, the matrix Λ should be interpreted as a *"pseudo kinetic energy matrix"*. This matrix is related to the joint space kinetic energy matrix by $\Lambda = [JA^{-1}J^T]^{-1}$.

Another important characteristic of redundant manipulator is concerned with the relationship between operational forces and joint forces. In the case of non-redundancy, an operational force vector F is produced by the joint force vector $J^T F$. The additional freedom of redundant mechanism results in infinities of possible joint force vectors Γ. However, for a given $\mathbf{F}$, all possible joint forces Γ satisfy the relation

$$\mathbf{F} = \overline{J}^T \mathbf{\Gamma}; \tag{6}$$

where

$$\overline{J}(\mathbf{q}) = A^{-1}(\mathbf{q}) J^T(\mathbf{q}) \Lambda(\mathbf{q}). \tag{7}$$

$\overline{J}(\mathbf{q})$ is actually a generalized inverse of the Jacobian matrix. A joint force vector Γ can then be decomposed into two terms: one contributes to the operational force vector, and the other only acts internally (in the null space associated with the Jacobian matrix)

$$\Gamma = J^T(\mathbf{q})\mathbf{F} + [I_n - J^T(\mathbf{q})\overline{J}^T(\mathbf{q})]\Gamma_o; \tag{8}$$

where I_n is the $n \times n$ identity matrix and Γ_o is an arbitrary joint force vector. It has been shown that a generalized inverse that is consistent with the system's dynamics is unique [5] and given by (eq. 7). This generalized inverse corresponds to the solution that minimizes the manipulator's instantaneous kinetic energy.

The relationships between the components of the operational space and joint space dynamic models are

$$\Lambda(\mathbf{q}) = [J(\mathbf{q})A^{-1}(\mathbf{q})J^T(\mathbf{q})]^{-1}; \tag{9}$$

$$\mu(\mathbf{q}, \dot{\mathbf{q}}) = \overline{J}^T(\mathbf{q})b(\mathbf{q}, \dot{\mathbf{q}}) - \Lambda(\mathbf{q})h(\mathbf{q}, \dot{\mathbf{q}}); \tag{10}$$

$$\mathbf{p}(\mathbf{q}) = \overline{J}^T(\mathbf{q})\mathbf{g}(\mathbf{q}); \tag{11}$$

where $h(\mathbf{q}, \dot{\mathbf{q}}) = \dot{J}(\mathbf{q})\dot{\mathbf{q}}$. The previous relationships are general. In particular, they still apply to non-redundant mechanisms. In this case of zero degree of redundancy, the matrix $\overline{J}$ reduces to J^{-1}.

Similar to the case of non-redundant manipulators, the dynamic decoupling and control of the end-effector can be achieved by selecting an operational command vector of the form (eq. 2). The manipulator joint motions produced by this command vector are those that minimize the instantaneous kinetic energy of the mechanism. Asymptotic stabilization is achieved by the addition of dissipative joint forces. In order to preclude any effect of the additional forces on the end-effector and maintain its dynamic decoupling, these forces are selected to act in the dynamically consistent nullspace associated with $\overline{J}(\mathbf{q})$. In the actual implementation, the control vector is developed in a form [5] that avoids the explicit evaluation of the expression of the generalized inverse of the Jacobian matrix.

End-Effector Dynamic Performance

The dynamic response of a mechanical system is determined by its inertial characteristics. Reducing the magnitude of inertias improves the system's dynamic response. The end-effector inertial characteristics at a configuration q are described by the kinetic energy matrix $\Lambda(\mathbf{q})$. It's effective inertia at a configuration q, when moving in a direction u is given by $\mathbf{u}^T \Lambda(\mathbf{q})\mathbf{u}$. The effective inertia varies with the configuration and direction. Isotropic and uniform inertial characteristics are therefore essential to provide isotropic and uniform end-effector's dynamic response.

The second characteristic is concerned with the acceleration characteristics at the end-effector. This is the minimum achievable acceleration given the bounds on actuator torques. Equivalently, this characteristic can be stated in terms of the bounds on the operational force vector $\mathbf{F}^*$, the input of the decoupled end-effector in (eq. 3). Let us examine the operational command vector $\mathbf{F}$ in (eq. 2), which achieves the dynamic decoupling and control of end-effector motion. Only a fraction of these operational forces, namely $\mathbf{F}^*$ the input of the decoupled end-effector, contributes to the end-effector acceleration. The end-effector dynamic performance is, therefore, dependent on the extent of the boundaries of $\mathbf{F}^*$, which determine the limitations on the magnitude of available end-effector acceleration.

The vector $\mathbf{F}$ of (eq. 2) is produced from the actuator joint force vector $\mathbf{\Gamma}$ by $\overline{J}^T(\mathbf{q})\mathbf{\Gamma}$, $\overline{J}(\mathbf{q})$ is equal to $J^{-1}(\mathbf{q})$ for a non-redundant manipulator. Substituting in (eq. 2) yields,

$$\overline{J}^T(\mathbf{q})\mathbf{\Gamma} = \Lambda(\mathbf{q})\mathbf{F}^* + \mu(\mathbf{q},\dot{\mathbf{q}}) + \mathbf{p}(\mathbf{q});$$

which, using (eq. 9- 11), can be written as

$$\mathbf{F}^* = E(\mathbf{q})[\mathbf{\Gamma} - \overline{\mathbf{b}}(\mathbf{q},\dot{\mathbf{q}}) - \overline{\mathbf{g}}(\mathbf{q})]; \tag{12}$$

where

$$E(\mathbf{q}) = J(\mathbf{q})A^{-1}(\mathbf{q}). \tag{13}$$

and

$$\overline{b}(q,\dot{q}) = [J^T(q)\overline{J}^T(q)]\,b(q,\dot{q}) - J^T(q)\Lambda(q)h(q,\dot{q}); \qquad (14)$$
$$\overline{g}(q) = [J^T(q)\overline{J}^T(q)]\,g(q). \qquad (15)$$

$\overline{b}(q,\dot{q})$ and $\overline{g}(q)$ are the joint force vectors corresponding to the end-effector Coriolis and centrifugal forces, and Gravity forces. For a non-redundant manipulator, $[J^T(q)\overline{J}^T(q)]$ reduces to the identity matrix and $\overline{g}(q)$ becomes identical to $g(q)$. For a redundant manipulator, $\overline{g}(q)$ reperesents the part of $g(q)$ that has a contribution at the end-effector, $\overline{b}(q,\dot{q})$ is similarly interpreted. Given (eq. 3), the matrix $E(q)$ also establishes the relationship between joint torques and accelerations.

$$\ddot{x} = E(q)\overline{\Gamma}; \qquad (16)$$

where

$$\overline{\Gamma} = \Gamma - \overline{b}(q,\dot{q}) - \overline{g}(q). \qquad (17)$$

$\overline{\Gamma}$ represents the vector of joint forces that contributes to the end-effector accelerations. These contributing forces are limited by the boundaries of actuator torques. At zero velocity the matrix $E(q)$ describes the bounds on the end-effector accelerations corresponding to the bounds on joint actuator torques corrected for the gravity. The bounds on $\overline{\Gamma}$ has been used [4] to construct a joint force normalization matrix $N_0(q)$. This matrix has been used to define

$$E_0(q) = WE(q)N_0(q); \qquad (18)$$

where W is a weighting matrix for the normalization of angular and linear accelerations. The matrix $E_0(q)$ can be interpreted as a joint force/acceleration transmission matrix at zero-velocity. Bounds on actuator torques are modified at non-zero velocities. Coriolis and centrifugal forces that arise at non-zero velocities also affect the bounds on $\overline{\Gamma}$. Similarly to $E_0(q)$, a matrix $E_v(q)$

$$E_v(q) = WE(q)N_v(q); \qquad (19)$$

has been constructed to describe the joint force/acceleration transmission at maximum operating velocities. At a given configuration q, the end-effector's acceleration characteristics will be described by the matrices $E_0(q)$ and $E_v(q)$.

Dynamic Optimization

The dynamic optimization is aimed at finding the design parameters under the various constraints to achieve the smallest, most isotropic, and most uniform end-effector inertial properties, while providing the largest, most isotropic, and most uniform bounds on the magnitude of end-effector acceleration, or equivalently, on the command vector $\mathbf{F}^*$ both at low and high velocities. The performance at high velocity is important for fast and gross motion, while performance at low velocity is particularly important for fast response in tasks with small range of motion, such as part-mating operations.

At a given configuration $\mathbf{q}$, the matrices $\Lambda(\mathbf{q})$, $E_0(\mathbf{q})$, and $E_v(\mathbf{q})$ are functions of the manipulator's geometric and motion parameters; e.g. link length, mass, moment of inertia, centers of mass, actuator mass, and bounds on actuator torques. Let η designate the set of these parameters.

The design process would typically start with an initial design based on workspace and geometric considerations. The various design parameters would be estimated within some range. These specifications and the dynamic and structural requirements form the set of design parameters η. Let $\{u_i(\eta); i = 1, \ldots, n_u\}$ and $\{v_i(\eta); i = 1, \ldots, n_v\}$ designate the sets of equality and inequality constraints on the manipulator design parameters η.

Expressed as a function of the manipulator configuration $\mathbf{q}$ and the design parameters η, the matrices $\Lambda(\mathbf{q})$, $E_0(\mathbf{q}, \eta)$ and $E_v(\mathbf{q}, \eta)$ constitute the basic components in this optimization problem. At a given configuration, the problem is to find the optimal design parameters η, under the constraints $\{u_i(\eta)\}$ and $\{v_i(\eta)\}$, that minimize some cost function based on the end-effector inertial and acceleration characteristics. This cost function is made up of three weighted components associated with the characterisitics of the matrices $\Lambda(\mathbf{q})$, $E_0(\mathbf{q})$, and $E_v(\mathbf{q})$,

$$\mathcal{C}(\mathbf{q}, \eta) = \sum_{i=1}^{3} w_i \mathcal{C}_i(\mathbf{q}, \eta);$$

subject to the equality and inequality constraints

$$u_i(\eta) = 0 \quad i = 1, \ldots, n_u;$$

$$v_i(\eta) \leq 0 \quad i = 1, \ldots, n_v;$$

where w_i are the weight coefficients. The cost function associated with the kinetic energy matrix is aimed at providing small and isotropic inertial properties at $\mathbf{q}$. The magnitude characteristics is described by the norm $\|\Lambda(\mathbf{q})\|$, and the isotropic properties are represented by the matrix condition number, i.e. $\kappa(\Lambda(\mathbf{q}, \eta))$. The first component becomes

$$\mathcal{C}_1(\mathbf{q}, \eta) = [\|\Lambda(\mathbf{q}, \eta)\| + \alpha_1 \kappa(\Lambda(\mathbf{q}, \eta))];$$

The cost functions associated with the end-effector accelerations at zero and maximum operating velocity are aimed at providing the largest and most isotropic properties at

q. This is

$$C_2(\mathbf{q},\eta) = [\frac{1}{\|E_0(\mathbf{q},\eta)\|} + \alpha_2 \kappa(E_0(\mathbf{q},\eta))];$$

$$C_3(\mathbf{q},\eta) = [\frac{1}{\|E_v(\mathbf{q},\eta)\|} + \alpha_3 \kappa(E_v(\mathbf{q},\eta))].$$

where $\alpha_1, \alpha_2, \alpha_3$. Finally, the problem of dynamic optimization over the manipulator work space $\mathcal{D}_\mathbf{q}$ can be expressed as

$$\text{minimize} \int_{\mathcal{D}_\mathbf{q}} \mathcal{C}(\mathbf{q},\eta)w(\mathbf{q})d\mathbf{q};$$

subject

$$u_i(\eta) = 0 \quad i = 1,\ldots,n_u;$$
$$v_i(\eta) \leq 0 \quad i = 1,\ldots,n_v;$$

where the function $w(\mathbf{q})$ is used to relax the weighting of the cost function $\mathcal{C}(\mathbf{q},\eta)$ in the vicinity of the work space boundaries and singularities.

Application to ARTISAN

Optimal dynamic characteristics at the end-effector has been one of the basic goals in the ARTISAN project [7]. These include high performance joint torque control ability, motion redundancy, micro-manipulation ability [10], light structure, and integrated sensing. The kinematic structure of the ARTISAN is divided into three subsystems: wrist positioning structure, wrist and micro-manipulator. The wrist positioning structure is the part of the manipulator composed of the first four joints. Joint 1 and joint 2 are intersecting, orthogonal revolutes. Joints 3 and 4 are revolutes with axes parallel to the axis of joint 2. This part of the system forms a redundant structure if we regard This part of the system forms a redundant structure with respect to the positioning of the wrist point. The dynamic optimization has been applied to the design of the redundant structure formed by the first four degrees of freedom of ARTISAN.

The design parameters consisted of the links' dimensions, masses, inertias, and motor parameters. The dynamic optimization was conducted in three main steps. Based on the preliminary design, the inertial characteristics were first optimized. This resulted in an initial selection of dimensions and mass distribution. This first set of design parameters is used to initialize, the second step which is aimed at providing optimal acceleration characteristics. Actuators are chosen in this second step. The overall optimization is achieved in the third step.

This procedure, illustrated in Fig. 1., has led to a significant reduction of the search space in steps 1 and 2 and provided a good initial estimate for the overall optimization in step 3. It is important to mention the impact of the various weights on the final solution.

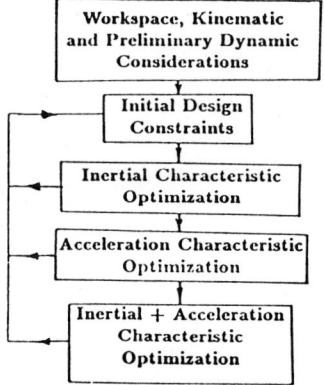

Fig. 1. The Three Step Optimization Procedure

The optimization was carried out using a sequential quadratic programming (SQP) algorithm. The results of this optimization for ARTISAN has been compared to a PUMA 560 arm. Fig. 2. shows the inertial characteristics of the PUMA arm (Fig. 2.a.) and ARTISAN (Fig. 2.b). At a given position of the end-effector, these figures show the projections of the ellipsoids associated with the three eigenvalues of Λ. Because of the redundancy, different ellipsoids would result at given end-effector poistion. The ellipsoids shown in Fig. 2.b. correspond to those that have the largest eigenvalues. Also, the scale used in Fig. 2.b. is twice that of Fig. 2.a. The average effective inertia of the PUMA is roughly three times that of ARTISAN.

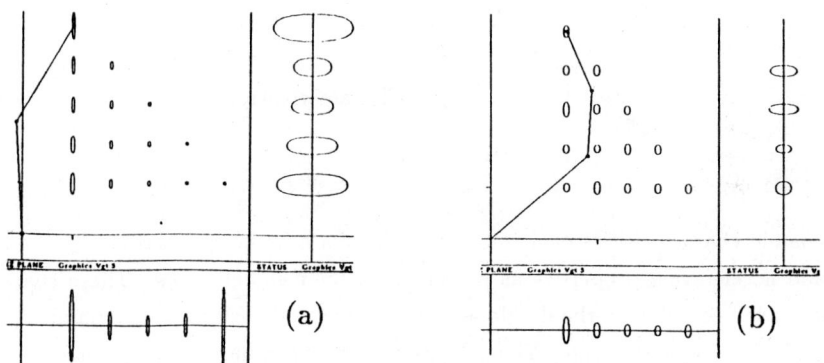

Fig. 2. The inertial Characteristics

Fig. 3. illustrates the minimum available end effector acceleration for the PUMA (Fig. 3.a.) and ARTISAN (Fig. 3.b) at zero joint velocity. The circles depict the minimum available accelerisitics at points in the workspace. On an average, the minimum available accelerations for ARTISAN is twice that of the PUMA arm for same joint torques.

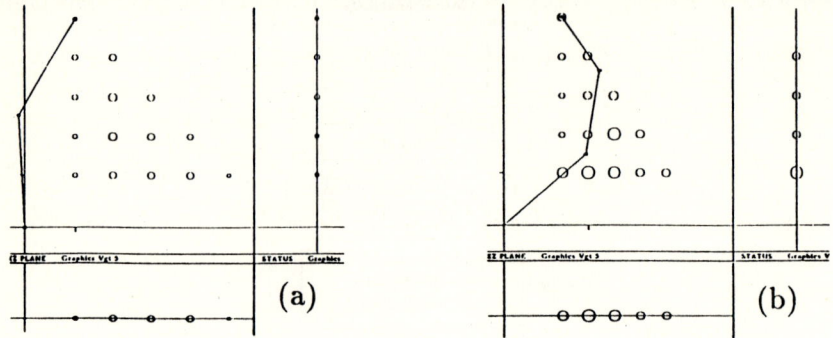

Fig. 3. Minimum Available End-Effector Accelerations

Fig. 4. shows the condition numbers of the acceleration characteristics at zero joint velocity for ARTISAN to be uniform over the workspace. These characteristics has been estimated to be roughly half of those computed for PUMA arm.

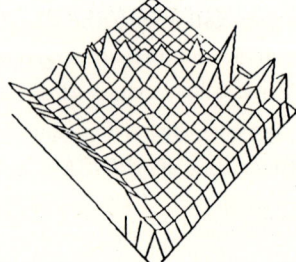

Fig. 4. Acceleration Characteristics

Conclusion

The dynamic characterisitics of manipulator systems have been described by the inertial and acceleration properties as perceived at their end-effectors. These characterisitics have been used in the developement of a methodology for the dynamic optimization in manipulator design. The optimization problem has been expressed as the minimization, with respect to the design parameters and constraints, of a cost function based on these characteristics. The small isotropic and uniform inertial characteristics will provide higher dynamic response at the end-effector. The large isotropic and uniform bounds on the end-effector accelerations will be translated into a large and well conditioned operational space command vector. The application to ARTISAN has demonstrated the effectiveness of this methodology to provide higher dynamic characteristics. With an optimal redistribution of masses, dimensions, and actuators, the resulting design has been shown to be significantly superior to conventional designs.

Acknowledgments

The financial support of the Systems Development Foundation and SIMA is gratefully acknowledged. We are thankful to professors Bernard Roth, Kenneth Waldron and Joel Burdick, who have made valuable contributions to the development of this work.

References

1. Asada, H.; A Geometrical Representation of Manipulator Dynamics and Its Application to Arm Design. Trans. of ASME, Journal of Dynamic Systems, Measurement, and Control, Vol. 105, No. 3,pp. 131-135. 1983.

2. Fournier, A.; Génération de Mouvements en Robotique. Application des Inverses Généralisées et des Pseudo Inverses. Thèse d'Etat, Mention Science, Université des Sciences et Techniques des Languedoc, Montpellier, France, 1980.

3. Khatib, O.; Commande Dynamique dans l'Espace Opérat- ionnel des Robots Manipulateurs en Présence d'Obstacles. Thése de Docteur-Ingénieur. École Nationale Supérieure de l'Aéro- nautique et de l'Espace (ENSAE). Toulouse, France, 1980.

4. Khatib, O. and Burdick, J.; Optimization of Dynamics in Manipulator Design: The Operational Space Formulation, Proceedings of the ASME Winter Annual Meeting, Miami, November 1985; also published in the International Journal of Robotics and Automation, vol. 2, no. 2, pp. 90-98, 1987.

5. Khatib, O.; A Unified Approach to Motion and Force Control of Robot Manipulators: The Operational Space Formulation," IEEE Journal on Robotics and Automation, vol. 3, no. 1, pp. 43-53, February 1987.

6. Paul, R.P and Stevenson, C.N.; Kinematics of Robot Wrists. International Journal of Robotics Research, vol. 2, No. 1, pp. 31-38, 1983.

7. Roth, B. et al.; The Design of the ARTISAN Research Manipulator System, submitted to the International Journal of Robotics Research.

8. Roth, B.; Performance Evaluation of Manipulators from a Kinematic Viewpoint. National Bureau of Standards Workshop on Performance Evaluation on Programmable Robots and Manipulators, National Bureau of Standards, NBS SP-459, pp. 39-61, 1976.

9. Shimano, B.; The Kinematic Design and Force Control of Computer Controlled Manipulators. Stanford A.I. Lab. Memo 313, 1978.

10. Waldron, K. J., Raghavan, M. and Roth, B.,; Kinematics of a Hybrid Series-Parallel Manipulation System (Part I and II). ASME Winter Annual Meeting. Boston, 1987.

11. Yoshikawa, T.; Analysis and Control of Robot Manipulators with Redundancy. Proc. of the 1st International Symposium of Robotics Research, MIT Press, Cambridge, MA, pp. 735-747, 1983.

12. Yoshikawa, T.; Dynamic Manipulability of Robot Manipulators. Proc. 1985 IEEE International Conference on Robotics and Automation, St. Louis, pp. 1033-1038, 1985.

Effect of Sampling Rates on the Performance of Model-Based Control Schemes

Pradeep K. Khosla

Department of Electrical and Computer Engineering
The Robotics Institute
Carnegie-Mellon University
Pittsburgh, PA 15213

Abstract

In our previous research, we experimentally implemented and evaluated the effect of dynamics compensation in model-based control algorithms. In this paper, we evaluate the effect of changing the control sampling period on the performance of the computed-torque and independent joint control schemes. While the former utilizes the complete dynamics model of the manipulator, the latter assumes a decoupled and linear model of the manipulator dynamics. We discuss the design of controller gains for both the computed-torque and the independent joint control schemes and establish a framework for comparing their trajectory tracking performance. Our experiments show that within each scheme the trajectory tracking accuracy varies slightly with the change of the sampling rate. However, at low sampling rates the computed-torque scheme outperforms the independent joint control scheme.

1. Introduction

Although many simulation results have been presented[1,2,3], the real-time implementation and performance of model-based control schemes with high control sampling rates had not been demonstrated on actual manipulators, until recently[4,5,6]. The main reasons for this have been the lack of a suitable manipulator system and the fact that it is difficult to evaluate the dynamics parameters for implementing model-based algorithms. One of the goals of the CMU Direct-Drive Arm II[7] project has been to overcome these difficulties and evaluate the effect of dynamics compensation on the real-time trajectory tracking of manipulators. For the real-time computation of the inverse dynamics, we have developed a high-speed and powerful computational environment. The computation of inverse dynamics has been customized for the CMU DD Arm II and a computation time of 1 ms has been achieved[8]. To obtain an accurate model we have computed and measured the various parameters from the engineering drawings of the CMU DD Arm II by modeling each link as a composite of hollow and solid cylinders, prisms, and rectangular parallelopipeds. We have also proposed an algorithm to identify the dynamics parameters[9] which has been implemented on the CMU DD Arm II. The results of the experimental implementation of our identification algorithm are presented in[10]. Finally, the negligible friction in our direct-drive arm makes it suitable to test the efficacy of the computed-torque scheme.

Based on the above contributions, in our previous research, we investigated the effect of high sampling rate dynamics compensation in model-based manipulator control methods.

Specifically, we compared the computed-torque scheme which utilizes the complete dynamics model of the manipulator with the independent joint control scheme[4] and the feedforward compensation method[11]. The control schemes were implemented on the CMU DD Arm II with a sampling period of 2 ms.

In this paper, we investigate the effect of reducing the sampling rate on the trajectory tracking performance of model-based manipulator control methods. We first compare the performance of each scheme as the sampling rate is changed. Next, we also compare the relative performance of both the computed-torque and the independent joint control schemes at different sampling rates. Our work represents the first experimental evaluation of the effect of the sampling rate on the performance of both the computed-torque and the independent joint control schemes. We discuss the design of the controller gains for both the independent joint control and the computed-torque schemes and establish a framework for the comparison of their trajectory tracking performance. Our experiments demonstrate that the computed-torque scheme exhibits a better performance than the independent joint control scheme. Our experiments also show that high sampling rates are important because they result in a stiffer system that is capable of effectively rejecting unknown external disturbances.

This paper is organized as follows: In Section 2, we describe previous research in manipulator control and provide a motivation for our work. Then in Section 3, we present an overview of the manipulator control schemes that have been implemented and evaluated on the CMU DD Arm II. The design of controllers is discussed in Section 4 and the real-time experimental results are presented and interpreted in Section 5. Finally, in Section 6 we summarize this paper. In the Appendix, we describe our experimental hardware set-up.

2. Past Work and Motivation

The robot control problem revolves around the computation of the actuating joint torques/forces to follow the desired trajectory. The dynamics of a manipulator are described by a set of highly nonlinear and coupled differential equations. The complete dynamic model of an N degrees-of-freedom manipulator is described by:

$$\tau = \mathbf{D}(\theta)\ddot{\theta} + \mathbf{h}(\theta,\dot{\theta}) + \mathbf{g}(\theta) \qquad (1)$$

where τ is the N-vector of the actuating torques; $\mathbf{D}(\theta)$ is the $N \times N$ position dependent manipulator inertia matrix; $\mathbf{h}(\theta,\dot{\theta})$ is the N-vector of Coriolis and centrifugal torques; $\mathbf{g}(\theta)$ is the N-vector of gravitational torques; and $\ddot{\theta}$, $\dot{\theta}$ and θ are N-vectors of the joint accelerations, velocities and positions, respectively.

This complex description of the system makes the design of controllers a difficult task. To circumvent the difficulties the control engineer often assumes a simplified model to proceed with the controller design. Industrial manipulators are usually controlled by conventional PID-type independent joint control structures designed under the assumption that the dynamics of the links are uncoupled and linear. The controllers based on such an overly simplified dynamics model result in low speeds of operation and overshoot of the end-effector.

To improve the performance of the PID controllers, researchers have investigated model-based control schemes which attempt to compensate for the nonlinearities and the mismatch in the dynamical description of the robot. One of the model-based techniques is the *feedforward dynamics compensation* method which computes the desired torques from the given trajectory and injects these torques as feedforward control signals. Independent joint feedback controllers are then added with the intention of compensating for the small coupling torques arising out of the mismatch in the dynamics of the model and the real arm[1]. More thorough compensation is achieved by the *computed-torque* technique in which the dynamics model is included in the feedback loop to decouple and linearize the manipulator dynamics. This technique has also been extended to operate in the Cartesian space and is called *resolved-acceleration* scheme[2].

One of the fundamental problems with real-time dynamics compensation has been the high computational requirements of the inverse dynamics formulations. This drawback led researchers to evaluate the significance of the terms in the dynamical model and compensate only for the significant terms. Another avenue that was explored involved the use of table look-up and interpolation techniques[12]. Recently there has been much work in reducing the computational requirements based on the structure of the dynamical equations[13, 14]. Further, high speed controller architectures have also been proposed and demonstrated[8]. These developments have made it possible to experimentally implement and evaluate the nonlinear model-based control schemes[4, 5].

Several researchers have followed an entirely different avenue of research that involves looking at alternate controllers that are computationally less expensive than the model-based schemes and at the same time robust. This has led to the development of alternate methods such as linear multivariable control[15], self-tuning and model-reference adaptive control[16, 17], sliding control[18], and prediction control[19].

Real-Time digital implementation of either model-based schemes or alternate control schemes requires the designer to make a choice of the sampling rate. Thus it is important to develop both theoretical and experimental methods to evaluate the effect of sampling rates on the performance of manipulator control methods. A theoretical investigation in this area is still an uncharted territory probably due to the nonlinear and coupled nature of the manipulator dynamics. In this paper, however, we present experimental results on evaluating the effect of changing the sampling rates on the performance of independent joint control and computed-torque schemes. A similar evaluation for alternate control schemes (as presented above) is beyond the scope of this paper and is a topic of current investigation. In the next section, we describe the control schemes that have been implemented on the CMU DD Arm II.

3. Schemes Implemented

We have implemented computed-torque and the independent joint control schemes and compared their real-time performance as a function of the control sampling rate. These schemes are described in the sequel.

<u>Independent Joint Control (IJC)</u>

In this scheme linear PD control laws were designed for each joint based on the assumption that the joints are decoupled and linear. The control torque τ applied to the joints at each sampling instant is:

$$\tau = \mathbf{J}\mathbf{u}_i \qquad (2)$$

where $\mathbf{J}$ is the constant $N \times N$ diagonal matrix of link inertias at a typical position, and $\mathbf{u}_i$ is the vector of commanded accelerations.

This scheme utilizes nonlinear feedback to decouple the manipulator. The control torque τ is computed by the inverse dynamics equation in (1), using the commanded acceleration $\mathbf{u}_i$ instead of the measured acceleration $\ddot{\theta}$:

$$\tau = \tilde{\mathbf{D}}(\theta)\mathbf{u}_i + \tilde{\mathbf{h}}(\theta,\dot{\theta}) + \tilde{\mathbf{g}}(\theta) \qquad (3)$$

where the " ~ " indicates that the estimated values of the dynamics parameters are used in the computation.

Before proceeding with a meaningful comparision of the performance of the computed-torque and the independent joint control schemes it is necessary to establish a common framework. In order to achieve this, we consider the control law in two steps; computation of the commanded acceleration and computation of the control torque. The commanded joint accelerations u_i can be computed in one of the following three ways:

$$u_1 = K_p(\theta_d - \theta) - K_v \dot{\theta} \tag{4}$$

$$u_2 = K_p(\theta_d - \theta) + K_v(\dot{\theta}_d - \dot{\theta}) \tag{5}$$

$$u_3 = K_p(\theta_d - \theta) + K_v(\dot{\theta}_d - \dot{\theta}) + \ddot{\theta}_d \tag{6}$$

where K_p and K_v are $N \times N$ diagonal position and velocity gain matrices, respectively. The N-vectors θ_d and θ are the desired and measured joint positions, respectively, and the " · " indicates the time derivative of the variables. Whereas only the position error and the velocity damping is used in (4), the commanded acceleration signal in (5) uses a velocity feedforward term, and the commanded acceleration signal in (6) uses both the velocity and acceleration feedforward terms. The idea is to increase the speed of response by incorporating a feedforward term.

The fundamental difference between the independent joint control schemes and the model-based schemes lies in the second step in the control law, i.e., the method of computing the applied control torque signals from the commanded acceleration signals. If the vector of actuating joint torques τ is computed from the commanded acceleration signal under the assumption that the joint inertias are constant, then we obtain an independent joint control scheme. On the other hand, if the actuating torques τ are computed from the *inverse dynamics* model in (1) then we obtain the computed-torque scheme.

We have performed real-time experiments and evaluated the effect of changing the sampling rates on the performance of the independent joint control and the computed-torque schemes. The experiments were performed on the CMU DD Arm II. In our experiments, we have used Equation 6 to compute the accelerations for both the computed-torque and the independent joint control schemes. In the next section, we explain our procedure to determine the gain matrices for both the computed-torque and the independent joint control schemes.

4. Controller Design

The performance of the nonlinear CT scheme and the linear IJC scheme can be compared only if the same criteria are used for design of the controller gain matrices. Fortunately, this is possible because the gain matrices K_p and K_v appear only in the commanded accelerations (Equations (4)-(6)) which are the same for both CT and IJC schemes. Thus, whether we implement the simplistic independent joint control scheme or the sophisticated computed-torque scheme, we are faced with the problem of designing the gain matrices K_p and K_v. These matrices are chosen to satisfy the specified output response criterion.

4.1. Design of Gain Matrices for Independent Joint Control

The closed loop transfer function relating the input θ_{jd} to the measured output θ_j for joint j is:

$$\frac{\theta_j}{\theta_{jd}} = \frac{s^2\delta + s\gamma k_{vj} + k_{pj}}{s^2 + k_{vj}s + k_{pj}} \tag{7}$$

where $\gamma=1$ if velocity feedforward is included and zero otherwise, and $\delta=1$ if acceleration feedforward is included and zero otherwise. The closed-loop characteristic equation in all the three cases is,

$$s^2 + k_{vj}s + k_{pj} = 0 \tag{8}$$

and its roots are specified to obtain a stable response. The complete closed-loop response of the system is governed by both the zeros and the poles of the system. In the absence of any feedforward terms, the response is governed by the poles of the transfer function.

Since it is desired that none of the joints overshoot the commanded position or the response be critically damped, our choice of the matrices K_p and K_v must be such that their elements satisfy the condition:

$$k_{vj} = 2\sqrt{k_{pj}} \quad \text{for } j=1,\dots,6 \tag{9}$$

Besides, in order to achieve a high disturbance rejection ratio or high stiffness it is also necessary to choose the position gain matrix K_p as large as possible which results in a large K_v.

4.2. Design of Gain Matrices for Computed-Torque Scheme

The basic idea behind the computed-torque scheme is to achieve dynamic decoupling of all the joints using nonlinear feedback. If the dynamic model of the manipulator is described by (1) and the applied control torque is computed according to (3), then the following closed-loop system is obtained:

$$\ddot{\theta} = u_i - [\tilde{D}]^{-1}\{[D - \tilde{D}]\ddot{\theta} + [h - \tilde{h}] + [g - \tilde{g}]\}$$

where the functional dependencies on θ and $\dot{\theta}$ have been omitted for the sake of clarity. If the dynamics are modeled exactly, that is, $\tilde{D}=D$, $\tilde{h}=h$ and $\tilde{g}=g$, then the decoupled closed loop system is described by

$$\ddot{\theta} = u_i.$$

Upon substituting the right hand side of either (4), (5) or (6) in the above equation, we obtain the closed-loop input-output transfer function of the system. The closed-loop characteristic equation in all the three cases is:

$$s^2 + k_{vj}s + k_{pj} = 0 \tag{10}$$

where k_{vj} and k_{pj} are the velocity and position gains for the j-th joint. Upon comparing (8) and (10), we obtain the relationships

$$k_{pj}^{[CT]} = k_{pj}^{[IJC]} \quad \text{and} \quad k_{vj}^{[CT]} = k_{vj}^{[IJC]}$$

which suggest that the gains of the IJC scheme are also the gains of the CT scheme. This equality must be expected because the closed-loop characteristic equation for both the independent joint control and the computed-torque scheme is the same.

4.3. Gain Selection

The gain matrices K_p and K_v are a function of the sampling rate of the control system[20]. The higher the sampling rate the larger the values of K_p and K_v that can be chosen. Since the stiffness (or disturbance rejection property) of the system is governed by the position gain matrix (K_p) a higher sampling rate implies higher stiffness also. In practice the choice of the velocity gain K_v is limited by the noise present in the velocity measurement. We determined the upper limit of the velocity gain experimentally: we set the position gain to zero and increased the velocity gain of each joint until the unmodeled high-frequency dynamics of the system were excited by the noise introduced in the velocity measurement. This value of K_v represents the maximum allowable velocity gain. We chose 80% of the maximum velocity gain in order to obtain as high value of the position gain as possible and still be well within the stability limits with respect to the unmodeled high frequency dynamics. The elements of the position gain matrix K_p were computed to satisfy the critical damping condition in (9) and also achieved the maximum disturbance rejection ratio. The elements of the velocity and position gain matrices (chosen for a sampling rate of 500 Hz) that were used in the implementation of the control schemes are listed in Table 1. The above procedure was repeated to select the gain matrices for sampling rates ranging from 500 Hz to 200 Hz.

5. Experiments and Results

In our experiments we implemented both the independent joint control scheme and the computed-torque scheme. We evaluated their individual and relative performances by changing the sampling rate but keeping both the position and the velocity gain matrices fixed. The maximum permissible velocity and position gains were chosen at a control sampling period of 5 ms (according to the method outlined in Section 4.3) and remained fixed even when the sampling period was changed. This allows us to determine the effect of the sampling rate on the trajectory tracking control performance. We have also evaluated the best performance of the CT method for a sampling period of 2 ms with its best performance for a sampling period of 5 ms. We conducted the evaluation experiments on a multitude of trajectories but due to space limitations we present our results for a simple but illustrative trajectory.

The first trajectory is chosen to be simple and relatively slow but capable of providing insight into the effect of dynamics compensation. In this trajectory only joint 2 moves while all the other joints are commanded to hold their zero positions and can be envisioned from the schematic diagram in Figure 1. Joint 2 is commanded to start from its zero position and to reach the position of 1.5 rad in 0.75 seconds; it remains at this position for an interval of 0.75 seconds after which it is required to return to its home position in 0.75 seconds. The points of discontinuity, in the trajectory, were joined by a fifth-order polynomial to maintain the continuity of position, velocity and acceleration along the three segments. The desired position, velocity and acceleration trajectories for joint 2 are depicted in Figure 2. The maximum velocity and acceleration to be attained by joint 2 are 2 rad/sec and 6 rad/sec^2, respectively.

The position tracking performance of joint 2 for both the CT and IJC schemes, for a control sampling rate of 200 Hz (corresponding to a control sampling period of 5 ms), is depicted in Figure 3. The corresponding position and velocity tracking errors are presented in Figures 4 and 5, respectively. We also depict the position tracking error of joint 1 in Figure 6 for both the CT and IJC schemes. We note that the CT scheme outperforms the IJC scheme. For example, in the case of joint 2 the maximum position tracking error for CT scheme is 0.03 rads while for the IJC scheme it is 0.45 rads, approximately. In an earlier paper[4], we had compared both the CT and IJC schemes with a control sampling period of 2 ms. It must be noted that in the earlier reported experiments[4] the gains were selected for a control sampling period of 2 ms whereas in the present experiments the gains have been selected for a control sampling period of 5 ms. To put the results in perspective, we recall that in the earlier experiment the maximum position tracking error for the CT method was 0.022 rads while for the IJC method it was 0.036 rads. From the above observations it may be deduced that increasing the control sampling period from 2 to 5 ms results in a noteworthy degradation of the performance of the IJC scheme. A similar increase in the sampling rate also improves the performance of the CT scheme.

In Figure 7, we depict the performance of the CT scheme as the sampling rate is increased from 200 Hz to 500 Hz. In this case the position and velocity gain matrices were determined for a sampling rate of 200 Hz and they remained fixed even when the sampling rate was increased to 500 Hz. Thus, Figure 7 presents the relative performance of the CT method as a function of the sampling rate only. We note that the trajectory tracking performance for both 200 Hz and 500 Hz sampling rates is comparable and has not changed in any appreciable manner with an increase in the sampling rate. Figure 8 depicts the results for the IJC method when a similar experiment was performed. In this case also we do not observe any appreciable change in performance when only the sampling rate is changed.

Thus, from the above set of experiments the following conclusions may be drawn:

1. If the gains are selected for a lower sampling rate and then if the sampling rate is increased, while keeping the gains fixed, there is no appreciable improvement in the performance of both the CT and the IJC schmes.

2. At lower sampling rates the CT scheme outperforms the IJC method. Even though the disturbance rejection ratio of both the schemes is diminished, it does not appreciably affect the CT method because of the compensation for the nonlinear and coupling terms. Whereas it affects the IJC method because the disturbance that is constituted by the nonlinear and the coupling terms is not rejected appreciably.

3. If the maximum possible gains are selected for the chosen sampling rates then the performance of CT at a higher sampling rate is better than its performance at a lower sampling rate. A similar conclusion is drawn for the IJC scheme also.

Our last conslusion is especially significant because it suggests that a higher sampling rate does not only imply improved performance but it also allows us to achieve high stiffness. It is desirable for a manipulator to have high stiffness so that the effect of unpredictable external disturbances on the trajectory tracking performance is significantly reduced.

6. Summary

In this paper, we have presented the first experimental evaluation of the effect of the sampling rate on the performance of both the computed-torque and the independent joint control schemes. We have discussed the design of the controller gains for both the independent joint control and the computed-torque schemes and established a framework for the comparison of their trajectory tracking performance. Based on our experiments we have demonstrated that the computed-torque scheme exhibits a better performance than the independent joint control scheme. Our experiments also show that high sampling rates are important because they result in a stiffer system that is capable of effectively rejecting unknown external disturbances.

7. Acknowledgements

This research was supported in part by the National Science Foundation under Grant ECS-8320364 and the Department of Electrical and Computer Engineering, Carnegie Mellon University. The author acknowledges the cooperation of Prof. Takeo Kanade (Head of Vision Laboratory, Carnegie Mellon University) throughout the course of this research.

I. The CMU DD Arm II

We have developed, at CMU, the concept of direct-drive robots in which the links are directly coupled to the motor shaft. This construction eliminates undesirable properties like friction and gear backlash. The CMU DD Arm II[7] is the second version of the CMU direct-drive manipulator and is designed to be faster, lighter and more accurate than its predecessor CMU DD Arm I[21]. We have used brushless rare-earth magnet DC torque motors driven by current controlled amplifiers to achieve a torque controlled joint drive system. The SCARA-type configuration of the arm reduces the the torque requirements of the first two joints and also simplifies the dynamic model of the arm. To achieve the desired accuracy, we use very high precision (16 bits/rotation) rotary absolute encoders. The arm weighs approximately 70 pounds and is designed to achieve maximum joint accelerations of 10 rad/sec^2.

The hardware of the DD Arm II control system consists of three integral components: the Motorola M68000 microcomputer, the Marinco processor and the TMS-320 microprocessor-based individual joint controllers. We have also developed the customized Newton-Euler equations for the CMU DD Arm II and achieved a computation time of 1 ms by implementing these on the Marinco processor. The details of the customized algorithm, hardware configuration and the numerical values of the dynamics parameters are presented in[8].

Joint (j)	Transfer Function ($\frac{1}{J_j s^2}$)	k_{pj}	k_{vj}
1	$\frac{1}{12.3s^2}$	2.75	3.33
2	$\frac{1}{2s^2}$	15.0	7.5
3	$\frac{1}{0.25s^2}$	256.0	32.0
4	$\frac{1}{0.007s^2}$	1285.0	71.5
5	$\frac{1}{0.006s^2}$	625.0	50.0
6	$\frac{1}{0.0003s^2}$	1110.0	50.0

Table 1: Transfer Functions and Gains of Individual Links

Figure 1: Schematic Diagram of 3 DOF DD Arm II

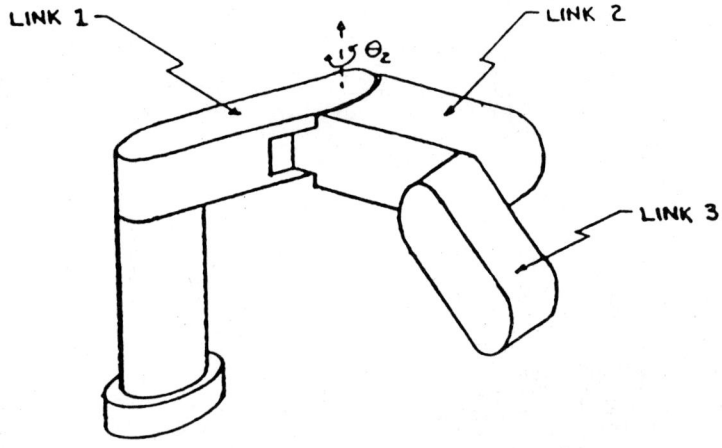

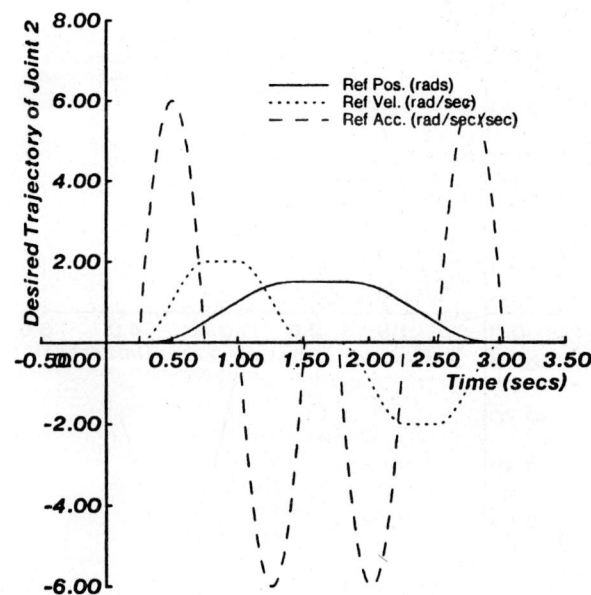

Figure 2: Desired Trajectories for Joint 2

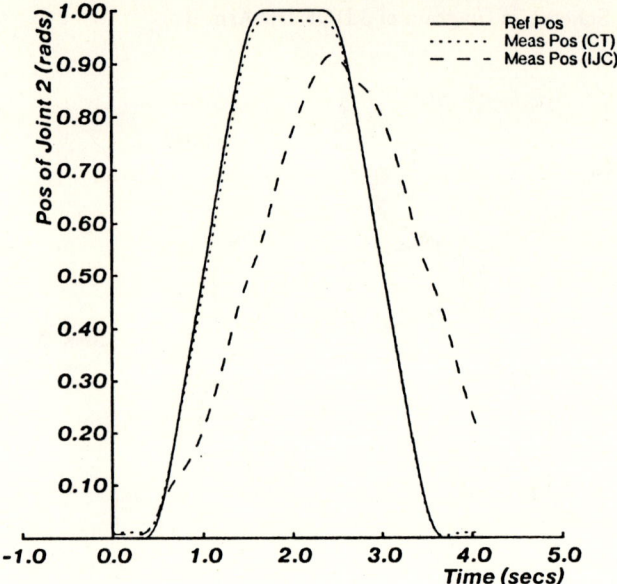

Figure 3: Position Tracking of CT and IJC at 5 ms Sampling

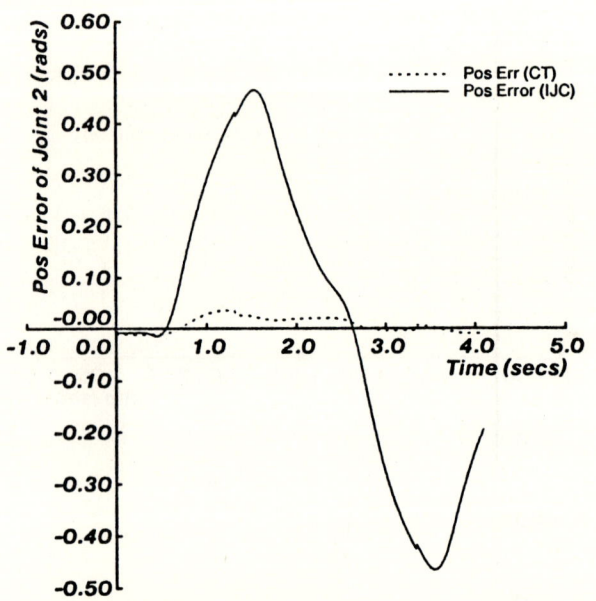

Figure 4: Position Tracking Error of Joint 2

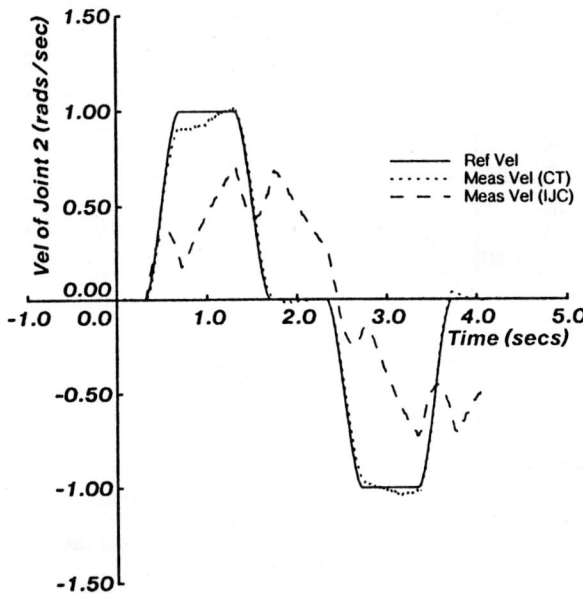

Figure 5: Velocity Tracking Errors of Joint 2

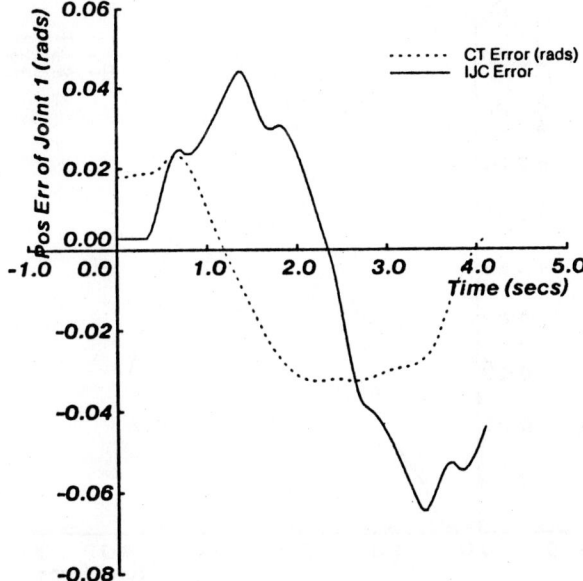

Figure 6: Position Tracking Errors of Joint 1

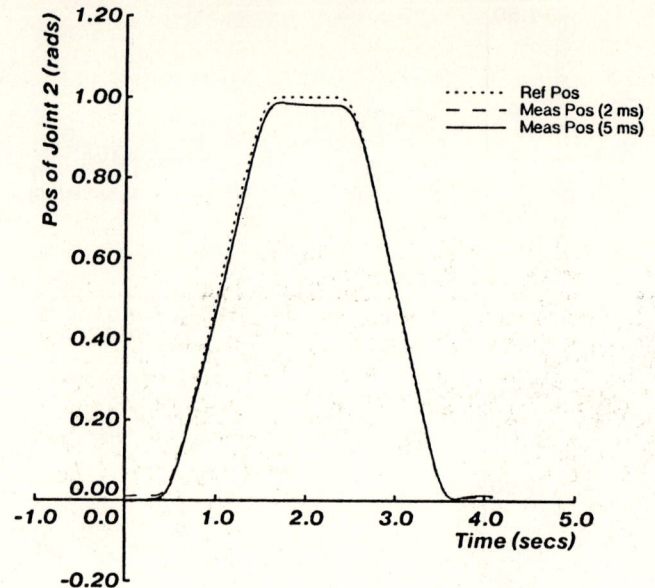

Figure 7: Performance of CT as a Function of Sampling Period

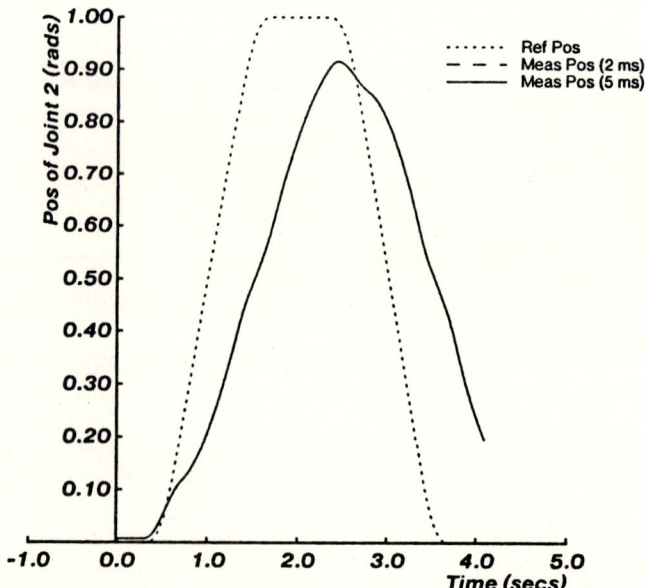

Figure 8: Performance of IJC as a Function of Sampling Period

References

1. Markiewicz, B. R., "Analysis of the Computed-Torque Drive Method and Comparision with the Conventional Position Servo for a Computer-Controlled Manipulator", Technical Memorandum 33-601, Jet Propulsion Laboratory, Pasadena, CA, March 1973.

2. Luh, J. Y. S., Walker, M. W. and Paul, R. P., "Resolved-Acceleration Control of Mechanical Manipulators", *IEEE Transactions on Automatic Control*, Vol. 25, No. 3, June 1980, pp. 468-474.

3. Bejczy A. K., "Robot Arm Dynamics and Control", Technical Memorandum 33-669, Jet Propulsion Laboratory, Pasadena, CA, February 1974.

4. Khosla, P. K. and Kanade, T., "Real-Time Implementation and Evaluation of Model-Based Controls on CMU DD ARM II", *1986 IEEE International Conference on Robotics and Automation*, Bejczy, A. K.,ed., IEEE, April 7-10 1986.

5. Leahy, M. B., Valavanis, K. P. and Saridis, G. N., "The Effects of Dynamics Models on Robot Control", *Proceedings of the 1986 IEEE Conference on Robotics and Automation*, IEEE, San Francisco, CA, April 1986.

6. An, C. H., Atkeson, C. G. and Hollerbach, J. M., "Experimental Determination of the Effect of Feedforward Control on Trajectory Tracking Errors", *Proceedings of 1986 IEEE Conference on Robotics and Automation*, Bejczy, A. K.,ed., IEEE, San Francisco, CA, April 7-10 1986, pp. 55-60.

7. Schmitz, D., Khosla, P. K. and Kanade, T., "Development of CMU Direct-Drive Arm II", *Proceedings of the 15-th International Symposium on Industrial Robotics*, Hasegawa, Yukio,ed., Tokyo, Japan, September, 11-13 1985.

8. Kanade, T., Khosla, P. K. and Tanaka, N., "Real-Time Control of the CMU Direct Drive Arm II Using Customized Inverse Dynamics", *Proceedings of the 23rd IEEE Conference on Decision and Control*, Polis, M. P.,ed., Las Vegas, NV, December 12-14, 1984, pp. 1345-1352.

9. Khosla, P. K. and Kanade, T., "Parameter Identification of Robot Dynamics", *Proceedings of the 24-th CDC*, Franklin, G. F.,ed., Florida, December 11-13 1985, pp. 1754-1760.

10. Khosla, P. K., "Estimation of Robot Dynamics Parameters: Theory and Application", *Proceedings of the Second International IASTED Conference on Applied Control and Identification*, ACTA Press, Los Angeles, CA, December 10-12 1986.

11. Khosla, P. K. and Kanade, T., "Experimental Evaluation of the Feedforward Compensation and Computed-Torque Control Schemes", *Proceedings of the 1986 ACC*, Stear, E. B.,ed., AAAC, Seattle, WA, June 18-20 1986.

12. Raibert, M. H., "Analytical Equations vs. Table Lookup for Manipulation: A Unifying Concept", *Proceedings of the IEEE Conference on Decision and Control*, New Orleans, La., December 1977, pp. 576-579.

13. Hollerbach, J. M. and Sahar, G., "Wrist Partitioned Inverse Kinematic Accelerations and Manipulator Dynamics", *Proceedings of the First International IEEE Conference on Robotics*, Paul, R. P.,ed., Atlanta, GA, March 13-15, 1984, pp. 152-161.

14. Khosla, P. K. and Neuman, C. P., "Computational Requirements of Customized Newton-Euler Algorithms", *Journal of Robotic Systems*, Vol. 2, No. 3, Fall 1985, pp. 309-327.

15. Seraji, H., "Linear Multivariable Control of Robot Manipulators", *Proceedings of the IEEE International Conference on Robotics and Automation*, IEEE, San Francisco, CA, April 1986, pp. 565-571.

16. Lee, C. S. G., Chung, M. J. and Lee, B. H., "Adaptive Control for Robot Manipulators in Joint and Cartesian Coordinates", *Proceedings of the First International IEEE Conference on Robotics*, Paul, R. P.,ed., Atlanta, GA, March 13-15, 1984, pp. 530-539.

17. Takesaki, M. and Arimoto, S., "Adaptive Trajectory Control of Manipulators", *International Journal of Control*, Vol. 34, 1981, pp. 201-217.

18. Slotine, J.-J. E and Coetsee, J. A., "Adaptive Sliding Controller Synthesis for Nonlinear Systems", *International Journal of Control*, 1986.

19. Tourassis, V. D., "Computer Control of Robotic Manipulators using Predictors", *Proceedings 1987 Symposium on Intelligent Control*, IEEE, Philadelphia, PA, January 1987, pp. 204-209.

20. Astrom, K. J. and Wittenmark, B., *Computer Controlled Systems: Theory and Design*, Prentice-Hall, Englewood Cliffs, N. J., Information and System Science Series, 1984.

21. Asada, H. and Kanade, T., "Design of Direct Drive Mechanical Arms", *Journal of Vibration, Stress, and Reliability in Design*, Vol. 105, No. 1, July 1983, pp. 312-316.

Modeling and Control of a Flexible Robot Link

L. KRUISE, J. VAN AMERONGEN and P. LÖHNBERG

Control Systems and Computer Engineering Laboratory
Department of Electrical Engineering
University of Twente, Enschede, the Netherlands,

and M.J.L TIERNEGO, Royal Militairy Academy, Breda, the Netherlands

SUMMMARY

When a flexible link is rotated around an axis, vibrations occur in the link. This paper describes a controller that is able to control the end-position of the link. The flexible link is modelled in state space. It is shown that the model is of infinite order. A method is given for reducing this model to a finite-order model, for which a controller can be designed. A number of experiments is carried out to demonstrate the performance of the controller.

1. INTRODUCTION

A flexible arm is studied, which can rotate around a vertical axis and is driven by a DC-motor via a gear transmission. In section 2 a model for the vibrations is derived. It is shown that the flexible arm is described in state space by a model with an infinite number of states. In section 3 a model reduction technique will be presented which yields a sufficient low-order model, to be used for the controller design. In section 4 the controller is discussed. State feedback is used to control the flexible arm. Because not all the states can be measured, an observer is used to estimate these states. After a series of simulations the designed controller has been tested on a flexible arm which was especially designed to demonstrate the problems and possible solutions in practice (section 5).

2. MODELLING

The experimental setup consists of a flexible arm, that can rotate in the horizontal plane. One end of the arm is clamped on a vertical gear shaft, which is driven by a DC-motor. The other end of the beam is free. Only the transversal vibrations in the horizontal plane are considered. This means that torsional and longitudinal vibrations are disregarded. Gravity effects may also be disregarded because the arm rotates in a horizontal plane.

2.1 MOTOR MODEL

Because the motor is fast with respect to the other components of the system, it is assumed that the DC-motor may be described by the following model:

$$\dot{\theta}_I = K_M \cdot u \tag{1}$$

where u : applied voltage to the amplifier ($-5V \leq u \leq 5V$)
K_M: motor constant (rad/s/V)
$\dot{\theta}_I$: angular position of the axis (rad/s)

The motor constant K_M depends on the motor parameters and on the gain of the power amplifier. The torque that the exerts on the axis is assumed to be negligibly small. K_M was measured to be 0.50 rad/s/V.

2.2 FLEXIBLE ARM MODEL

Only small motions of the link about the equilibrium state are considered. This implies that only linear terms are taken into account. Following the approach described by Sakawa et al. [1], the motion of the link can be described by the following partial differential equation (see figure 1 for definitions of the symbols)

$$\frac{E \cdot I}{\rho \cdot a} \frac{\partial^4 W}{\partial r^4} + 2 \cdot \delta \cdot \frac{E \cdot I}{\rho \cdot a} \cdot \frac{\partial^5 W}{\partial r^4 \partial t} + \frac{\partial^2 W}{\partial t^2} = -r \cdot \ddot{\theta}_I , \tag{2}$$

where E : elasticity or Young modulus
I : area moment of inertia
ρ : specific weight
a : cross section area
r : position along the beam
W(r,t): displacement at postion r, at time t
δ : damping constant
L : length of the link
θ_I : angular position of the axis
θ_U : angular position of the tip

The boundary conditions for the beam are

$$W(0,t) = \left.\frac{\partial W(r,t)}{\partial r}\right|_{r=0} = \left.\frac{\partial^2 W(r,t)}{\partial r^2}\right|_{r=L} = \left.\frac{\partial^3 W(r,t)}{\partial r^3}\right|_{r=L} = 0 \qquad (3)$$

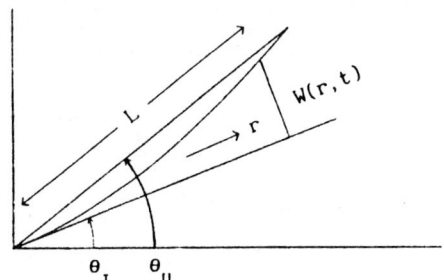

Figure 1. A flexible link

Fukunda and Kuribayashi [2] have shown that the solution of equation (2) can be written as:

$$W(r,t) = \sum_{n=1}^{\infty} Y_n(r) \cdot T_n(t) \qquad (4)$$

where $T_n(t)$, the modal motion, is a function that only depends on time, and $Y_n(r)$ is a mode shape function that only depends on the position r on the beam. A combination $Y_n(r) \cdot T_n(t)$ is called a mode. The shape functions can be found by solving the eigenvalue problem (see ref. [1])

$$\frac{E.I}{\rho.a} \cdot \frac{d^4 Y_n(r)}{dr^4} = \omega_n^2 \cdot Y_n(r), \qquad (5)$$

where ω_n is the resonance frequency of the nth mode. These shape functions have the following important property

$$\int_0^L Y_i(r) \cdot Y_j(r) dr = 0 \qquad i \neq j \qquad (6)$$

The shape functions are scaled in such a way that

$$Y_n(L) = (-1)^{n+1} . \tag{7}$$

The modal motion $T_m(t)$ can be found by substituting eq. (4) into eq. (2). Using eq. (5) this yields

$$\sum_{n=1}^{\infty} \omega_n^2 . Y_n . T_n + 2.\delta \sum_{n=1}^{\infty} \omega_n^2 . Y_n . \dot{T}_n + \sum_{n=1}^{\infty} Y_n . \ddot{T}_n = -r.\ddot{\theta}_I . \tag{8}$$

The next step is to multiply all the terms by Y_m and integrate the resulting terms over r. Using eq. (6) this yields

$$\omega_m^2 . T_m + 2.z_m . \omega_m . \dot{T}_m + \ddot{T}_m = -A_m . \ddot{\theta}_I \tag{9}$$

with $z_m = \delta . \omega_m$ and $A_m = \dfrac{\int_0^L r.Y_m(r)\,dr}{\int_0^L Y_m^2(r)\,dr}$.

The result of these calculations is that a partial differential equation (eq. 2) is split up into an infinite number of second order systems. The two states describing each second order syste are chosen as: T_m integrated once (for brevity $\int T_m$) and T_m integrated twice ($\int\int T_m$).

2.3 COMPLETE MODEL

The complete model, consisting of motor and flexible link, with input the applied voltage and output θ_U (see figure 1) defined a:

$$\theta_U(t) = \theta_I(t) + W(L,t)/L,$$

can be transformed into the following state space model:

$$\begin{aligned}\dot{\underline{x}} &= A.\underline{x} + \underline{b}.u, \\ y &= \underline{c}.\underline{x}\end{aligned} \tag{10}$$

with $\underline{x}^T = [\theta_I, \int T_1, \int\int T_1, \ldots, \int T_n, \int\int T_n, \ldots]$,

$y = \theta$,

$$A = \begin{bmatrix} 0 & 0 & 0 & \cdots\cdots\cdots & \\ \begin{bmatrix} -A_1 \\ 0 \end{bmatrix} & \begin{bmatrix} -2.z_1.\omega_1 & -\omega_1^2 \\ 1 & 0 \end{bmatrix} & & \underline{0} & \\ \vdots & & \ddots & & \\ \begin{bmatrix} -A_n \\ 0 \end{bmatrix} & \underline{0} & & \begin{bmatrix} -2.z_n.\omega_n & -\omega_n^2 \\ 1 & 0 \end{bmatrix} & \\ \vdots & & & & \ddots \end{bmatrix}$$

$\underline{b}^T = [K_M, 0, 0, 0, \ldots]$,

$\underline{c} = [0, -2z_1\omega_1/L, -\omega_1^2/L, \ldots, (-1)^n.2z_n\omega_n/L, (-1)^n.\omega_n^2/L, \ldots]$

3. MODEL REDUCTION

The description of the transverse vibrations in a flexible arm by a model of infinite order has no practical use. The order of the model has to be reduced. This can be achieved by assuming that the bandwidth of the motor is limited at ω_M. This implies that excitation of frequencies higher than ω_M may be disregarded. Assume further that the bandwidth of the controlled system ω_c (and thus also the related performance indices, such as settling time) is less than ω_M. This allows that indeed the motor may be considered as an integrator. Modes with a resonance frequency higher than ω_c may be disregarded. The number of modes n in the reduced model can thus be determined by

$$\omega_{n+1} > \omega_c. \tag{11}$$

By selecting a higher value of ω_c more modes must be taken into account. The reduced model can be used to find a control law.

4. CONTROL STRATEGY

State feedback is used to control the flexible arm. The method that is used to find the feedback gains is pole placement. The controller was first tested in simulations.

TABLE 1. poles and zeros of the reduced order model, for different values of the number of modes n.

number of modes	poles	zeros
1	0.0 $-0.055 \pm 5.21.i$	-13.4 14.2
2	0.0 $-0.055 \pm 5.21.i$ $-0.30 \pm 32.4.i$	-15.7 17.3 -49.4 46.7
3	0.0 $-0.055 \pm 5.21.i$ $-0.30 \pm 32.4.i$ $-0.5 \pm 91.7.i$	-15.5 17.0 $-73.3 \pm 34.7.i$ $73.7 \pm 31.0.i$

4.1 SIMULATIONS

The parameters used in the simulations, for example the resonance frequencies, have been determined from experiments with the real system. The open loop poles and zeros for the reduced-order model are presented in table 1. From table 1 it can be seen that only the zeros shift when the number of modes in the reduced-order model increases.

In figure 2 three simulations are shown. In each simulation three modes are simulated. In the first two simulations only the first mode is controlled; in the last one also the second mode is controlled (see figure 2). The poles are chosen such that the responses have no overshoot. In figure 2a the bandwidth ω_c is that small that the dynamics of the second mode may be neglected. In figure 2b this bandwidth is increased. This results in a response where the second mode can be recognized clearly. In figure 2c the second mode is controlled also. This results in a response with no overshoot and a properly damped second mode. The third mode gives almost no contribution. From this it may be concluded that the fourth and higher modes will also be negligibly small. From the last two simulations it may be concluded that making the

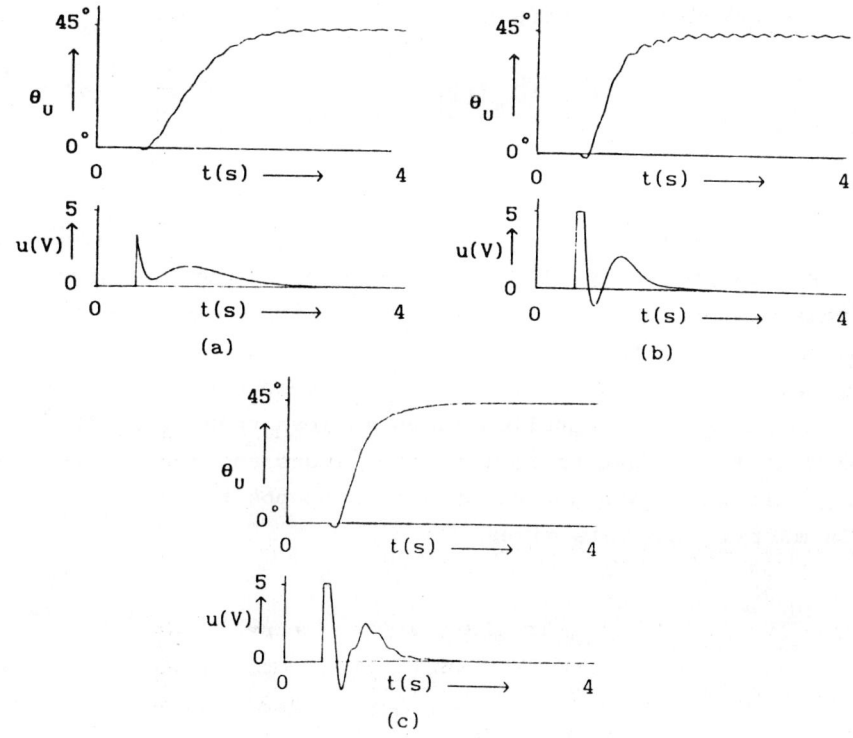

Figure 2. Simulated responses for
a) poles placed at -3, -4, -5
b) " " " -3, -5 ± 7i
c) " " " -3, -5 ± 7i, -3 ± 22i

bandwidth larger will have hardly any effect, because in these responses the input voltage is already maximal for a relatively long period.

4.2 STATE ESTIMATOR

State feedback requires knowledge of all the states. The states are: the angle θ_I, and for every mode two states. The angle θ_I is measured by a resolver. With the aid of a Resolver-to-Digital

Converter (RDC) the angle is converted into a 14 bits digital number. The states which describe the vibration modes cannot be measured directly. Strain gauges are used to measure the vibrations as suggested by van Vugt [3] and Hastings and Book [4]. The signal measured with a strain gauge is:

$$u_{sg}(r_0,t) = K_c \sum_{n=1}^{\infty} \left. \frac{d^2 Y_n(r)}{dr^2} \right|_{r=r_0} T_n(t), \qquad (12)$$

where r_0 is the position of the strain gauge and K_c is a constant that depends on the specific resistance of the strain gauge and the thickness of the beam. From eq. (12) it can be seen that the signal measured with a strain gauge depends on its position on the beam. From the simulations in the previous section it follows that the number of controlled modes is less than four. It is assumed at first that there are three modes and that there are three pairs of strain gauges at different positions $r = r_1$, r_2, r_3. In matrix form this gives

$$\underline{U}_{sg} = \underline{S} \cdot \underline{T} \qquad (13)$$

with
$$\underline{U}_{sg} = \begin{bmatrix} u_{sg}(r_1,t) \\ u_{sg}(r_2,t) \\ u_{sg}(r_3,t) \end{bmatrix}, \quad \underline{T} = \begin{bmatrix} T_1(t) \\ T_2(t) \\ T_3(t) \end{bmatrix}$$

and $\underline{S}$ a 3*3 matrix with elements

$$S_{ij} = K_c \cdot \left. \frac{d^2 Y_i(r)}{dr^2} \right|_{r=r_j}$$

By calculating the inverse matrix S^{-1} of S, the modal motions T_n can be calculated by

$$\underline{T} = \underline{S}^{-1} \cdot \underline{U}_{sg}.$$

Each of the modal motions can thus be calculated from the three measurements, as long as there is no influence of the higher modes. The first mode that disturbs this decoupling of the strain gauge signals is the fourth mode. To reduce this effect, the strain gauges are placed at positions where

$$\frac{d^2 Y_4(r)}{dr^2} = 0.$$

This equation has roots at $r_1 = 0.0944L$, $r_2 = 0.356L$ and $r_3 = 0.642L$. From the simulations in the previous section it is clear that the effect of the fifth and higher modes are negligibly small. To check the decoupling, impulse responses were measured. From these reponses a number of parameters was obtained. In table 2 the theoretical and experimental values for the resonance frequencies are compared. The theoretical values differ from the experimental values. This is due to the fact that the Young modulus E depends on how the profile of the flexible link is made. Therefore the resonance frequencies are scaled such that for the first mode the scaled value equals the experimental value. From table 2 it can be seen that the values for the other modes agree well.

TABEL 2. experimental, theoretical and scaled resonance frequencies

ω experimental	ω theoretical	ω scaled
5.21 (rad/s)	5.72	5.21
32.4	35.9	32.7
91.7	100.5	91.5

The modal motions T_n are linear combinations of $\int T_n$, $\int\int T_n$ and θ_I (see eq. (9)). This implies that the unknown states $\int T_n$ and $\int\int T_n$ cannot be measured but have to be estimated. Without the decoupling of the strain gauge signals an observer for a SIMO system (three pairs of strain gauges) has to be designed. This means that 18 elements of an observer matrix have to be calculated. With the decoupling three independent observers for SISO systems have to be designed, one for each mode. This implies that only 6 elements of an observer matrix have to be calculated. The observer gains are calculated using pole placement.

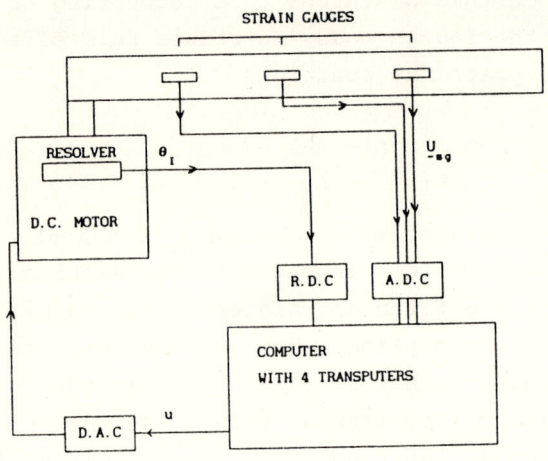

Figure 3. Experimental setup

5 EXPERIMENTAL SETUP

After a number of simulations the designed controller and observer were tested on the real system. A sketch of the setup is given in figure 3. The arm parameters are

material: aluminum
width d: 4.0 mm
height h: 60.0 mm
lenght L: 1.90 m

With the computer setup used for the control of the flexible arm a sampling frequency of about 1.5 kHz could be reached. This high frequency could be obtained by the use of transputers. These transputers make it possible to do a large part of the calculations in parallel. A detailed description of the implementation of the control algorithm is given by ter Reehorst [5], a student who worked at our group. The experimental results are presented in figure 4. In the experiments the same pole locations are used as in the simulations presented in section 4.1. This means that figures 2a-c can be compared with the figures 4a-c. In figure 4 there remains a small vibration. This is due to the fact that t motor has smallsignal non-linearities, for example backlash and static friction. In figure 4b the contribution of the second mode is smaller than in the related simulation (see figure 2b). In figure 4c the second mode is controlled also. It can be seen that the second mode is damped properly. The response has a small overshoot. This is probably due to the fact that a number of assumptions are not completely valid, for example that the torque that the beam exerts on the motor axis is negligible small. Fi-

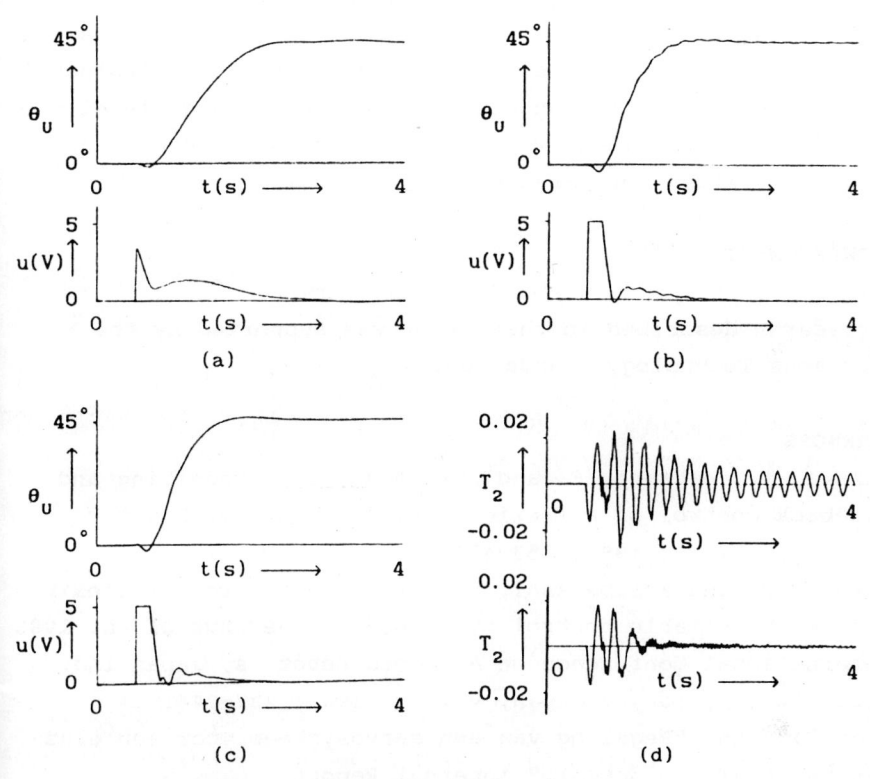

Figure 4. Measured responses for
    a) poles placed at $-3, -4, -5$,
    b)   "   "   " $-3, -5 \pm 7i$,
    c)   "   "   " $-3, -5 \pm 7i, -3 \pm 22i$,
    d) measured second modes for figures 4b and 4c.

nally in figure 4d the measured second mode in figure 4b is compared with the second mode in figure 4c. This figure shows that the effect controlling the second mode gives a significant reduction of the ouput error.

6. CONCLUSIONS

A model with an infinite number of modes can be reduced to a model with a number of modes that depends on the desired bandwidth ω_c. The maximal bandwidth is determined by the bandwidth of the motor. The use of strain gauges for the estimation of the states describing the transversal vibrations works well. The design of observers was simplified by the decoupling of the strain gauge signals. The results can probably be improved by using a more extensive motor model and by taking into account that the torque that the link exerts on the motor axis is not negligible small. With the aid of transputers it was easy to obtain a high-sample frequency. All the implementations were done in high-level programming languages. These were C and OCCAM (a computer language especially written for parallel programming).

ACKNOWLEDGMENT

The research described in this paper was supported by the Netherlands Technology Foundation.

REFERENCES

1. Sakawa, Y., Matsuno, F. and Fukushima, S., "Modelling and feedback control of a flexible arm," Journal of Robotic Systems, vol. 2, 1985, 453-472.
2. Fukuda, T. and Kuribayashi, Y., "Precise control of flexible arms with reliable control systems," Proceedings of the 1983 International Conference on Advanced Robotics, Japan Ind. Robot Assoc., Tokyo, Japan, vol. 1, 1983, 237-244.
3. Vugt, J. van, "Regeling van een servosysteem voor een elastische staaf (in dutch)," Internal Report, Nr 065.1708/86-007R, University of Twente, 1986.
4. Hastings, G.G. and Book, W.J., "A linear dynamic model for flexible robotic manipulators," IEEE Control Systems Magazine February 1987, 61-64.
5. Ter Reehorst, G., "Control of a flexible link by using transputers," Internal Report, University of Twente, 1988 (not yet finished).

Decomposed Parameter Identification Approach of Robot Dynamics

D. LU, Z. Y. QIAN, Z. J. ZHANG

Department of Automatic Control
Shanghai Jiao Tong University
Shanghai 200030, P. R. China

Summary

A new method is proposed in this paper for the parameter identification of robot dynamics. Different from existing methods, the identification of 10n dynamic parameters for a robot with n joints is firstly decomposed into n subproblems, each of which deals with the task of identifying 10 parameters. Based on the decomposed models, both sequential off-line and recursive on-line identification algorithms are then developed. These new algorithms reduce the computational burden greatly and make the parallel computation possible. The determination and processing of unidentifiable and combined identifiable parameters are particularly convenient by using the decomposed models. Algorithm for this purpose is also given in this paper. Finally, simulation results of identifying the dynamic parameters for the first three links of PUMA-560 by using the proposed method are presented to show the effectiveness of this new method.

Introduction

It is well known that dynamic parameter uncertainties of a robot, such as inaccuracies on inertias, location of mass center for each link, mass of the load and its exact position in the end-effector, will result in the degeneration of control performance. To solve this problem, two strategies can be used. The one is parameter identification by which the estimation of dynamic parameters being accurate enough for control purpose is obtained. The other is robust control by which the effect of parameter uncertainties on the control performance will be reduced to an acceptable extent. Both of these two methods have attracted much research attention. In this paper, only the parameter identification will be considered.

It has been pointed out in many literatures [2]-[5] that the generalized force τ is a linear function of the dynamic para-

---

* Supported by National Natural Science Foundation of China

meters which are expressed with reference to the local D - H coordinate systems, i.e.,

$$\tau = K \theta \quad (1)$$

where $\theta = [\theta_1{}^t, \theta_2{}^t, \ldots, \theta_n{}^t]^t$, $\tau = [\tau_1, \tau_2, \ldots, \tau_n]^t$

$\theta_i = [m_i, m_i c_{xi}, m_i c_{yi}, m_i c_{zi}, I_{xxi}, I_{xyi}, I_{xzi}, I_{yyi}, I_{yzi}, I_{zzi}]^t$

The m_i the mass of the ith link, and $C_i = [c_{xi}, c_{yi}, c_{zi}]^t$ is the position vector of the mass center of the ith link. The last 6 elements of θ_i are the elements of the inertia tensor I_i of the ith link.

$$I_i = \begin{bmatrix} I_{xxi} & I_{xyi} & I_{xzi} \\ I_{xyi} & I_{yyi} & I_{yzi} \\ I_{xzi} & I_{yzi} & I_{zzi} \end{bmatrix}, \quad K = \begin{bmatrix} K_{11} & K_{12} & \cdots & K_{1n} \\ & K_{22} & \cdots & K_{2n} \\ & & \ddots & \vdots \\ 0 & & & K_{nn} \end{bmatrix} \quad (2)$$

Each k_{ij} is an 1 x 10 row vector. The matrix K can be computed according to the measurement of $q, \dot{q}, \ddot{q}$ and τ [2]-[5].

The equation (1) can be used as an identification model. It can be seen, however, in equation (1) there are 10n parameters to be identified. Hence, the efficiency of existing schemes which solve the identification problem by treating these 10n parameters as a whole is limited due to the computational burden caused by the high dimension of the problem. Nevertheless, the equation (1) has an obvious characteristic that K is an upper triangular matrix. This observation motivated us to think about the possibility of decomposing the original problem of 10n parameters into n subproblems of 10 parameters. As a result, we contrive the decomposed identification approach of robot dynamics.

In this paper, the decomposed identification models for robot dynamic parameter estimation are firstly set up. Based on these models, the off-line sequential identification algorithms and on-line decentralized recursive algorithm are then developed. Compared with the existing algorithms, the proposed ones require far less computation and allow to be implemented in parallel fashion, which improves the efficiency of robot dynamic parameter identification greatly. Moreover, with the decomposed models it is easy to determine and process the unidentifiable or combined identifiable parameters. The algorithm for this purpose is also presented.

Decomposed Identification Models

After N times of sampling, the equation (1) can be augmented into

$$\tau_N = K_N \theta \qquad (3)$$

where $\tau_N = [\tau(1)^t, \tau(2)^t, \ldots, \tau(N)^t]^t$,
$K_N = [K(1)^t, K(2)^t, \ldots, K(N)^t]^t$; $\tau(i) = [\tau_1(i), \tau_2(i), \ldots, \tau_N(i)]^t$,

$$K(i) = \begin{bmatrix} K_{11}(i) & K_{12}(i) & \cdots & K_{1n}(i) \\ & K_{22}(i) & \cdots & K_{2n}(i) \\ & & \ddots & \vdots \\ & & & K_{nn}(i) \end{bmatrix} \qquad i=1,2,\ldots,n$$

The $\tau_j(i)$, $i=1,2,\ldots,N$; $j=1,2,\ldots,n$, is the ith measured input torque to the jth joint. B^t means the transpose of B.

Suppose that all the parameters of θ are individually identifiable, or equivalently, K_N is of full column rank, then the least squares estimation of θ is

$$\theta = (K_N^t K_N)^{-1} K_N^t \tau_N \qquad (4)$$

Since $K_N^t K_N$ is a 10n x 10n matrix, the computation of its inverse is very complex. In order to reduce the computational complexity, the decomposed identification models which are equivalent to the original one in the sense of least squares estimation are derived in this section.

It can be seen that every $K(i)$ in K_N is an upper triangular submatrix. Therefore, by collecting together the N sampling values of the generalized force $\tau_j(i)$, $i=1,2,\ldots,N$, applied to the same joint j to form a new subvector τ_j^*, the equation (3) can be rewritten into the following form:

$$\tau^* = K^* \theta \qquad (5)$$

where

$$K^* = \begin{bmatrix} K_{11}^* & K_{12}^* & \cdots & K_{1n}^* \\ & K_{22}^* & \cdots & K_{2n}^* \\ & & \ddots & \vdots \\ & & & K_{nn}^* \end{bmatrix}, \quad \tau^{*t} = [\tau_1^{*t}, \tau_2^{*t}, \ldots, \tau_n^{*t}]$$

and $K_{ij}^* = [K_{ij}^t(1), K_{ij}^t(2), \ldots, K_{ij}^t(N)]^t$

$\tau_j^* = [\tau_j(1), \tau_j(2), \ldots, \tau_j(N)]^t$

The equation (5) can be further rewritten into a set of decomposed sequential identification models:

$$\tau_i^* - \sum_{j=i+1}^{n} K_{ij}^* \theta_j = K_{ii}^* \theta_i , \quad i = n, n-1, \ldots, 1 \quad (6)$$

and we have the following theorem:

Theorem 1: The least squares estimation of θ obtained from the decomposed models (6) is the same as that obtained from the model (3).

The proof of this theorem is very easy and hence is omitted.

Estimation Algorithms Based on Decomposed Models

Based on the decomposed models (6), the effective off-line and on-line estimation algorithms will be developed in this section.

I. Off-line Algorithms

In the case of off-line estimation, τ_N and K_N in the equation (3) are obtained in advance. It is very easy to arrange these data in the form of equation (5) and derive the following off-line sequential algorithm.

Algorithm 1

step 1. set $i = n$;
step 2. $\theta_i = (K_{ii}^{*t} K_{ii}^*)^{-1} K_{ii}^{*t} (\tau_i^* - \sum_{j=i+1}^{n} K_{ij}^* \theta_j)$ (7)
step 3. if $i = 1$ then stop otherwise goto step 4;
step 4. $i = i - 1$, goto step 2.

Another estimation approach with better data stability is the following sequential recursive least squares estimation based also on the decomposed models (6).

Algorithm 2

step 1. set $i = n$;
step 2. set $m = 0$, $\theta_i^0 = 0$, $P_i^0 = \mu_i I$, where μ_i is a sufficiently large positive number;
step 3. compute the $(m+1)$th estimation of θ :

$$\theta_i^{m+1} = \theta_i^m + L_i^{m+1} (\tau_i(m+1) - K_{ii}(m+1) \theta_i^m \\ - \sum_{j=i+1}^{n} K_{ij}(m+1) \theta_j) \quad (8)$$

$$L_i^{m+1} = P_i^m K_{ii}^t(m+1) / (\alpha + K_{ii}(m+1) P_i^m K_{ii}^t(m+1)) \quad (9)$$

$$P_1^{m+1} = (P_1^m - L_1^{m+1} K_{11}(m+1) P_1^m)/\alpha \qquad (10)$$

step 4. if $m = N - 1$ then goto step 6;
step 5. $m = m + 1$, goto step 3;
step 6. if $i = 1$ then stop otherwise goto step 7;
step 7. $i = i - 1$, goto step 2.

It has already been shown [8] that the solution determined by algorithm 2 is uniformly convergent to the least squares solution of equation (5). Thus, according to theorem 1, it convergences also uniformly to the least squares solution of the original problem (3).

Since the existing methods treat the 10n parameters as a whole, we call them the centralized estimation methods. The comparison of computational complexity between the above two algorithms and the existing ones is shown in the following tables.

		multiplication	addition
	centra-lized	$(\frac{1000}{6}+200N)n^3-22+$	$(\frac{1000}{6}+200N)n^3-11$
*		$(7550+10N)n^2+1143\frac{1}{3}n$	$+5400n^2+818\frac{1}{3}n$
*	algorithm 1	$5Nn^2+(205N+8838)n$	$5Nn^2+(195N+6374)n$
*	centra-lized	$(300n^3+30n^2+n)N$	$(300n^3+10n^2)N$
‡	algorithm 2	$5Nn^2+326nN$	$5Nn^2+305Nn$

*------nonrecursive; ‡------recursive

Table 1. Comparison of computational complexity

		multiplication	addition
*	centra-lized	4070648	4555299
*	algorithm 1	194028	173244
‡	centra-lized	6588600	6516000
‡	algorithm 2	213600	201000

*------nonrecursive
‡------recursive

Table 2. Comparison of computational complexity as $N=100$, $n=6$.

From these two tables, it can be seen that the new algorithms require far less computation than the centralized methods.

II. On-line algorithm

In the situation of on-line estimation, the data sampling and the parameter estimation are carried out simultaneously, it is impossible to arrange the data in the form of equation (5). Therefore, the algorithms proposed above can not be used in real-time. In this case, one can use the existing on-line estimation methods which are derived from the centralized model (3). The computation burden, however, will result in difficulty in real-time implementation. The algorithm 2 is a sequential off-line recursive algorithm. In order to estimate θ_1, the estimated values of θ_j's $(j>1)$ must be used. In the situation of on-line estimation, θ_1 and θ_j's $(j>1)$ are estimated simultaneously. This sequential recursive algorithm seems to be useless in this case. However, since all the dynamic parameters of a robot manipulator are time-invariant, when m is large enough the mth estimation $\hat{\theta}_j^m$ of θ_j will be accurate enough and can be viewed as its real value. From this point of view, the algorithm 2 can be modified into the following on-line decentralized recursive one.

Algorithm 3

step 1. set $m=0$, $\hat{\theta}_1^0=0$, $P_1^0=\mathcal{U}_1 I$, where $\mathcal{U}_1$ is a sufficiently large positive number.

step 2. compute

$$\hat{\theta}_1^{m+1}=\hat{\theta}_1^m+L_1^{m+1}\left[\tau_1(m+1)-\sum_{j=1}^{n}K_{1j}(m+1)\hat{\theta}_j^m\right] \quad (11)$$

$$L_1^{m+1}=P_1^m K_{11}^t(m+1)/(\alpha+K_{11}(m+1)P_1^m K_{11}^t(m+1)) \quad (12)$$

$$P_1^{m+1}=\left[P_1^m - L_1^{m+1} K_{11}(m+1) P_1^m\right]/\alpha \quad (13)$$

Similar to the algorithm 2, the forgetting factor α is also used here.

step 3. $m=m+1$, goto step 2.

The advantage of this algorithm is that the computation is completely decentralized. The following figure shows that this algorithm can be implemented in a parallel fashion.

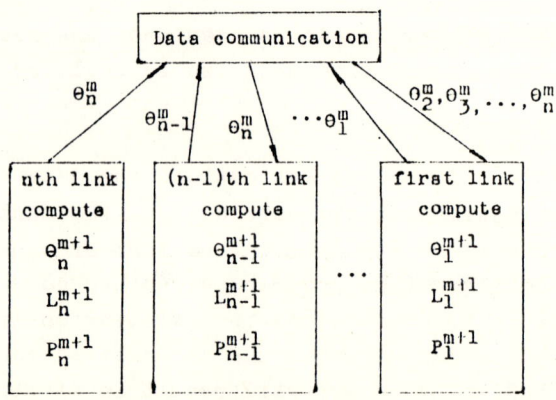

Fig.1. Parallel implementation of the on-line recursive algorithm

As to the convergency of algorithm 3, we have the following:

Theorem 2; The estimation $\hat{\theta}_i$ of θ_i obtained by using algorithm 3 is uniformly convergent to its real value θ_i ($i=1, 2,\ldots,n$), if the forgetting factor $\alpha < 1$.

Proof: It can be easily seen from equations (9) and (12) that the estimation $\hat{\theta}_n^m$ obtained by using algorithm 3 is the same as the estimation θ_n^m obtained by using algorithm 2. It is well-known that the solution of algorithm 2 is uniformly convergent to its real value [10]. Hence, $\hat{\theta}_n^m$ also converges uniformly to its real value θ_n, i.e., there exists a sufficiently large positive number s, when $m \geqslant s$, we have $\hat{\theta}_n^m = \theta_n$.

For $m \geqslant s$, equation (12) can be written as

$$\hat{\theta}_i^{m+1} = \hat{\theta}_i^m + L_i^{m+1}(\tau_i(m+1) - K_{in}(m+1)\theta_n$$
$$- \sum_{j=i}^{n-1} K_{ij}(m+1)\hat{\theta}_j^m), \quad i=n-1, n-2,\ldots,1$$

For the (n-1)th link of a robot manipulator, its dynamic parameters can be estimated by algorithm 3:

$$\hat{\theta}_{n-1}^{m+1} = \hat{\theta}_{n-1}^m + L_{n-1}^{m+1}(\tau_{n-1}(m+1) - K_{n-1,n}(m+1)\theta_n$$
$$- K_{n-1,n-1}(m+1)\hat{\theta}_{n-1}^m)$$

When $m = s + r$, we have

$$\hat{\theta}_{n-1}^{s+r+1} = P_{n-1}^{s+r+1}(\alpha^{r+1} K_{n-1,n-1}^s y_{n-1}^s + \alpha^r K_{n-1,n-1}(s+1)y_{n-1}(s+1)$$
$$\ldots + \alpha^0 K_{n-1,n-1}(s+r+1)y_{n-1}(s+r+1))$$

where $P_{n-1}^{s+r+1} = (\alpha^{r+1} P_{n-1}^s + \alpha^r K_{n-1,n-1}^t(s+1)K_{n-1,n-1}(s+1) + \ldots$
$$+ K_{n-1,n-1}^t(s+r+1)K_{n-1,n-1}(s+r+1))^{-1}$$

$$K_{n-1,n-1}^s = (\alpha^{\frac{s-1}{2}} K_{n-1,n-1}^t(1), \alpha^{\frac{s-2}{2}} K_{n-1,n-1}^t(2),\ldots,$$
$$\alpha^0 K_{n-1,n-1}^t(s))$$

$$y_{n-1}^s = (\alpha^{\frac{s-1}{2}} y_{n-1}(1), \alpha^{\frac{s-2}{2}} y_{n-1}(2),\ldots,\alpha^0 y_{n-1}(s))^t$$

$$y_{n-1}(j) = \begin{cases} \tau_{n-1}(j) - K_{n-1,n}(j)\hat{\theta}_n^j & j \leq s \\ \tau_{n-1}(j) - K_{n-1,n}(j)\theta_n & j > s \end{cases}$$

Since $\alpha < 1, \alpha^{r+1}$ will tends to zero, if r is sufficiently large. Then, we have P_{n-1}^{s+r+1} tends to P_{n-1}^{r+1}. Similarly

$$\hat{\theta}_{n-1}^{s+r+1} \longrightarrow P_{n-1}^{r+1}(\alpha^r K_{n-1,n-1}(s+1)y_{n-1}(s+1) + \cdots \cdots +$$
$$K_{n-1,n-1}(s+r+1)y_{n-1}(s+r+1)) = \theta_{n-1}^{r+1}$$

That is, the estimation $\hat{\theta}_{n-1}^{m+1}$ obtained by algorithm 3 will tend to the estimation θ_{n-1}^{m+1} obtained by using algorithm 2, and in turn, will converge to its real value θ_{n-1}.

Similar method can be used to prove the convergence of $\hat{\theta}_j^{m+1}$ to its real value θ_j for $j = n-2, n-3, \ldots, 1$. This completes the proof.

Identifiability of Robot Dynamic Parameters

In the previous sections, it was assumed that all the dynamic parameters of a robot manipulator are identifiable. That is, the matrix K in equation (3), or equivalently, the matrices K_{11}^* for $i = 1, 2, \ldots, n$ in equation (6) are all of full column rank. However, because of the restriction of robot motion and limitations on measurement, some parameters are unidentifiable and some are identifiable only in linear combination. Hence, we have to determine and process these unidentifiable parameters before the decomposed algorithm is used. However, the high dimension of the overall identification model (3) makes it very complex to determine and process the unidentifiable parameters. Fortunately, with the help of decomposed models, this work can be simplified a lot. This is because we can analyse the lower dimensional K_{11}^* rather than K. The following algorithm based on the principle of singular value decomposition is given for this purpose.

Algorithm 4

step 1. set $i = 1$
step 2. singular value decomposition of K_{11}^*.
$K_{11}^* = U_1 \Sigma_1 V_1^t$
where $U_1 = [U_1^1 \ U_1^2 \ \ldots \ U_1^N]$ and $\Sigma_1 = \begin{bmatrix} \sigma_i^1 & & & \\ & \sigma_i^2 & & \\ & & \ddots & \\ & & & \sigma_i^{10} \end{bmatrix}$
$V_1 = [V_1^1 \ V_1^2 \ \ldots \ V_1^{10}]$

(1) If the jth singular value equalssto zero, then θ_{ij} is an unidenfifiable parameter. Then, set $\theta_{ij} = 0$

and delete θ_{1j} in θ_1 and the jth column in the submatrix $[K_{11}^{*t}\ K_{21}^{*t}\ \ldots\ K_{11}^{*t}]^t$.

(2) If the jth singular value is nonzero, $v_i^{jt}\theta_1 = a_1\theta_{1k_1} + a_2\theta_{1k_2} \cdots + a_{kj}\theta_{1k_j}$ is identifiable only in linear combination. Then set $\theta_{1k_2} = \theta_{1k_3} = \cdots = \theta_{1k_j} = 0$, and delete these elements in θ and the k_2th, k_3th,..., k_jth columns in $[K_{11}^{*t}\ K_{21}^{*t}\ \ldots\ K_{11}^{*t}]^t$.

step 3. if i=1, stop; otherwise
step 4. i=i - 1, goto step 2.

After applying this algorithm to the decomposed identification models, all the K_{ij}^* in equation (6) are changed into column reduced $\widetilde{K}_{ij}$. The equation (6) can be rewritten into the following one:

$$\tau_i - \sum_{j=i+1}^{n} \widetilde{K}_{ij} \widetilde{\theta}_j = \widetilde{K}_{ii} \widetilde{\theta}_i \qquad i=n, n-1, \ldots, 1 \qquad (14)$$

and then all the proposed algorithms can be used to solve the problem (14).

Simulation Results

To verify the effectiveness of the proposed methods, digital simulation experiments have been made on DPS-8 by Fortran programming language. In the simulations, the dynamic parameters of the first three links of PUMA-560 were estimated by using all the algorithms proposed in this paper. The robot dynamic equation that we used can be seen from [11]. The simulation was designed as follows: the first three joints of PUMA-560 move from $(0°, 90°, -60°)$ to $(100°, -80°, 170°)$ in one second, the number of sampling points is 200, sampling period is 5 miliseconds.

By using algorithm 4, we obtained the following 11 parameters or combinations of parameters which are identifialle:

$\theta(1,1) = I_{yy1} + I_{xx2} + I_{xx3} + m_3 d_3^2$; $\theta(2,1) = m_2 a_2 + m_2 c_{x2} + m_3 a_2$

$\theta(2,2) = a_2 m_2 c_{x2} + I_{zz2}$; $\theta(2,3) = m_2 c_{y2}$

$\theta(2,4) = m_2 c_{z2} + m_3 d_3$; $\theta(2,5) = m_2 c_{x2} a_2 - I_{xx2} + I_{yy2}$

$\theta(3,1) = m_3 a_3 + m_3 c_{x3}$; $\theta(3,2) = m_3 c_{x3} a_3 + I_{yy3}$

$\theta(3,3) = m_3 c_{y3}$; $\theta(3,4) = m_3 c_{z3}$

$\theta(3,5) = a_3 m_3 c_{x3} - I_{xx3} + I_{zz3}$

The estimation results obtained by using the first three algorithms in this paper are listed in table 3. The convergent procedure of the algorithm 3 is shown in figure 2. From these results, one can see that the estimation accuracy of all these three algorithms is rather high, and that the convergent rate of the algorithm 3 is very fast.

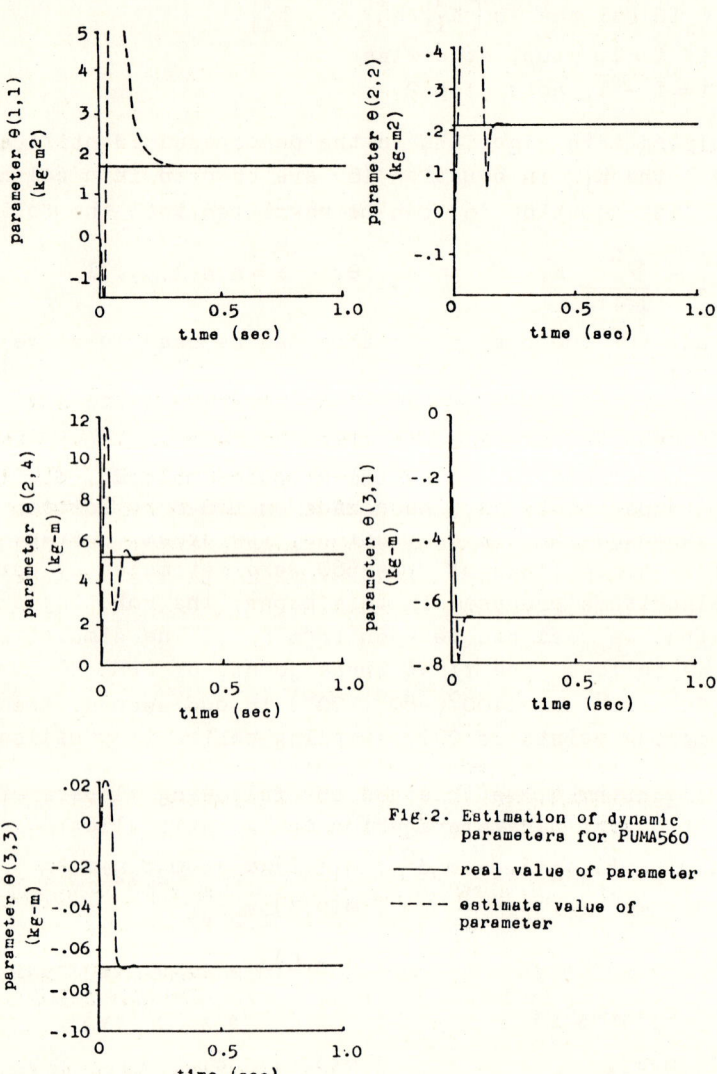

Fig.2. Estimation of dynamic parameters for PUMA560
——— real value of parameter
- - - - estimate value of parameter

parameter	real value	algorithm 1	algorithm 2	algorithm 3
$\theta(1,1)$ (kg-m^2)	1.7171	1.7170	1.7171	1.7171
$\theta(2,1)$ (kg-m)	4.4619	4.4664	4.4619	4.4619
$\theta(2,2)$ (kg-m^2)	0.2088	0.2074	0.2088	0.2088
$\theta(2,3)$ (kg-m)	0.1117	0.1117	0.1117	0.1117
$\theta(2,4)$ (kg-m)	5.3077	5.3077	5.3077	5.3077
$\theta(2,5)$ (kg-m^2)	-1.4754	-1.4771	-1.4754	-1.4754
$\theta(3,1)$ (kg-m)	-0.0065	-0.0065	-0.0065	-0.0065
$\theta(3,2)$ (kg-m^2)	0.0758	0.0758	0.0758	0.0758
$\theta(3,3)$ (kg-m)	-0.0686	-0.0686	-0.0686	-0.0686
$\theta(3,4)$ (kg-m)	0.0185	0.0185	0.0185	0.0185
$\theta(3,5)$ (kg-m^2)	-0.0675	-0.0675	-0.0675	-0.0675

Table 3. Estimation results of the dynamic parameters for the first three links of PUMA-560

Conclusions

Because of the characteristic of the robot dynamics, the identification of its dynamic parameters can be decomposed into n subproblems corresponding to individual links. The efficiency of parameter estimation is then improved greatly. Furthermore, the determination and processing of unidentifiable or combined identifiable parameters are simplified. In this paper, based on the decomposed identification models, several effective estimation algorithms are proposed, which require far less computation than the existing ones.

References

1. Paul, R.P.; Robot manipulators: mathematics, progrmming and control, MIT Press, 1981.

2. Khalil, W., et al: Automatic generation of identification models of robotics, IASTED Int. J. Robotics and Automation, Vol.1, No.1, 1986.

3. An, C.H., et al: Estimation of inertial parameters of rigid-body links of manipulators, Proc. 24th IEEE Conf. Decision and Contr., 1985.

4. ibid, Estimation of inertial parameters of manipulator loads and links, Int. J. Robotics Research, Vol.5, No.3, 1985.

5. Khosla, P.K., et al: Parameter identification of robot dynamics, Proc. 24th IEEE Conf. Decision and Contr., 1985.

6. Hollerbach, J.M.: A recursive lagrangian formulation of manipulator dynamics and a comparative study of dynamics formulation complexity, IEEE Trans. Syst. Man. Cyber. Vol. 10, No.11, 1980.

7. Luh, J.Y.S.,et al: On-line computational scheme for mechanical manipulators, ASME J. Dyn. Syst. Meas. Contr., Vol. 102, 1980.

8. Ljung, L. et al: Theory and practice of recursive identification, MIT Press, 1981.

9. Golub, G.H.,et al: Matrix computations, John Hopkins University Press, 1983.

10. Goodwin, G.C., et al: Dynamic system identification: experiment design and data analysis, Academy Press, 1977.

11. Bejczy, A. K., et al: Nonlinear feedback control of PUMA-560 robot arm by computer, Proc. 24th IEEE Conf. Decision and Contr., 1985.

Dynamic Behavior of a Flexible Robotic Manipulator

ERIC WEHRLI
Research Assistant, Ph. D. Candidate

THEODORE KOKKINIS
Assistant Professor

Department of Mechanical Engineering & Center for Robotic Systems in Microelectronics
University of California, Santa Barbara, California 93106, U.S.A.

The problem of modelling a revolute-joint robotic manipulator with either open or closed chain and flexibility in both the links and the joints is discussed. A kinematic formulation utilizing the Denavit-Hartenberg formalism, modal representation of the link flexible motion and infinitesimal elastically restrained joint rigid motion is developed and used to obtain the dynamic model by the Lagrangian approach. Details of the formulation are shown for a one-link structure. A direct scheme for the control of the motion of the end effector in Cartesian space based on nonlinear inversion of the system is developed. As this scheme requires simulation of the left inverse system, it has to be implemented off-line and to be supplemented by regulator terms to ensure convergence to the desired trajectory in the presence of disturbances and uncertainties.

1. Introduction

The present work is the first step in the development of a tool for simulation of the dynamic behavior of robotic arms, taking into account the flexibility of the links and the joints. The final goal is to use this for the control of the motion of the end effector of the arm in Cartesian space. There are many other factors that must be accounted for in studying the dynamic behavior of general robotic arms, such as friction in the joints, and backlash and flexibility of the mechanical transmissions. This work is not concerned with these factors; the reason is that the object of attention is a special class of robotic arms, namely direct-drive arms with either open- or closed-chain mechanisms. For such arms the aforementioned factors are negligible. It is, however, expected that for some of these arms possessing relatively stiff, lightweight links [1] the flexibility of the joints will be a significant contributor to the overall compliance of the arm. For this reason it is an essential part of this development.

The modelling and control of flexible-link robotic arms (with a view to space applications) has developed very rapidly over the past ten years. For simulation, notable is the work of Sunada [2]. He used finite element analysis to obtain the modal characteristics of the flexible links, and a time-varying component-mode synthesis approach to reduce the size of the problem. His model is semi-nonlinear in the sense that a nominal rigid motion is prescribed. Yoo and Haug [3] presented a full nonlinear formulation, with each link considered separately as a free body; constraints due to contact between the links were formulated for different types of joints. Flexibility was represented using a combination of normal modes and attachment modes. Kim and Haug [4] addressed the problem of mode selection for this model. Very interesting is Book's approach [5]. He derived a comprehensive nonlinear model for simulation of the dynamic behavior of open-chain arms with flexible links, using a modal representation for the flexibility. Particular attention was payed to the computational efficiency of the formulation, employing recursive computation. The equations as presented are valid for links with zero twist and offset, and therefore for the planar or special antropomorphic case only.

Dynamic modelling of closed-chain arms has been attempted mainly for rigid links. The most important special feature of the problem is the formulation of the loop constraint. Luh and Zheng [6], and in parallel Kleinfinger and Khalil [7] used a tree structure, a Newton-Euler formulation for the branches and a Lagrange multiplier approach for the loop constraints to derive the active joint generalized forces. One aspect of the problem of a closed-chain flexible arm was partially addressed by Kiedrzynski and Becquet [8]. They formulated simplified beam-link and rod-link models of the flexible links in the closed loop.

Modelling the flexibility of the joints has received less attention and then mostly in connection with the flexibility of the transmission, rather then the compliance of the bearings and/or the contact surfaces. Ahmad and Widman [9] and Dado and Soni [10] considered dynamic models of two-link (rigid) planar arms with flexible shafts and gear trains in the joints. Rivin [11] studied the relative importance of the compliance of the links and the joints/transmissions and concluded that the latter is dominant. Shih and Frank [12] developed spatial dynamic compliant joint models considering the stiffness of the shaft as well as that of the bearings and the supporting structure. These models included frictional effects but the links of the mechanism were considered rigid.

There are many contributions for the control of one-link arms; as such systems are represented by a linear model these contributions will not be mentioned here. Singh and Schy [13] presented a technique for joint-space control of multi-link flexible arms which introduces an elastic mode stabilizer requiring feedback of the elastic states. Gebler [14] developed a feedforward control strategy for a two-link robot with flexible links and joint shafts. Finally Bayo [15] proposed an open-loop technique based on the linearization of the system about the desired Cartesian trajectory for a two-link planar manipulator.

2. Flexible Arm Kinematic Model and Lagrangian Dynamics

A kinematic model is developed for the motion of robotic arms with flexibility in both the links and the (revolute) joints. Links consist of two rigid hubs connected with a flexible midsection (figure 1). To link (i) we assign two frames R_{i1} and R_{i2} bound to the hubs, with their z-axes coinciding with the joints' axes. The geometry of the undeformed link is described with three of the Denavit-Hartenberg parameters (length a_i, twist α_i, offset d_{i+1}); the fourth is the joint angle θ_{i+1}. Thus the transformation R_{i1} to R_{i2} (undeformed) is:

$$T_u = \begin{bmatrix} 1 & 0 & 0 & a_i \\ 0 & C\alpha_i & -S\alpha_i & -d_{i+1}S\alpha_i \\ 0 & S\alpha_i & C\alpha_i & d_{i+1}C\alpha_i \\ 0 & 0 & 0 & 1 \end{bmatrix} \quad \ldots\ldots\ldots\ldots(1)$$

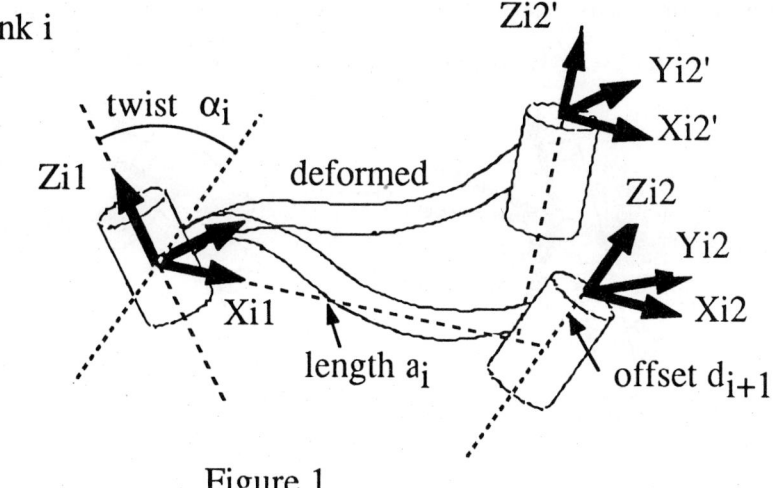

Figure 1

When the midsection of link i is deformed R_{i2} moves to a new position $R_{i2'}$; we write the transformation from R_{i2} to $R_{i2'}$ using a modal summation:

$$T_f = I_{4x4} + \sum_j q_{ij}(t) \, A_{f,ij} = I + A_f \quad \ldots\ldots(2)$$

where $q_{ij}(t)$, $j=1,2,\ldots$ are modal coordinates for link i (all considered infinitesimal quantities of order ε), and $A_{f,ij}$ is the the time-invariant mode-shape transformation matrix for mode j of link i, having the form:

$$A_{f,ij} = \begin{bmatrix} 0 & -\delta_{ij} & \gamma_{ij} & u_{ij} \\ \delta_{ij} & 0 & -\beta_{ij} & v_{ij} \\ -\gamma_{ij} & \beta_{ij} & 0 & w_{ij} \\ 0 & 0 & 0 & 0 \end{bmatrix} \quad \ldots\ldots(3)$$

The transformation from $R_{i2'}$ to $R_{i+1,1}$ (the frame bound to the proximal hub of the next link), in the absence of joint deformation, is a rotation about the z-axis of $R_{i2'}$ and the relevant matrix is given by:

$$T_r = \begin{bmatrix} C\theta_{i+1} & -S\theta_{i+1} & 0 & 0 \\ S\theta_{i+1} & C\theta_{i+1} & 0 & 0 \\ 0 & 0 & 1 & 0 \\ 0 & 0 & 0 & 1 \end{bmatrix} \quad \ldots\ldots(4)$$

Due to the flexibility of the contact surfaces and other elements of the joint the joint rotation brings us to an intermediate frame $R_{i+1,1'}$, and an infinitesimal elastically restrained rigid motion representing the joint deflection takes us to $R_{i+1,1}$ (figure 2).

Joint i+1

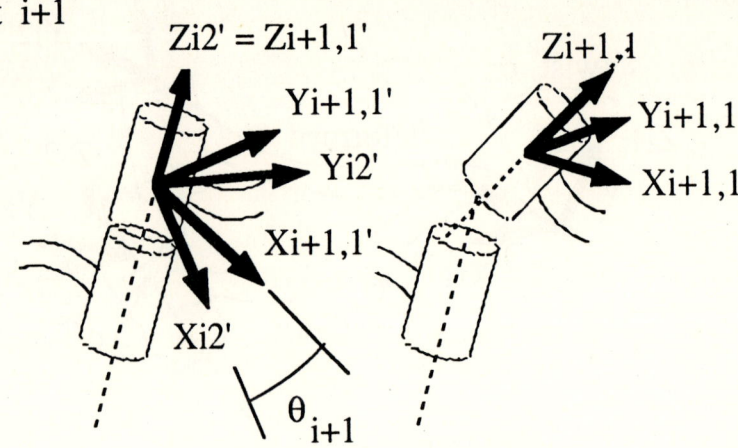

Figure 2

The transformation for the latter is written as:

$$T_a = I_{4\times 4} + \begin{bmatrix} 0 & -\omega_{i+1} & \psi_{i+1} & b_{i+1} \\ \omega_{i+1} & 0 & -\varphi_{i+1} & e_{i+1} \\ -\psi_{i+1} & \varphi_{i+1} & 0 & h_{i+1} \\ 0 & 0 & 0 & 0 \end{bmatrix} \quad \dots\dots(5)$$

T_a can be written in a summation form similar to that used for T_f in (2) as follows:

$$T_a = I_{4\times 4} + \sum_{j=1}^{6} p_{i+1,j}(t)\, A_{a,i+1,j} = I + A_a \quad \dots\dots(6)$$

where $p_{i+1,j}(t)$, j=1,2,...,6 are the six deformation variables for joint i+1 (all infinitesimal quantities of order ε), and $A_{a,i+1,j}$ are appropriate constant matrices. Following these considerations the transformation from R_{i1} at the base of link i to $R_{i+1,1}$ at the base of link i+1 is written as:

$${}^{i,1}_{i+1,1}T_d = T_u\, T_f(q_{ij})\, T_r(\theta_{i+1})\, T_a(p_{i+1,k}) \quad \dots\dots(7)$$

with T_u, T_f, T_r and T_a given by (2), (3), (4) and (6). An approximation for T_d correct to order ε and linear in the variables q and p is:

$$T_d \approx T_u T_r + T_u A_f T_r + T_u T_r A_a \quad \dots\dots(8)$$

The first term is the rigid transformation, the second is due to the link flexibility, and the third is due to the joint flexibility. If R_0 is a frame fixed in the ground, the transformation from R_0 to R_{i1} is given by the recursive relationship:

$${}^0_iT_d = {}^0_{i-1}T_d \; {}^{i-1}_iT_d \qquad \text{...............(9)}$$

The instantaneous position of a material point on link i relative to R_0 is given by:

$${}^0P = {}^0_iT_d \; {}^iP \qquad \text{...............(10)}$$

Here ${}^iP$ is the instantaneous position of the point relative to R_{i1} (figure 3), and is

Link i

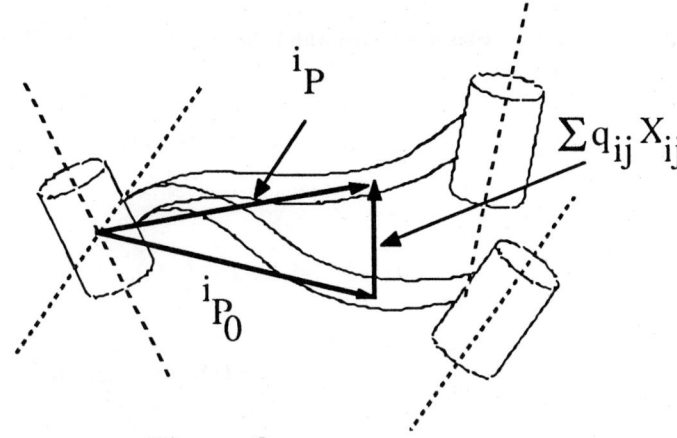

Figure 3

expressed using the same modal summation as in (2):

$${}^iP = {}^iP_0(\text{undeformed}) + \sum_j q_{ij}(t) \; X_{ij}({}^iP_0) \qquad \text{...............(11)}$$

This kinematic model forms the basis for obtaining the equations of motion of the system by the Lagrangian approach. The recursive nature of the position equations (9) and (10) makes the calculation of velocities and therefore the formulation of the kinetic energy easier. Velocities must be calculated for the three constituent elements of each link, namely the proximal and distal hubs (rigid bodies) and the flexible midsection (deformable solid). Considering link i we have that for the proximal hub the velocity, written as a twist or screw, is:

$$\xi_{i1} = {}^0_i\dot{T}_d \; {}^0_iT_d^{-1} \qquad \text{...............(12)}$$

For the distal hub of link i the velocity is:

$$\xi_{i2} = \frac{d}{dt}({}^0_iT_d \, T_u \, T_f)({}^0_iT_d \, T_u \, T_f)^{-1} \qquad \text{...............(13)}$$

Finally for a point on the flexible midsection, the velocity is:

$${}^0\dot{P} = {}^0_i\dot{T}_d \; {}^iP + {}^0_iT_d \; {}^i\dot{P} \qquad \text{...............(14)}$$

The following two recursive velocity equations, derived from (8) and (9) are necessary for velocity calculations using (12)-(14):

$${}^0_i\dot{T}_d = {}^0_{i-1}\dot{T}_d \; {}^{i-1}_iT_d + {}^0_{i-1}T_d \; {}^{i-1}_i\dot{T}_d \qquad \text{...............(15)}$$

$${}^{i-1}_{i}\dot{T}_d = T_u \dot{T}_r + T_u \dot{A}_f T_r + T_u A_f \dot{T}_r + T_u \dot{T}_r A_a + T_u T_r \dot{A}_a \quad \ldots\ldots(16)$$

The Lagrangian is formulated as a function of the joint angles θ, the link modal coordinates q and the joint deformation variables p, using (8)-(16):

$$L(\theta, q, p, \dot{\theta}, \dot{q}, \dot{p}) = KE(\theta, q, p, \dot{\theta}, \dot{q}, \dot{p}) - PE(\theta, q, p) + \sum_{n=1}^{6} \lambda_n C_n \quad \ldots\ldots(17)$$

6 scalar constraints have been included with Lagrange multipliers λ_n for the case of a manipulator with one closed mechanical loop of links and joints in its structure. The equations of motion take the form:

$$\frac{d}{dt}(\frac{\partial L}{\partial \dot{\theta}_i}) - \frac{\partial L}{\partial \theta_i} = \begin{cases} T_i \text{ (torque, if joint is active)} \\ 0 \text{ (if joint is passive)} \end{cases} \quad i=1,2,\ldots,I$$

$$\frac{d}{dt}(\frac{\partial L}{\partial \dot{q}_{jk}}) - \frac{\partial L}{\partial q_{jk}} = 0 \quad j=1,2,\ldots,I \quad k=1,2,\ldots,K_j$$

$$\frac{d}{dt}(\frac{\partial L}{\partial \dot{p}_{lm}}) - \frac{\partial L}{\partial p_{lm}} = 0 \quad l=1,2,\ldots,I+1 \quad m=1,2,\ldots,6$$

$$C_n = 0 \quad n = 1,2,\ldots,6 \quad \text{(constraints from a possible closed loop)} \quad (18)$$

I is the number of links of the manipulator. This is equal to the number of joints for an open-chain structure; for a structure with a closed chain (one closed loop of links and joints) there is one additional joint, but the joint angle for it is not included in the formulation. This joint is chosen to be a passive one. The deformation variables for it are included in the formulation as they contribute to the system's potential energy.

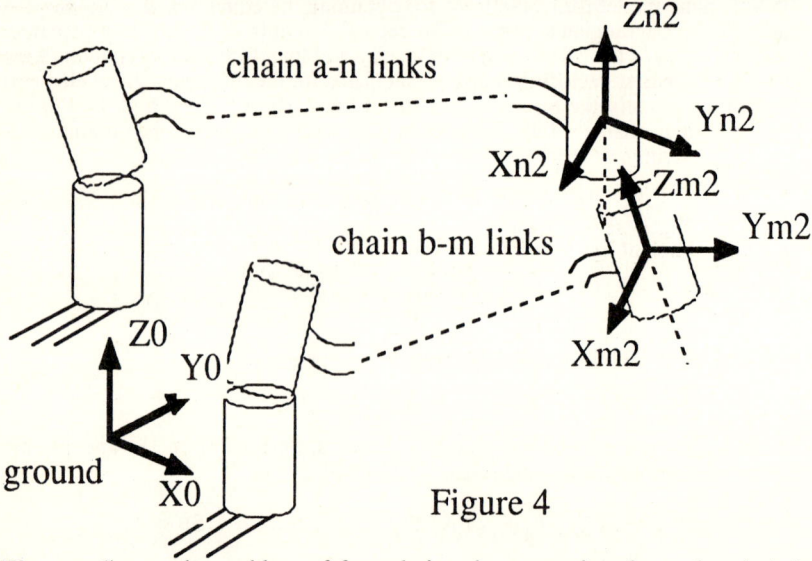

Figure 4

We now discuss the problem of formulating the constraints for a closed-chain structure. The structure is viewed as two open chains with their terminal links in contact via a revolute joint. The joint variables, link modal coordinates and joint deformation variables for the two chains are not independent, but have to satisfy constraint equations arising from the loop closure. Let the closed-chain manipulator consist of chain A with N links and chain B with M links. The condition for chain closure is the matching of the frames R_{N2} and R_{M2} bound to the distal hubs of the terminal links of the two chains modulo a rotation in the passive revolute joint connecting the hubs (figure 4). Depending on the geometry of the structure an offset may have to be included. If the connecting joint is flexible, we must take into account its deformation. The constraint equation is:

$$^{0}_{N1}T_A \, ^{N1}_{N2}T_A \, (^{N2}_{M2}T_{AB}) = ^{0}_{M1}T_B \, ^{M1}_{M2}T_B \qquad \ldots\ldots(19)$$

The transformation in parentheses must include depending on the case rotation, deformation and/or offset. Equation (19) yields 12 scalar constraints obtained by equating the elements of the matrices on its right and left hand side (the last row is trivial). From these 12 nonindependent conditions we must select 6 to include in (18); these must be such that the angle of the connecting joint does not appear.

3. One-Link Model: Inertial Coefficients and Computational Considerations

The details of the formulation are now considered for a one-link structure. In reality we have to consider two bodies: the ground and the link. Thus the transformation leading to the proximal hub of the link is:

$$^{0}_{1}T_d = T_{u,0} \, T_{f,0} \, T_{r,1} \, T_{a,1} \qquad \ldots\ldots(20)$$

We may select a frame in the ground so that $T_{u,0}=I$. Further the flexibility of the ground is negligible, therefore $T_{f,0}=I$ as well. Thus:

$$^{0}_{1}T_d = T_{r,1} \, T_{a,1} = T(\theta) \, (I + \sum_{i=1}^{6} P_{1,i} \, A_{a,1,i}) \qquad \ldots\ldots(21)$$

The instantaneous position of a point of the flexible midsection of the link (which we model as a thin 3D flexible rod) relative to R_0 is given by:

$$^{0}P = T(\theta) \, (I + \sum p_i A_{ai}) \, (^{1}P + \sum q_j X_j) \qquad \ldots\ldots(22)$$

q_j are the modal coordinates used to represent the elastic deflection of the midsection. The constants X_j are obtained from the mode shapes, which can correspond to normal modes or static correction modes. Carrying out the multiplication in (22) and deleting second-order terms we find:

$$^{0}P \approx T(\theta) \, ^{1}P + \sum q_j \, T(\theta) \, X_j + \sum p_i \, T(\theta) \, A_{ai} \, ^{1}P \qquad \ldots\ldots(23)$$

By differentiation we find the velocity of the point:

$$^{0}\dot{P} = \dot{\theta} \, (T_\theta \, ^{1}P + \sum q_j \, T_\theta \, X_j + \sum p_i \, T_\theta \, A_{ai} \, ^{1}P)$$

$$+ \sum \dot{q}_j \, T(\theta) \, X_j + \sum \dot{p}_i \, T(\theta) \, A_{ai} \, ^{1}P \qquad \ldots\ldots(24)$$

Here T_θ is the derivative of $T(\theta)$ with respect to $\theta=\theta_1$, the joint angle. The kinetic energy of the link (neglecting the two hubs) is:

$$KE = \int_0^L \frac{1}{2} \mu \, ds \, \text{Trace} \, (^0\dot{P} \, ^0\dot{P}^T) \quad \quad \quad (25)$$

Equation (25) can be manipulated to obtain the expression of the kinetic energy as a quadratic form of the generalized coordinates and extract the expressions for the elements of the inertia tensor of the system. These expressions are shown here for the case without joint flexibility:

$$I(\theta,\theta) = \text{Trace} \, (T_\theta \int_0^L \mu \, ds \, (^1P^1P^T + \sum q_j \, ^1P \, X_j^T +$$

$$+ \sum q_j X_j \, ^1P^T + \sum\sum q_j q_k X_j X_k^T) T_\theta^T) \quad \quad (26a)$$

$$I(\theta,\dot{q}_j) = \text{Trace} \, (T_\theta \int_0^L \mu \, ds \, (^1P \, X_j^T + \sum q_k X_j X_k^T) T^T(\theta)) \quad \quad (26b)$$

$$I(\dot{q}_j,\dot{q}_k) = \text{Trace} \, (T(\theta) \int_0^L \mu \, ds \, (X_j X_k^T) \, T^T(\theta)) \quad \quad (26c)$$

Once the elements of the elements of the inertia tensor have been determined the calculation of the centrifugal and Coriolis forces can be automated by exploiting known identities connecting these to the derivatives of the elements of the inertia tensor. Let φ denote the vector of generalized coordinates. The nonlinear inertial forces can be written in the form:

$$F_c(\varphi,\dot{\varphi}) = C(\varphi,\dot{\varphi}) \, \dot{\varphi} \quad \quad \quad (27)$$

The elements of the C matrix are given by:

$$C(\varphi,\dot{\varphi}) = \sum_n \{ [e_n \dot{\varphi}^T I_n(\varphi)]^T - \frac{1}{2} [e_n \dot{\varphi}^T I_n(\varphi)] \} \quad \quad (28)$$

where e_n is the nth unit vector and I_n is the derivative of the inertia tensor with respect to the nth generalized coordinate. The calculation of the gravitational forces is not shown here. It is based on equation (20) which gives the instantaneous position of a point of the flexible link; from this we can easily formulate the expression for the gravitational potential energy. The elastic potential energy is a positive semidefinite quadratic form of the generalized coordinates. Thus the elastic forces can be found easily if we know the stiffness matrix utilizing the equation:

$$F_e = \begin{bmatrix} 0 & 0 & 0 \\ 0 & K_{qq} & 0 \\ 0 & 0 & K_{pp} \end{bmatrix} \begin{bmatrix} \theta \\ q \\ p \end{bmatrix} \quad \quad (29)$$

The submatrix K_{qq} is directly obtained from the finite element program that is usually employed to obtain the mode shapes for the link. It is diagonal if exclusively normal modes (which are orthogonal) are used. K_{pp} is diagonal and is obtained from modelling the joint flexibility with linear and rotational springs. Equations (26)-(29) are written for a very simple case and yet are sufficient to demonstrate the computational complexity of the problem and the difficulties existing in implementing model-based control schemes for flexible manipulators. It must be mentioned that in the absence of gravitational forces, we are able to write linear equations of motion for the one-link structure; this linear model is the one employed in control studies for such arms.

4. Solution of the Control Problem by Direct Nonlinear Inversion

The model for multi-link arms is always nonlinear; thus the methods developed for control of the motion of one-link arms cannot be directly used for their control, as they are based on linear models. Here we discuss the solution of the inverse dynamic problem for manipulators with several flexible links (and possibly flexibility in the joints). We limit the discussion to open-chain structures for which there are no constraint equations and all joints are active. Further we consider a nonredundant arm, so that the number of degrees of freedom of the end effector, n, is equal to the number of joints. Thus the equations of motion (18) take the general form:

$$I(\varphi)\ddot{\varphi} + C(\varphi,\dot{\varphi})\dot{\varphi} + G(\varphi) = B_1 T \quad \ldots\ldots(30)$$

where φ is the vector of generalized coordinates, whose first n components are the joint angles; the other are the modal coordinates of the links and possibly the joint deformation variables. T is the vector of torques applied at the joints. The right-hand-side of (30) was written using the assignment matrix:

$$B_1 = \begin{bmatrix} I_{nxn} \\ 0 \end{bmatrix} \quad \ldots\ldots(30a)$$

to emphasize the fact that there are no applied forces for the flexible coordinates. For the solution of the control problem we write the equation in state-space form, with state vector $x = [\varphi^T\ d\varphi/dt^T]^T$:

$$\dot{x} = a(x) + B(x) T \quad \ldots\ldots(31a)$$

where:

$$a(x) = \begin{bmatrix} \dot{\varphi} \\ -I^{-1}(\varphi)[C(\varphi,\dot{\varphi})\dot{\varphi} + G(\varphi)] \end{bmatrix} \quad \ldots\ldots(31b)$$

$$B(x) = \begin{bmatrix} 0 \\ I^{-1}(\varphi) B_1 \end{bmatrix} \quad \ldots\ldots(31c)$$

The vector of joint torques T (nx1) is viewed as the control vector of the system. Important for the solution developed here is the output equation; as we are interested in the position of the end effector in Cartesian space the output vector is the vector (nx1) of Cartesian coordinates of the end effector, y, and the output equation is nothing but the forward kinematic map of the arm with flexibility:

$$y = f(\varphi) \quad \dots\dots(31d)$$

We can now state the inverse dynamic problem as follows: "given the Cartesian trajectory of the end effector y(t), find the torques that were applied at the joints T(t)." Rephrasing this using control terminology we have: "given a desired Cartesian trajectory of the end effector $y_d(t)$, find the open-loop torques that must be applied at the joints to achieve it." A solution to this problem is obtained by applying a direct nonlinear inversion procedure, as suggested by Hirschorn [16] for general nonlinear systems. The procedure is based on differentiation of the output equation and construction of a series of systems having as output the derivatives of the output of the original system, until one obtains a system which is invertible. Applying the procedure we obtain:

$$\text{system 0 output:} \quad z_0 = y = f(\varphi) \quad \dots\dots(32a)$$

$$\text{system 1 output:} \quad z_1 = \dot{y} = f_\varphi \dot{\varphi} = c_1(x) \quad \dots\dots(32b)$$

$$\text{system 2 output:} \quad z_2 = \ddot{y} = \frac{\partial c_1}{\partial x} \dot{x} = c_2(x) + D_2(x) T \quad \dots\dots(32c)$$

where:

$$c_2(x) = \frac{\partial c_1}{\partial x} a(x) \quad \dots\dots(33a)$$

$$D_2(x) = \frac{\partial c_1}{\partial x} B(x) = f_\varphi I^{-1}(\varphi) B_1 \quad \dots\dots(33b)$$

(32c) can be solved for T in terms of the Cartesian acceleration of the end effector, provided the nxn matrix $D_2(x)$ is nonsingular:

$$T = D_2^{-1}(x) [\ddot{y} - c_2(x)] \quad \dots\dots(34)$$

In such a case we say that the second system is invertible, or, the relative order of the original system is 2. Elsewhere [17] it is shown that $D_2(x)$ is nonsingular everywhere except at the singularities of the forward kinematic map of the flexible arm. (34) involves the state vector of the system as well; thus knowledge of the desired Cartesian acceleration of the end effector is not sufficient; we need the trajectory of the system in the state space. To obtain this we construct the following left-inverse system:

$$\dot{\hat{x}} = \hat{a}(x) + \hat{B}(x) T \quad \dots\dots(35a)$$

$$\dot{\hat{y}} = \hat{c}(x) + \hat{D}(x) T \quad \dots\dots(35b)$$

Here:

$$\hat{a}(x) = a(x) - B(x) D_2^{-1}(x) c_2(x)$$

$$\hat{B}(x) = B(x) D_2^{-1}(x)$$

$$\hat{c}(x) = -D_2^{-1}(x) c_2(x)$$

$$\hat{D}(x) = D_2^{-1}(x)$$

$$\hat{T} = \ddot{y} \qquad \hat{y} = T \qquad \ldots\ldots(36)$$

The input or control for the left-inverse system is the desired Cartesian acceleration, and the output is the vector of open-loop joint torques. Of course the design of the control is not complete, and these torques will produce the desired trajectory only with the right initial conditions and in the absence of disturbances. Regulator terms must be added to produce a stable control. Work is in progress in this direction. The stiffness and the material damping of the links are used to advantage. As (35a) cannot be solved in closed form, the determination of the open-loop joint torques requires numerical simulation of the inverse system. This is not feasible with the computing means of today. The only way to overcome this problem is to construct a library of torque time histories by off-line simulation of the inverse system, corresponding to a number of frequently executed trajectories, and store them in memory for use by the control system. Other techniques suggested for the control of multi-link flexible robots are based on ad hoc perturbation of the nonlinear model. To be able to construct stable controllers the perturbation approach must be formalized.

Conclusion

The problems of dynamic modelling and control of robotic arms with flexibility in both the links and the joints were discussed. Details of the dynamic model obtained by Lagrangian method were shown for a one-link structure. For the control nonlinear inversion was used to solve the open-loop problem; it was found that the utilization of the full nonlinear model is not at present feasible, because we cannot obtain closed-form solutions for the left-inverse system.

References

[1] Kokkinis, T., et al., 1987, "Kinematics and Design of a Direct-Drive Robot Arm," Proceedings, IASTED International Conference on Robotics and Automation, Santa Barbara, California.

[2] Sunada, W. H., 1982, "Dynamic Analysis of Flexible Spatial Mechanisms and Robot Manipulators," Ph.D. Thesis, University of California, Los Angeles.

[3] Yoo, W.S., and Haug, E.J., 1986, "Dynamics of Articulated Structures, Part I: Theory," Journal of Structural Mechanics, 14(1): 105-126, and, "Dynamics of Articulated Structures, Part II: Computer Implementation and Applications," Journal of Structural Mechanics, 14(2): 177-189.

[4] Kim, S.S., and Haug, E.J., 1987, "Selection of Deformation Modes for Flexible Multi-Body Dynamics," Proceedings, ASME Design Automation Conference, Boston, Massachussetts.

[5] Book, W.J., 1984, "Recursive Lagrangian Dynamics of Flexible Manipulator Arms," International Journal of Robotics Research, 3(3): 87-101.

[6] Luh, J.Y.S., and Zheng, Y.F., 1985, "Computation of Input Generalized Forces for Robots with Closed Kinematic Chain Mechanisms," IEEE Journal of Robotics and Automation, RA-1(2): 95-103.

[7] Kleinfinger, J.F., and Khalil, W., 1986, "Dynamic Modelling of Closed-Loop Robots," Proceedings, 15th International Symposium on Industrial Robotics.

[8] Kiedrzynski, A., and Becquet, M., 1985, "Modelisation of the Elastic Links of Closed-Loop Robots," Proceedings, International Conference on Automation and Robotics.

[9] Ahmad, S., and Widman, G.R., 1987, "Control of Industrial Robots with Flexible Joints," Proceedings, IEEE International Conference on Robotics and Automation, Raleigh, N. Carolina.

[10] Dado, M.H.F., and Soni, A.H., 1987, "Dynamic Response Analysis of 2-R Robot with Flexible Joints," Proceedings, IEEE International Conference on Robotics and Automation, Raleigh, N. Carolina.

[11] Rivin, E.I., 1985, "Effective Rigidity of Robot Structures: Analysis and Enhancement," Proceedings, American Control Conference, Boston, Massachussetts.

[12] Shih, L., and Frank, A.A., 1987, "Dynamic Modelling and Analysis of General Linked Mechanisms with Compliance," Proceedings, ASME Design Automation Conference, Boston, Massachussetts.

[13] Singh, S. N., and Schy, A. A., 1986, "Control of Elastic Robotic Systems by Nonlinear Inversion and Modal Damping," ASME Transactions, Journal of Dynamic Systems, Measurement and Control, 108(3):180-189.

[14] Gebler, B., 1987, "Feed-Forward Control Strategy for an Industrial Robot with Elastic Links and Joints," Proceedings, IEEE International Conference on Robotics and Automation, Raleigh, N. Carolina.

[15] Bayo, E., 1988, "Computed Torque for the Position Control of Open-Chain Flexible Robots," Proceedings, IEEE International Conference on Robotics and Automation, Philadelphia, Pennsylvannia.

[16] Hirchorn, R.N., 1979, "Invertibility of Multivariable Nonlinear Control Systems," IEEE Transactions of Automatic Control, AC-26(6): 855-865.

[17] Kokkinis, T., and Sahraian, M., 1988, "Cartesian Space Control for Flexible Manipulators with Nonlinear Inversion," Proceedings, IASTED Conference on Automation and Robotics, Santa Barbara, California.

Vehicles

Control of an Active Suspension System for a Wheeled Vehicle

E. HORIUCHI, S. USUI, K. TANI, N. SHIRAI

Namiki 1-2, Tsukuba-city, Ibaraki 305 Japan
Mechanics Division,
Robotics Department
Mechanical Engineering Laboratory

Abstract
An active suspension system with an actuator in parallel with a spring and a shock absorber is designed for a wheeled vehicle. The passive elements in the mechanism reduce high frequency vibration. A possible implementation with a DC torque motor and a pseudo-straight line mechanism is illustrated. A simple analysis of the kinematics of a wheeled vehicle with an active suspension is also discussed. A control method based on sliding mode control is developed for a two-wheeled two degree of freedom model of the vehicle. Simulation results show that, in the sense of root-mean-square vibration reduction, active suspension is superior to passive suspension and that proposed approaches are effective when applied to a nonlinear system.

1 Introduction

Recently, the use of wheeled vehicles in inaccessible and dangerous environments is increasing. The transport of wood or planting work in forests, rescue activities in the area hit by a disaster, and maintenance and inspection work in nuclear plants are examples of possible practical applications of wheeled vehicles. A wheeled vehicle which aims at traveling over irregular terrains needs a specific suspension system in place of conventional passive devices, a spring or a shock absorber, employed as suspensions for ordinary vehicles. The suspension system required for such a vehicle must be provided with actuators to have the accommodation to terrains and realize the suspension properties suited for the vehicle function.

Energy consumption by active suspension systems is an important problem because practicable wheeled vehicles should be self-contained. Semi-active suspensions [1],[2] are alternatives for this problem. They use an active damper in parallel with a passive spring. In this paper, a concept of an active suspension system (AS) composed of passive elements and an actuator as well as a suspension controller design is presented. So far, several optimum control strategies based on linear models have been studied: optimum output feedback control [3] and preview control [4] (a control scheme in which an input is sensed before it reaches the controlled plant). This paper employs a control strategy based on sliding mode control [5] for AS.

2 Active suspension system

2.1 Concept of an active suspension system (AS)

The concept of AS is shown in Fig. 1. This system has an actuator generating a force f(t) in the vertical direction, a spring with stiffness K, and a shock absorber with damping ratio C in parallel between the vehicle body with mass M and the wheel axle. An unknown terrain elevation r(t) at the wheel is transmitted to the body through the suspension mechanism and causes the body displacement x(t) in the vertical direction. Both x(t) and r(t) refer to some absolute frame. Only the vibration in the vertical direction is considered in the rest of this paper.

2.2 Effect of passive elements

The actuator in AS only has to control the low frequency body vibration because the spring sustains the static load and the choice of spring stiffness and damping ratio to the given body mass determines the upper bound of the frequency of the vibration to be actively controlled. These passive elements highly reduce the high frequency vibration. To make the effect of them clear, the dynamics of the system shown in Fig.1 is investigated.

$$M\ddot{x}(t) + C\{\dot{x}(t) - \dot{r}(t)\} + K\{x(t) - r(t)\} = f(t) \quad \ldots\ldots(1)$$

From (1) with f(t) = 0, a transfer function G(s) with input r(t) and output x(t) is obtained.

$$G(s) = \frac{Cs + K}{Ms^2 + Cs + K} \quad \ldots\ldots\ldots\ldots\ldots\ldots\ldots\ldots\ldots\ldots\ldots\ldots(2)$$

The gain diagram of G(s) is shown in Fig. 2. The values of C and K are shown in Table 1, and M = 25 kg. A peak observed at about 2 Hz indicates that the upper bound of the frequency of the vibration to be actively controlled is 2 Hz or so.

2.3 An implementation of the active suspension system

An implementation of AS utilizing the Chebyshev's four-bar link to transform the actuator rotational motion into the pseudo-straight line motion at the wheel axle is shown in Fig. 3. Figure 5 shows the pseudo-straight line motion generated by this mechanism. The actuator torque is also transformed into the force applying at the wheel axle by way of this structure although the force-torque relationship is dependent on the actuator angle. The force at the axle generated by 1 Nm actuator torque is shown in Fig. 6. A DC torque motor is employed as the actuator of AS since precise open loop torque control is possible because of its small friction, although active suspensions often use hydraulic or pneumatic actuators [6],[7].

3 A two-wheeled two degree of freedom model

3.1 Formulation

A two-wheeled two degree of freedom model of the wheeled vehicle is derived on assumptions that (a) the tires are rigid and (b) the wheels keep contact with the ground. Since the upper bound frequency of the vibration to be actively controlled decreases with decreasing K if C and M are fixed, spring stiffness is desired to be small in spite of the drawback of the large body sink by the gravity. The tires are much more rigid than the spring in AS. The locomotion speed of the vehicle in the scope of this paper is rather slow, so that the vehicle presumedly will not spring free from the ground.

The model shown in Fig. 4 has two degrees of freedom: the body height and the body tilt. If the wheels correspond to the front and rear wheels, $\theta(t)$ is the pitch angle; and if they represent the right and left wheels, $\theta(t)$ is the roll angle. The height of the center of mass of the body, $x_G(t)$, refers to some absolute frame. Subscripts, F and R, indicate that the variable is related to either the F-wheel or the R-wheel. And M, I, and 2L mean the body mass, the moment of inertia of the body, and the body length (wheelbase or tread), respectively. The body length is presumed to be independent of the body tilt and unchanged.

$$y_i(t) = x_i(t) - r_i(t) \; ; \; i = F, R \quad \ldots\ldots\ldots\ldots\ldots\ldots(3)$$

where $y_i(t)$ is the distance between the body and the ground. Formulation should be done in terms of $y_i(t)$ since precise measurement of $x(t)$ and $r(t)$ is difficult. Dynamic equations based on $x(t)$ and $r(t)$ are rewritten by measurable variables. Suppose that the distances between the body and the ground at the suspension positions are available, $y_G(t)$ defined in (4) is introduced as a variable which represents the body height.

$$y_G(t) = \{y_F(t) + y_R(t)\}/2 \quad \ldots\ldots\ldots\ldots\ldots\ldots(4)$$

Let the equilibrium position of the vehicle body the origin, and the dynamic equations concerning the body height and the body tilt based on measurable variables are obtained as follows.

$$M\ddot{y}_G(t) = f_Y(t) - 2C\dot{y}_G(t) - 2Ky_G(t) + d_Y(t) \quad \ldots\ldots\ldots (5)$$

$$I\ddot{\theta}(t) = L\cos\theta(t)\{f_\theta(t) - 2CL\dot{\theta}(t)\cos\theta(t) \quad \ldots\ldots\ldots\ldots (6)$$
$$- 2KL\sin\theta(t)\} + d_\theta(t)$$

where

$$f_Y(t) = f_F(t) + f_R(t), \quad f_\theta(t) = f_F(t) - f_R(t)$$

$d_Y(t)$ and $d_\theta(t)$ are disturbance forces caused by unknown terrain elevations to the body height and to the body tilt, respectively It is assumed that only maximum absolute values of $d_Y(t)$ and $d_\theta(t)$ can be estimated. Note that these equations are decoupled between $y_G(t)$ and $\theta(t)$ by letting $K_F = K_R$ and $C_F = C_R$ and that (6) is nonlinear.

3.2 Kinematics of the wheeled vehicle

The use of $y_G(t)$ in stead of $x_G(t)$ avoids difficult problems in the locomotion over unknown terrains. If $x_G(t)$ is kept unchanged while traveling over a long slope, for example, the limits of the suspension stroke will make it impossible to continue the locomotion. It is necessary to analyze the kinematics of the wheeled vehicle with AS under the condition that $y_G(t)$ and $\theta(t)$ are kept constant.

When $y_G(t)$ and $\theta(t)$ are fixed, the body of the vehicle traveling over a long flat slope follows the slope profile. Consider terrain elevations defined by a sinusoidal function.

$$r_F(t) = \sin 2\pi\omega t, \quad r_R(t) = \sin 2\pi\omega(t - B/V)$$

where V is a constant locomotion speed, B is the wheelbase, and is the frequency of terrain elevations. If $y_G(t)$ is controlled to be a constant H and $\theta(t)$ is kept zero, then,

$$x_G(t) = \{x_F(t) + x_R(t)\}/2$$
$$= H + \{\sin 2\pi\omega t + \sin 2\pi\omega(t - B/V)\}/2 \quad \ldots\ldots\ldots (7)$$

Consider two cases: (a) $B\omega/V = n$, (b) $B\omega/V = n + 1/2$. In case (a), (7) is reduced to $x_G(t) = H + \sin 2\pi\omega t$. This means that the vehicle body tracks the same trajectory as $r_F(t)$. While in case (b), (7) becomes $x_G(t) = H$ which shows that the height of the vehicle is held fixed with respect to some absolute frame. Considering general cases, the path of $x_G(t)$ becomes a trajectory with an intermediate amplitude between those two trajectories because the apparent frequency of terrain elevation depends on the locomotion speed and actual terrain profiles cannot be modeled by a simple sinusoidal curve.

4 Sliding mode control

4.1 Purposes of sliding mode control

The nonlinearity in the dynamics of the vehicle with AS should be paid attention to. The fact that, in physical suspension systems, spring stiffness and damping ratio are prone to involve nonlinear and time-varying properties which bring about parameter errors and parameter variations should also be considered. Sliding mode control is one solution to these problems because it is capable of dealing directly with nonlinear systems and needs only estimation values about nonlinearity of the system.

In addition, the vehicle discussed here is planned to be equipped with a vision sensor system which detects obstacle surfaces. Sliding mode control is a simple model following control; the state variables of the system in sliding mode are constrained on defined switching lines. Thus, the correction of sensor data based on the vibration model is possible.

4.2 Design of a sliding mode controller

The design procedure of a controller which regulates $y_G(t)$ and $\theta(t)$ is as follows. First, on the basis of the premise that state variables are available, switching lines for $y_G(t)$ and $\theta(t)$ with negative inclinations are defined in phase plain.

$$s_Y = \dot{y}_G(t) + \alpha_Y y_G(t) = 0 \; (\alpha_Y > 0) \quad \ldots\ldots\ldots\ldots\ldots\ldots (8)$$
$$s_\theta = \dot{\theta}(t) + \alpha_\theta \theta(t) = 0 \; (\alpha_\theta > 0) \quad \ldots\ldots\ldots\ldots\ldots\ldots (9)$$

Second, the control structure is defined. In this approach, a control strategy which switches between two values according to the sign of s_Y and s_θ is introduced.

$$f_i = \begin{cases} f_i^+ & (s_i > 0) \\ f_i^- & (s_i < 0) \end{cases} \; ; \; i = Y, \theta \; \ldots\ldots\ldots\ldots (10)$$

Third, control signals must be determined so that sliding mode exists in the neighborhood of the switching line. The occurrence of sliding mode is assured by global asymptotic stability of an equilibrium point, $s_i = 0$; $i = Y, \theta$, which is proven by the second method of Lyapunov. After the certification of $V(t) = s_i^2$; $i = Y, \theta$ being Lyapunov functions, which is omitted here, next conditions are derived.

$$\dot{s}_i s_i < 0 \; ; \; i = Y, \theta \quad\quad\quad (11)$$

These are sufficient conditions for the occurrence of sliding mode. As for the pitch angle, $\theta(t)$,

$$\dot{s}_\theta = \ddot{\theta} + \alpha_\theta \dot{\theta}$$
$$= L\cos\theta\{f_\theta - 2CL\dot{\theta}\cos\theta - 2KL\sin\theta\}/I + d_\theta/I + \alpha_\theta \dot{\theta} \quad\quad (12)$$

In the case that $s_\theta > 0$, from the condition (11), $\dot{s}_\theta < 0$. From (12), the condition to be satisfied by f_θ^+ is obtained.

$$f_\theta^+ < \{2CL\cos\theta - I\alpha_\theta/(L\cos\theta)\}\dot{\theta} + 2KL\sin\theta - d_\theta/(L\cos\theta) \quad\quad (13)$$

Each term on the right hand of (13) is minimized with respect to $\theta(t)$. For the simplicity, the parameters, M, I, C, and K, are assumed to be time-invariant and perfectly identified in advance If they should involve parameter errors or parameter variations, the minimization would have to take their influence into consideration. In this formulation, only the information about maximum and minimum values of parameter errors or parameter variations would be required.

From the premise that maximum absolute values of $d_y(t)$ and $d_\theta(t)$ are known, minimum values of the second and the third term on th right hand of (13) is obtained with $-\pi/4 \le \theta \le \pi/4$.

The second term $\ge -\sqrt{2}KL$

The third term $\ge -\sqrt{2}|d_\theta|_{max}/L$

If $1/\sqrt{2} \le x \le 1$ for a function, $f(x) = ax - b/x$ $(a, b > 0)$, then

$$(a - 2b)/\sqrt{2} \le f(x) \le a - b$$

Therefore,

$$|f(x)| \le \beta = \max\{|2CL - I\alpha_\theta/L|, \sqrt{2}|CL - I\alpha_\theta/L|\}$$

The first term $\ge -|f(\cos\theta)||\dot{\theta}| \ge -\beta|\dot{\theta}|$

The desired control law is obtained by combining these terms.

$$f_\theta^+ = -\beta|\dot{\theta}| - \sqrt{2}KL - \sqrt{2}|d_\theta|_{max}/I \quad\quad\quad (14)$$

When $s_\theta < 0$, control law f_θ^- is derived in the same way.

$$f_\theta^- = \beta|\dot{\theta}| + \sqrt{2}KL + \sqrt{2}|d_\theta|_{max}/I \quad\quad\quad (15)$$

As for $y_G(t)$, next control laws are introduced after the similar discussion as above.

$$f_Y^+ = -|2C - M\alpha_Y||\dot{y}_G| - 2K|y_G| - |d_Y|_{max} \quad \cdots\cdots\cdots(16)$$
$$(s_Y > 0)$$

$$f_Y^- = |2C - M\alpha_Y||\dot{y}_G| + 2K|y_G| + |d_Y|_{max} \quad \cdots\cdots\cdots(17)$$
$$(s_Y < 0)$$

5 Hybrid control

One demerit of sliding mode control is chattering caused by the delay in physical systems. Suction control [8] which employs continuous control laws to approximate switched control is an approach for rejecting the chattering. A hybrid method which combines sliding mode control with state feedback control, is introduced to overcome this problem.

Consider a system whose dynamics is expressed in the form that

$$M\ddot{x}(t) = u(t) - A\dot{x}(t) - Bx(t) \quad \cdots\cdots\cdots\cdots\cdots\cdots\cdots(18)$$

where A, B, and M are positive constants and u(t) is a control signal. For a hybrid controller design, a new parameter, z(t), is introduced.

$$z(t) = \dot{x}(t)/x(t) \quad \cdots\cdots\cdots\cdots\cdots\cdots\cdots\cdots\cdots\cdots\cdots(19)$$

This parameter indicates the direction in which the point corresponding to the state variables in phase plain converges on the origin of phase plain. When this system is controlled by a state feedback controller,

$$u(t) = -k_1 x(t) - k_2 \dot{x}(t) \quad \cdots\cdots\cdots\cdots\cdots\cdots\cdots(20)$$

with z(t), (18) is reduced to

$$\dot{z}(t) = -z^2(t) - Pz(t) - Q \quad \cdots\cdots\cdots\cdots\cdots\cdots(21)$$

where

$$P = (k_2 + A)/M > 0, \quad Q = (k_1 + B)/M > 0$$

Let z_1 and z_2 ($0 > z_1 > z_2$) be solutions of the equation

$$z^2 + Pz + Q = 0 \quad \cdots\cdots\cdots\cdots\cdots\cdots\cdots\cdots\cdots\cdots(22)$$

If the next condition (23) is satisfied, z_1 and z_2 are negative.

$$P^2 - 4Q > 0 \quad \cdots\cdots\cdots\cdots\cdots\cdots\cdots\cdots\cdots\cdots\cdots\cdots(23)$$

Both z_1 and z_2 can be interpreted as equilibrium points in the space of z of the system described by (21). After a simple analysis, it is proven that the equilibrium point z_1 is locally asymptotically stable for $z > z_2$.

The basic idea on which hybrid control is based is that (a) when the system is subject to disturbances, sliding mode controller draws the state of the system into the neighborhood of z_1 which satisfies $z > z_2$, and (b) once the state enters the domain, the control is switched to state feedback controller which assures that the state remains within the domain until it converges on the origin of phase plain. A division of phase plain is illustrated in Fig. 7. Both $-\alpha_1$ and $-\alpha_2$ are inclinations of boundaries of the domain S^f, where state feedback controller is on. Both S^+ and S^- represent domains in which sliding mode controller is on. The magnitudes of hybrid control parameters, α, α_1, α_2, z_1, and z_2 must follow the next inequality.

$$0 > -\alpha_1 > -\alpha = z_1 > -\alpha_2 > z_2 \quad \ldots\ldots\ldots\ldots\ldots\ldots (24)$$

These parameters provide more intuitive guidelines for controller design than the index performance utilized in optimum control. Let $z_1 = -\alpha$, and feedback gains, k_1 and k_2, are determined from (25) under the condition (23).

$$\alpha^2 - P\alpha + Q = 0 \quad \ldots\ldots\ldots\ldots\ldots\ldots\ldots\ldots\ldots\ldots\ldots (25)$$

Obtained hybrid control laws are

$$\begin{aligned}
&\text{if } (y_G, \dot{y}_G) \in S_Y^+ \text{ then } f_Y = f_Y^+ \\
&\text{if } (y_G, \dot{y}_G) \in S_Y^- \text{ then } f_Y = f_Y^- \quad \ldots\ldots\ldots\ldots\ldots (26) \\
&\text{if } (y_G, \dot{y}_G) \in S_Y^f \text{ then } f_Y = -k_{Y1} y_G - k_{Y2} \dot{y}_G \\
&\text{if } (\theta, \dot{\theta}) \in S_\theta^+ \text{ then } f_\theta = f_\theta^+ \\
&\text{if } (\theta, \dot{\theta}) \in S_\theta^- \text{ then } f_\theta = f_\theta^- \quad \ldots\ldots\ldots\ldots\ldots (27) \\
&\text{if } (\theta, \dot{\theta}) \in S_\theta^f \text{ then } f_\theta = -k_{\theta 1} \theta - k_{\theta 2} \dot{\theta}
\end{aligned}$$

As for $\theta(t)$, a linearization of (6) is used. In the vicinity of the origin, control signals by the sliding mode controller and the hybrid controller are set zero values.

6. Simulation results and discussion

6.1 Test terrain

The ability of the proposed controllers is verified by the computer simulation of a two-wheeled vehicle with a front and a rear wheel traveling over test terrains. Parameters used in the simulation are listed in Table 1. The test terrain profile, $r_i(t)$, a function of time, is given by an output of a second order system

$$\ddot{r}_i(t) + a_1 \dot{r}_i(t) + a_2 r_i(t) = w(t); \quad i = F, R \quad \ldots\ldots (28)$$

where $w(t)$ denotes a signal to form the test terrain, and both a and a_2 are parameters to specify the smoothness and the amplitude of the test terrain. Test terrains generated by a trapezoid-form signal are shown in Fig. 8.

6.2 Simulation results

The suspension performance is verified among a passive suspension composed of a spring and a shock absorber only and AS with three different control strategies: (1) the optimum regulator, whose diagonal weights of the performance index, Q, are listed in Table 1, (2) the sliding mode controller, and (3) the hybrid controller. The control cycle is 20 msec and control signals are updated when 10 msec is past after the acquisition of new state variables, which takes the computation time and the servo delay into consideration.

Figure 9 shows $y_G(t)$'s and Fig. 10 shows $\theta(t)$'s. Root mean square values of $y_G(t)$'s and $\theta(t)$'s are presented in these figures. It is shown that the body vibration in $y_G(t)$ and $\theta(t)$ is highly reduced by AS and, especially in the case of the body tilt $\theta(t)$ which include nonlinear dynamics, both the sliding mode controller and the hybrid controller achieved better results than those by the optimum regulator. In Fig. 10, the chattering rejection by hybrid control appears in the curve of $\theta(t)$ approaching zero.

Figure 11 depicts the F-wheel actuator forces during the locomotion. Although the sliding mode controller and the hybrid controller generate greater signals than the optimum regulator, the hybrid controller succeeds in smoothing control signals to a certain extent, which leads to the chattering rejection.

7. Conclusion

The main purpose of this paper is the control system design of an active suspension system for a wheeled vehicle for applications in hazardous environments. In the sense of root-mean-square vibration reduction, both sliding mode controller and hybrid controller which generate nonlinear and discontinuous signals achieved comparable suspension performance to optimum regulator based on linear model. In addition, proposed hybrid method which gives intuitive control specifications succeeded in reducing the chattering. Future work will treat the design of a new active suspension system for a four-wheeled vehicle with six degrees of freedom and the extension of the proposed control strategies to the six-degree-of-freedom model.

References

[1] D.Karnopp, M.J.Crosby, R.A.Harwood: Vibration Control Using Semi-Active Force Generators, ASME Journal of Engineering for Industry, pp.619/626 (1974)
[2] J.Alanoly, S.Sankar: A New Concept in Semi-Active Vibration Isolation, ASME Journal of Mechanisms, Transmissions, and Automations in Design, Vol.109 pp.242/247 (1987)
[3] A.G.Thompson, B.R.Davis, F.J.M.Salzborn: Active Suspensions with Vibration Absorbers and Optimal Output Feedback Control, SAE Technical Paper No.841253 (1984)
[4] E.K.Bender: Optimum Linear Preview Control With Application to Vehicle Suspension, ASME Journal of Basic Engineering pp.213/221 (1968)
[5] V.I.Utkin: Variable Structure Systems with Sliding Modes, IEEE Transactions, Vol.AC-22, No2 pp.212/222 (1977)
[6] D.Cho, J.K.Hendrick: Pneumatic Actuators for Vehicle Active Suspension Applications, ASME Journal of Dynamic Systems, Measurement, and Control, Vol.107 pp.67/72 (1985)
[7] J.Dominy, D.N.Bulman: An Active Suspension for a Formula One Grand Prix Racing Car, ASME Journal of Dynamic Systems, Measurement, and Control, vol.107 pp.73/78 (1985)
[8] J-J.E.Slotine: The Robust Control of Robot Manipulators, The International Journal of Robotics Research, Vol.4 pp.49/64 (1985)

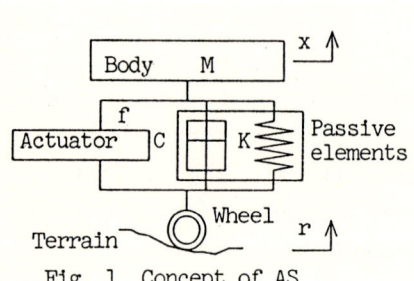

Fig. 1 Concept of AS

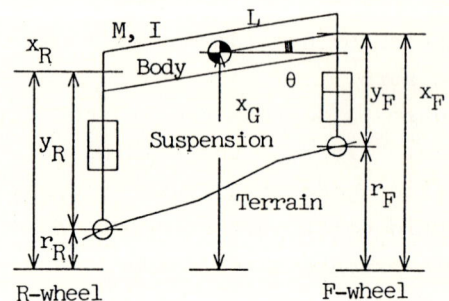

Fig. 4 Two-wheeled two d.o.f. model

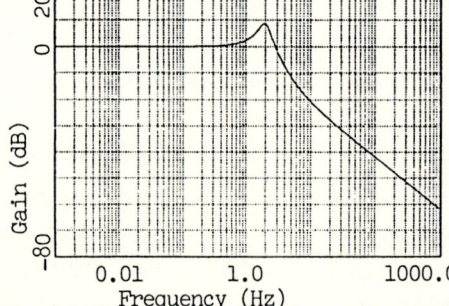

Fig. 2 Gain diagram of passive elements

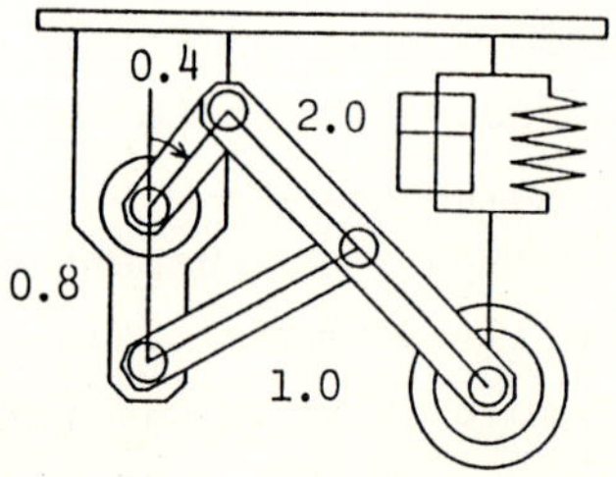

Fig. 3 An implementation of AS

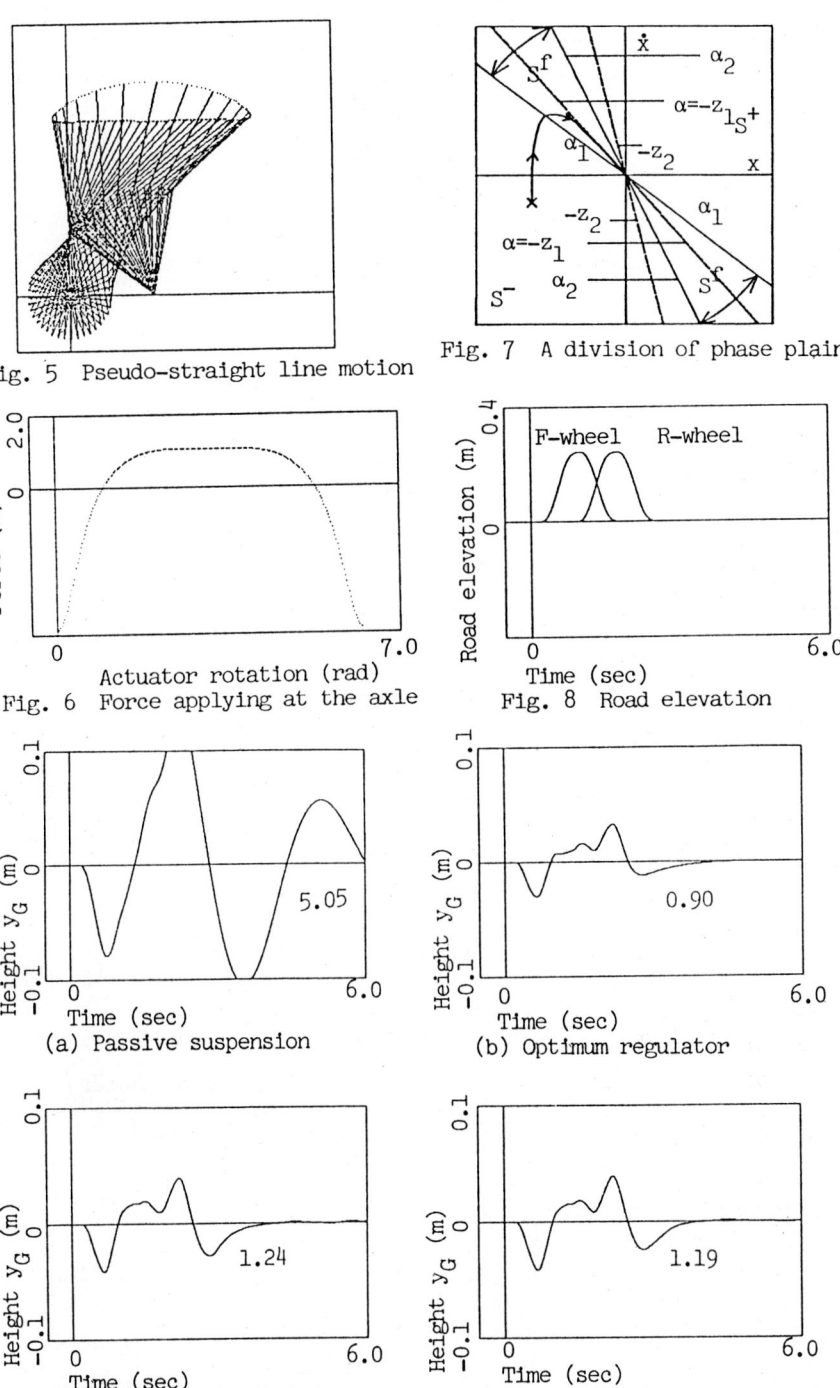

Fig. 5 Pseudo-straight line motion

Fig. 6 Force applying at the axle

Fig. 7 A division of phase plain

Fig. 8 Road elevation

(a) Passive suspension

(b) Optimum regulator

(c) Sliding mode control

(d) Hybrid control

Fig. 9 Body height

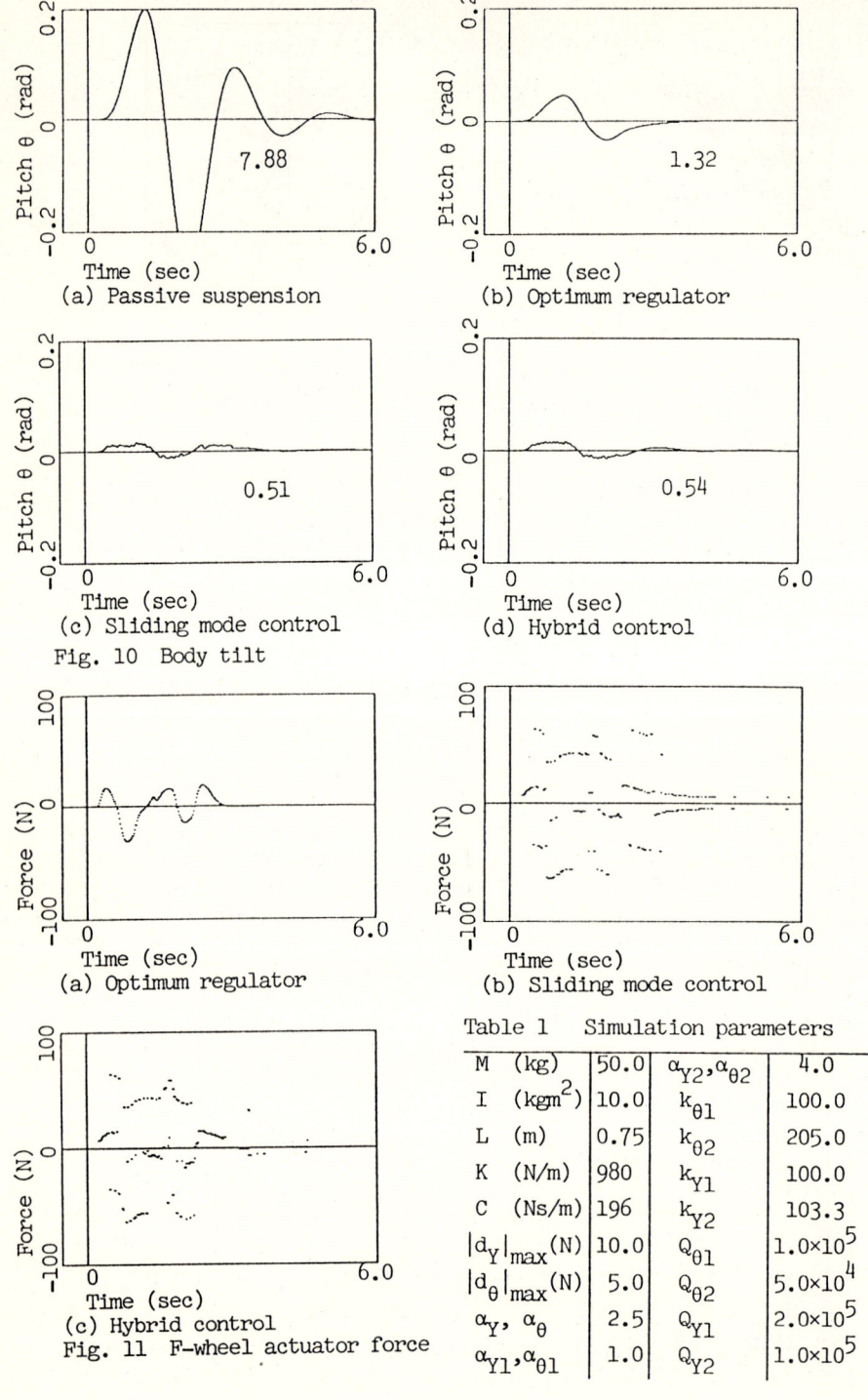

Fig. 10 Body tilt
(a) Passive suspension
(b) Optimum regulator
(c) Sliding mode control
(d) Hybrid control

Fig. 11 F-wheel actuator force
(a) Optimum regulator
(b) Sliding mode control
(c) Hybrid control

Table 1 Simulation parameters

M	(kg)	50.0	$\alpha_{Y2}, \alpha_{\theta 2}$	4.0
I	(kgm^2)	10.0	$k_{\theta 1}$	100.0
L	(m)	0.75	$k_{\theta 2}$	205.0
K	(N/m)	980	k_{Y1}	100.0
C	(Ns/m)	196	k_{Y2}	103.3
$|d_Y|_{max}$	(N)	10.0	$Q_{\theta 1}$	1.0×10^5
$|d_\theta|_{max}$	(N)	5.0	$Q_{\theta 2}$	5.0×10^4
α_Y, α_θ		2.5	Q_{Y1}	2.0×10^5
$\alpha_{Y1}, \alpha_{\theta 1}$		1.0	Q_{Y2}	1.0×10^5

Dynamic Modeling of High Speed Ground Transportation Vehicles for Control Design and Performance Evaluation

W. Kortüm, W. Schwartz, I. Fayé

DFVLR-Institute for Flight Systems Dynamics,
D-8031 Wessling

Summary

Interest in high speed ground transportation systems results in correspondingly increased interest in modeling and simulation of the dynamic behavior of vehicle systems. MEDYNA is a program especially well suited for the analysis of rail-guided vehicles: conventional railways, as well as magnetically levitated vehicles. This paper demonstrates the application of MEDYNA to the control law design and to the performance evaluation of such advanced vehicles. A brief description is made of the modeling requirements of magnetically levitated systems, along with a summary of some of the related capabilities of MEDYNA. As a case study, analysis of a vehicle based on the German TRANSRAPID system is presented. System matrices of a simplified vehicle model are established, and control design is performed with the aid of MATLAB. Finally, performance evaluation is studied with a complex model of the TRANSRAPID vehicle and elastic guideways.

1. Introduction

High speed ground transportation systems are of current interest because of saturation of air traffic and the limited capacity of road traffic with automobiles. Magnetically levitated (MAGLEV) vehicles provide one alternative for improving conventional railways. With the most advanced concept, the German TRANSRAPID system, basic levitation and guidance are provided by the attractive forces of electromagnets (EMS). The use of such MAGLEV vehicles together with elevated guideways allows for travel over greater topography variations, as well as optimal use of available land space.

Considering the full complexity of such advanced HSGT systems results in very complicated and high order mathematical models. Thus, attention must be paid to the proper handling of these models during development analysis. Dynamic modeling, computational analysis, and design modifications are desirable at an early stage of development because of the high costs, risks, and development time involved in designing and testing new concepts. Existing software tools aid the modern dynamicist and make it possible to avoid the costly endeavor of developing personal computer codes. Especially useful in this context are general purpose vehicle system dynamics software packages based on multibody formalisms, [1].

The analysis in this paper relies mainly on the general purpose program MEDYNA. This name stands for "Mehrkörper-Dynamik", the German expression for Multibody Dynamics. The operation of MEDYNA and its general functional attributes have been presented in a number of previous papers, [2], [3], [4], and thus only a brief review is made of the relevant features of MEDYNA. Earlier results on simulating MAGLEV vehicles using MEDYNA have been reported in [5].

2. Dynamic Modeling of HSGT-Systems and MEDYNA

MEDYNA contains a linear formalism in terms of the kinematic relations and thus treats "small" rigid and elastic body motions relative to "large" prescribed motions of the global reference frames. The motion of the global reference frame(s) (i.e. one for a single vehicle, several for a train) are guideway oriented functions.

A vehicle or a train is made up of a number of physical bodies consisting of car bodies, bogies or trucks, wheelsets etc. The number of bodies and degrees of freedom of the system depend on the vehicle type and the dynamic problem under consideration (e. g. stability, curving, or ride comfort). The individual bodies are interconnected by joints and linkages constraining their relative motion, as well as by coupling or compliant elements resulting in interaction forces between contiguous bodies.

HSGT systems, restricted by appropriate guidance forces, are designed to move along a guideway, and thus encounter only small deviations relative to the guide rails. This allows the equations of motion for all parts of the vehicle to be linearized with respect to guideway oriented moving reference frames. In these equations all terms which are linear and time-invariant are incorporated into the system matrices, [6]. Thus only the terms resulting from nonlinear force laws of suspension elements and those resulting from time-varying Coriolis or centrifugal forces have to be evaluated during numerical integration.

Structural flexibility must often be taken into account for modern ground transportation systems, because of the substantial dynamic loads and efforts being undertaken to arrive at light-weight vehicle and guideway constructions.

For HSGT vehicles there are two areas where elastic deformations may be of importance:

- Elasticity of vehicle bodies (car body, chassis etc.) for evaluating load and stress on certain parts including problems of material fatigue. Also certain vibrational and noise problems are caused by elastic vehicle deformations.

- Elasticity of the guidway (bridges, elevated guideways) and the dynamic interaction of vehicle and guideway may have a significant influence on the dynamic stability of rail-guided vehicles as well as on the maximal deflection and dynamic loads on the guideway structures, [7].

In considering the flexibility of elevated guideways for MAGLEV vehicles the following situations are of interest:

- The vehicle is hovering over a certain position on the guideway (i. e. it has zero speed). The hovering case may be the the most critical case for MAGLEV vehicles, since in this situation instabilities associated with vehicle and guideway interactions can build up. This stationary situation can be readily computed with MEDYNA because the vehicle together with the flexible guideway resembles *one* large elastic multibody system. If, in addition, the vehicle model is comprised of only linear coupling elements (e.g. linear suspensions, etc.), then it is possible to compute the system matrices for the elastic multibody system. Thus, eigenvalues can be computed and stability conclusions drawn.

- The vehicle is traveling with a certain speed, $V \neq 0$, along the guideway. No instabilities can build up because the vehicle is only instantaneously on any particular section of the guideway and with no sustained loads on this part, guideway deflections diminish in the form of (lightly) damped oscillations.

Finally, *suspension* systems are of special importance for modeling multi-body systems since they dominantly determine the basic dynamic and vibrational behavior of the vehicle. One distinguishes between primary suspensions as those providing levitation and guidance along the guideway and secondary suspensions providing support and cushion between the vehicle bodies. Both types of suspensions may result in interaction forces and/or can cause kinematic constraints. The force laws may be rather complex, described by nonlinear characteristics or additional differential equations. The inclusion of sensors, feedback control laws, and actuators as active components could be used to enhance the performance of suspensions in advanced vehicle concepts. With such active components it may be possible to achieve vehicle characteristics which are impossible to obtain using purely passive systems, [8].

For multibody systems a suspension can be modeled as coupling elements between bodies. These connections are made between specified interconnection points by selecting desired coupling elements from a library containing both passive and *active elements*. With a general *user-specified element*, the user can implement any force law for a coupling element not contained in the menu. Although the existing coupling elements in MEDYNA readily allow the implementation of non-contacting force laws for electromagnets, any special operating conditions of the electromagnets require the use of a user-specified element (e.g. magnet failure).

Various possibilities for primary suspensions are steel wheels on steel rails, air cushions, magnetic levitation based on attractive electromagnets (EMS), or based on superconducting repulsive magnets (EDS). In these investigations the magnet levitation is based on the EMS concept, and active coupling elements are used for modeling the electromagnets. Feedback control, however, is necessary for the stabilization of EMS vehicles because the action of the uncontrolled attractive electromagnets is unstable.

3. Case Study: Controller Design and Performance Evaluation for a Transrapid Vehicle

In this section a presentation is made of the controller design studies and performance evaluation that have been carried out with MEDYNA.

The vehicle data chosen for the performance evaluation closely resembles that of the TRANSRAPID (TR06) vehicle, [9], which is presently tested at high speeds (so far up to 412 km/hr) at the Emsland test site near Lathen, FRG. With the proper modeling assumptions the TR06 vehicle has been reduced to a vehicle model for use with MEDYNA. The complexity of this modeled vehicle, however, remains primarily unsimplified in the performance analysis in order to achieve the most realistic simulation possible. The following design and analysis studies have been performed:

1. Controller design based on a simplified vertical model of the vehicle (henceforth referred to as design-model). The development of two different control strategies is presented:
   a. measurement vector feedback with pole-assignment, and
   b. measurement vector feedback with the Riccati design.
2. Performance analysis of control strategies for a planar model of the complete vehicle (one section), Figure 1, hovering over an elastic guideway section, as well as traveling over a guideway modeled as a half-sine wave to account for elastic deformation of the guideway spans. This analysis includes both normal operation and a magnet break-down.
3. Performance of the three-dimensional model of the vehicle while traveling into a curve. Again the deflections of the guideway are included, as well as the magnet break-down case.

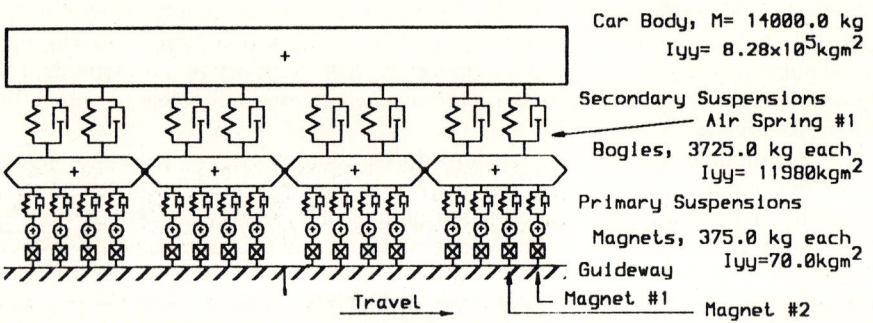

Figure 1. Planar Model of the Complete Vehicle (one section)

3.1 Control Law Design Studies

1. Design-model

The simplified vehicle model shown in Figure 2 was used to provide a less complicated model for the design process. In the interest of formulating a low order controller, it was necessary to use a low order design model.

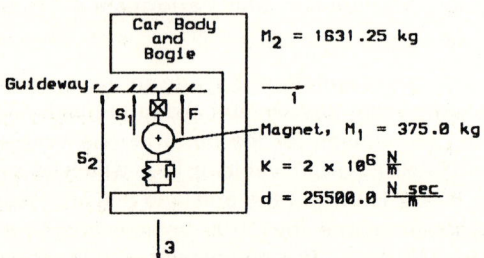

Figure 2. Simplified Model with Two Bodies

For this model, as in the most general cases, MEDYNA establishes the dynamic equations in the state-space form. The system matrices generated are very useful for further analysis and design since the majority of modern methods rest on the state-space description. Modern controller design strategies can be carried out with a number of software packages, including the program MATLAB. This program has access to the MEDYNA storage file and hence direct access to the system matrices.

To maintain proper characteristics of the vehicle, the linearized force law

$$\dot{F} = -\frac{R}{L} \cdot F - \frac{Rc_2}{L} \cdot s - c_2 \cdot \dot{s} + \frac{c_1}{L} \cdot u \qquad (1)$$

for a single magnet unit is used, along with two system bodies. In this relation F and u are the force and corresponding voltage of the magnet. The terms s and $\dot{s}$ are the measure of gap and gap separation velocity between the magnet and the guideway. The coefficients R, L, c_1 and c_2 represent electrical properties of the magnet (i.e. resistance, inductance, etc.) and the sensitivity of sensors.

The first body is the magnet itself, and the second body being a combination of one fourth of a bogie (four magnets per bogie) and the mass of one sixteenth of the car body (there are sixteen magnets for a half section of the car body). This model is restricted to only vertical motion and possesses two degrees of freedom. The parameters for this design-model are all given in Figure 2. With the inclusion of the magnet force as an active coupling element with dynamics, the equations of motion for this design-model are fifth-order, and the eigenvalues of the open-loop case are computed:

$\lambda_1 = 9.664522$ $rad/$ sec, $\qquad \lambda_2 = -1.942857$ $rad/$ sec
$\lambda_3 = -9.753468$ $rad/$ sec, $\qquad \lambda_{4,5} = -40.64405 \pm 65.4544i$ $rad/$ sec

Clearly the positive eigenvalue indicates that the open-loop system is unstable. Thus, closed-loop (feedback) control is necessary for stabilization of the system.

2. Pole Assignment

The pole placement method involves selecting a set of desired pole locations and calculating the gains required to achieve such a pole configuration and hence desired system response. The desired closed-loop eigenvalues for this system have been selected to be:

$\lambda_{1,2} = -40.64405 \pm 65.4544i$ $rad/$ sec
$\lambda_3 = \lambda_4 = -50.0$ $rad/$ sec
$\lambda_5 = -30.0$ $rad/$ sec

The complex pair of eigenvalues was attributed to the rigid body motion of the bogie/car body combined mass and the spring/damper between the bogie and the magnet. Thus, no attempt was made to modify these eigenvalues. The selection of the other eigenvalues was made in an attempt to produce a vehicle with a high apparent suspension "stiffness" and thus less susceptible to system misalignments due to magnet failure.

Although computation of the open-loop system matrix, and its corresponding eigenvalues are readily computed with MEDYNA, the feedback control gains required for closed loop control are computed with Ackermann's formula using the MATLAB package.

The control law of interest is dependent on measurement vector feedback and is given:

$$u = -\underline{c}^T \underline{z}_M, \qquad \underline{z}_M = [s_1 \; s_2 \; \dot{s}_1 \; \dot{s}_2 \; F]^T \tag{2}$$

where $s_{1,2}$ are the measured gap displacements between the magnet and guideway, and bogie and guideway, and $\dot{s}_{1,2}$ are the measured gap velocities between magnet and guideway and bogie and guideway. For the control law in equation 2, the pole placement method results in the following gains:

$$\underline{c}^T = [\;-23118.94 \quad -200206.9 \quad -2930.92 \quad -13896.63 \quad 0.179155\;]$$

3. Riccati Design

The Riccati design involves minimizing the cost function given in the equation:

$$J = \int_0^\infty (\underline{x}^T Q \underline{x} + \underline{u}^T R \underline{u})\, dt \tag{3}$$

based on a set of weighting values, Q and R, corresponding to allowable system tolerances. A first approximation of the values of Q and R is made by taking the inverse of the square of the maximum allowable deviation of the state variables, or, respectively, the input variable. In this analysis only the diagonals of the Q and R matrices are filled, such that the maximum allowable deflection is ± 5.0 mm, the maximum vertical velocity is ± 1.0 m/sec, the maximum allowable magnet force is ± 1500 N and the maximum allowable voltage is ± 100 volts. These represent only target values used in the solution of the Riccati relation. In actual performance it is possible that these values will be exceeded.

The model used in this analysis is that shown in Figure 2. Again the feedback control gains for this "optimal" closed-loop control are computed with the MATLAB package. In MATLAB the Hamiltonian is formed and the method of reduced eigenspace is used to compute the following gains [10]:

$$\underline{c}^T = [2466.367 \quad -31299.87 \quad -683.796 \quad -3036.359 \quad 0.09532628]$$

With the feedback control law given in equations (2), the following eigenvalues have been computed for the closed-loop system with Riccati gains:

$$\lambda_1 = -53.32269 \; rad/sec$$
$$\lambda_{2,3} = -9.030262 \pm 9.750922i \; rad/sec$$
$$\lambda_{4,5} = -40.00444 \pm 66.30884i \; rad/sec$$

Not only do the Riccati gains result in a stable system, but it is interesting to note that the design method has made no attempt to modify the pole placement of the two eigenvalues associated with the rigid body motion of the bogie attached to the magnet by a spring/damper interconnection.

4. Time Simulation

Numerical integration methods available in MEDYNA include a Runge-Kutta-Bettis code with error control and variable step size, and two multistep codes. For the analysis in this paper the Runge-Kutta-Bettis method was used.

The simplified vehicle model shown in Figure 2 has been extended to analyze the case where the bogie and the car body are two bodies separated by a spring/damper interconnection. This new design-model is still restricted to vertical motion and is shown in Figure 3a. Analysis has been performed to show that gains computed with a simplified model can be applied to a model of higher order.

For all of the simulations it assumed that the MAGLEV vehicle hovers at a nominal air gap of 1.0 cm above the guideway, and the maximum gap variation should be no greater than $\pm$ 5.0 mm. To maintain this air gap each magnet requires a nominal voltage of approximately 60.0 volts to support the vehicle. The maximum allowable voltage for a single magnet is approximately $\pm$ 400.0 volts.

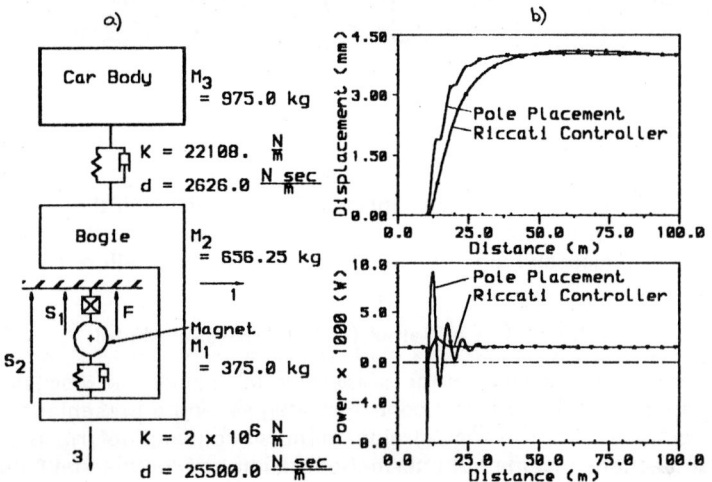

Figure 3. a) Simplified Model with Three Bodies, b) System Responses for Step Inputs

Figure 3b shows the system response for a step input of 4mm. The figure shows a time history of the displacement of the magnet, as well as the power required by the magnet. Although this pole placement "controller" will perform better than this Riccati controller during magnet failure, the power required by the latter controller is much smaller.

5. Standard TR06 Controller

The final control strategy of interest in this report is one based on the former TRANSRAPID TR06 controller shown in Figure 4. Careful approximation and proper reduction of the controller shown in this figure leads to a scheme that can be implemented in MEDYNA as an actuator possessing dynamics. Although this TR06-C controller ("C" used to denote controller approximation) is very similar to the TRANSRAPID controller, it is linear and thus its implementation is possible without the need of any user-specified coupling elements.

A direct comparison between the two previously mentioned controllers and this TR06-C controller is not intended due to fundamental differences. For both the Riccati and pole placement controllers, measurements are made of the gap between magnet and guideway, and between bogie and guideway, as well as the corresponding velocities of these separations. These measurements combined with magnet force, F, form the measurement feedback vector. In the case of the TR06-C controller a measurement is made of the relative acceleration of the magnet and the gap. Velocity and position terms required for feedback are obtained in

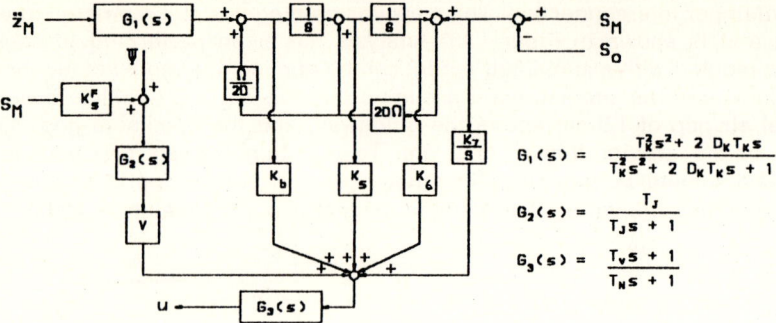

Figure 4. TR06 Controller Layout

the controller itself with the help of internal integration. With the inclusion of an integral term acting on the feedback of measured gap, the TR06-C controller is understandably successful at set-point control (e.g. magnet failure case).

3.2 Performance Analysis with a Planar Two-dimensional Model

Performance analysis for the planar model includes a discussion of the eigenvalues for the three different control strategies, and a presentation of a time simulation of the vehicle traveling straight along smooth and deformed guideways. The time simulations include both normal operation and the magnet failure case.

Eigenvalue/Stability Analysis

Stability of the stationary hovering case over an elastic guideway section has been investigated by combining the planar vehicle model with an elastic body as the supporting track. This elastic body has been generated with three eigenmodes (the modal shape functions are given in Appendix A); thus, the addition of this elastic guideway to the multibody system adds six eigenvalues to the overall system. The combination of vehicle, elastic guideway, and TR06-C controller yields 214 eigenvalues. The same system with the pole placement and Riccati controllers has 70 eigenvalues. The large number of eigenvalues for the TR06-C controller arises from the complexity of that controller and its implementation in MEDYNA.

Although stability conclusions for the overall system are drawn from the computed eigenvalues, a complete listing is not practical due to the large number of eigenvalues. Table 1, however, is a listing of the eigenvalues relevant to the stability of the system. The first column of positive eigenvalues are those eigenvalues associated with a guideway section that has been modeled with no inherent damping. They are generally the "fastest" and least damped eigenvalues of the system and indicate an unstable condition. Although the actual inherent structural damping of steel construction is relatively low (typical modal damping ranges from $\zeta = 0.008$ to 0.05), it is not zero. The second column in Table 1 represents further eigenvalue analysis with a damping factor of $\zeta = 0.015$ for the steel guideway section.

Control Strategy	Eigenvalues $\zeta=0.000$	Eigenvalues $\zeta=0.015$	Damping	f (Hz)
TR06-C Controller	0.6886 ±171.194i 0.3307 ±382.105i	-1.83894 ±171.213i -5.39604 ±382.058i	0.0107 0.0141	27.25 60.81
Pole Placement	0.069555 ±42.6625i 0.155358 ±169.596i 0.036515 ±381.919i	-0.57359 ±42.662i -2.33024 ±169.575i -5.6937 ±381.875i	0.0134 0.0141 0.0149	6.79 26.99 60.78
Riccati Method	0.74469 ±42.4897i 0.090388 ±169.5564i 0.020033 ±381.9134i	0.12328 ±42.498i -2.45583 ±169.536i -5.71033 ±381.870i	--- 0.0145 0.0150	-- 26.98 60.78

Table 1.   Selected Eigenvalues for Two-dimensional Planar Model

The unstable eigenvalues are all characterized by natural frequencies that are nearly identical to the first three eigenmodes of the elastic guideway. If these eigenvalues reflect directly the addition of the elastic eigenmodes to the multibody system, then the eigenvalues for the first eigenmode are missing in the case of the TR06-C controller. These eigenvalues are in fact present as a stable pole pair with a very low damping ratio of 0.0126:

$$\lambda_{181,182} = -.6239226 \pm 49.55016i \text{ rad/sec}$$

The undamped elastic guideway section alone should result in three pole pairs along the imaginary axis (with increasing frequency according to to the modal shape functions). It is the interaction of the multibody system with its various controllers that moves these poles off the imaginary axis, resulting in unstable systems. Inspite of the vast differences between controller structures (especially between the measurement feedback controllers and the TR06-C controller), the location of the critical eigenvalues is nearly unchanged. One might then conclude that these instabilities due to elastic guideway interactions are only marginally dependent on the controller structure.

With addition of light damping to the elastic guideway, the system becomes stable in two of the three cases. The damping ratio of the "fastest" set of eigenvalues reflects almost directly the damping that has been added to the elastic guideway. In the case of the TR06-C controller a further increase in the inherent damping is shown in Table 2. It is clear that the eigenvalues are increasingly more stable and that the damping ratios of these now stable eigenvalues are nearly the same as the added inherent guideway damping.

Control Strategy	Eigenvalues $\zeta=0.015$	Eigenvalues $\zeta=0.030$	Damping	f (Hz)
TR06-C Controller	-1.83894 ±171.213i -5.39604 ±382.058i	-4.36511 ±171.197i -11.1228 ±381.925i	0.0255 0.0291	27.25 60.81

Table 2.   Additional Eigenvalue Analysis for Two-dimensioal Planar Model

Time Simulation

For time simulations of both the two-dimensional and three-dimensional models the normal operation of the vehicle is of interest, as well as an emergency condition. The emergency state in these simulations consists of a magnet failure (magnet 2) after 0.50 seconds of simulation, then after 2.50 seconds, a complete blow out (lasting 0.50 seconds) of a secondary suspension air spring (air spring 1). This emergency situation for the two-dimensional planar model is shown in two different simulations, namely, travel on a straight perfectly smooth guideway, see Figure 5, and straight travel on a guideway with a series of half-sine waves in the vertical plane used to approximate the elastic deformation of the track, see Figure 6. The speed of the vehicle is 80.0 m/sec (288 km/h) and the total simulation time was 5.0 seconds. The analysis was first performed on a smooth guideway to gain insight into the system behavior without the added complexity of the vertical deformations.

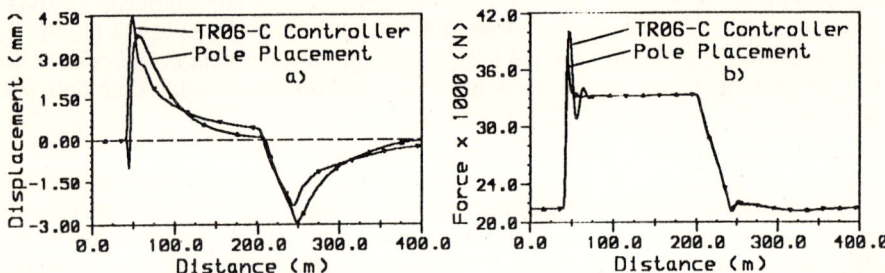

Figure 5. Planar Model Traveling Over a Smooth Guideway a) Vertical Displacement of Lift Magnet 1, b) Lift Force of Magnet 1

With the failure of magnet 2, Figure 5a shows that the greatest displacement of magnet 1 is seen in is seen in the case of the TR06-C controller. However, for the corresponding air spring blow out, the largest displacement occurs in the case of the pole placement controller. In both cases the maximum allowable gap deflection of $\pm$ 5 mm has not been exceeded and the system response is qualitatively the same, so that no operational differences exist between the controllers. With the pole placement controller, however, the initial displacement of the magnet in the negative direction indicates the possible presence of a transmission zero. Figure 5b shows a plot of the resulting lift forces for the two cases. The TR06-C controller shows greater oscillations, but behavior again qualitatively the same.

The addition of vertical excitations (input as kinematic excitations), see Figure 6a, to model track elasticity is combined with both normal and emergency operating conditions. In both situations Figure 6 shows that both the TR06-C and the pole placement controllers result in displacements of magnet 1 that do not exceed specified displacement tolerances. Magnet 1 is of interest here because it shows the greatest deflections in the emergency case. Although the oscillations due to the deformed guideway are smaller for the pole placement controller than for the TR06-C controller, the emergency condition relative to normal operation is more noticeable in the pole placement case.

Another difference between the two control strategies is most clearly seen in the plots of magnet force for magnet 1, Figure 7. Sharp peaks in the lower half of the oscillations for the pole placement case are not present in the correspond-

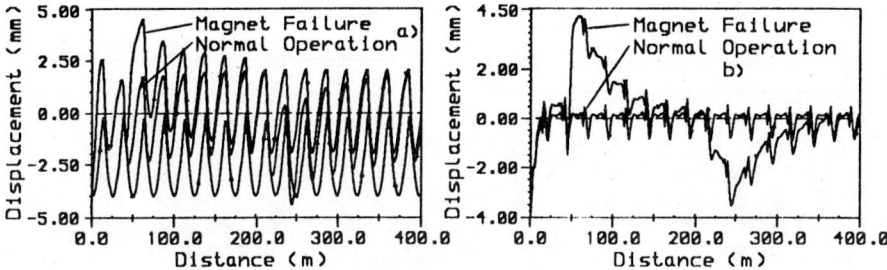

Figure 6. Displacement of Magnet 1 for Planar Model Traveling Over Elastic Track: a) TR06-C Controller, b) Pole Placement Controller

ing results for the TR06-C controller. These peaks result from abrupt transitions in the half-sine waves and the lack of filters in the pole placement controller. The force plots do show, however, that behavior of each control strategy, in terms of magnet force, is qualitatively nearly the same: the double peak at the top of each oscillation, and the return to the normal operating force after the air spring blow out. This was done to bring the magnet force back to the nominal operating force.

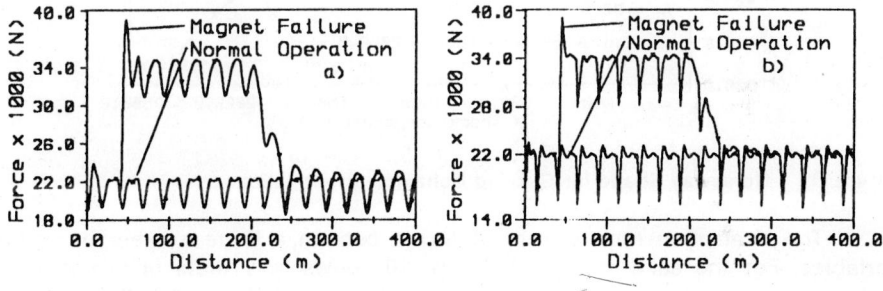

Figure 7. Lift Force of Magnet 1 for Planar Model Traveling Over Elastic Track: a) TR06-C Controller, b) Pole Placement Controller

3.3 Curving Behavior Analysis with Three-Dimensional Model

Curving behavior of the "complete" three-dimensional model is performed here as a final analysis of the MAGLEV vehicle. The model used in these simulations is shown in Figure 8. Figure 8a shows of a side view of the MAGLEV vehicle and Figure 8b shows of a top view of the same vehicle. The speed of the vehicle is again 80.0 m/sec (288 km/h), but the total simulation time has been increased to 30 seconds. As in the case of the planar model simulation, the curving simulations include both normal and emergency operation. Again, failure of magnet 2 occurs after 0.5 seconds of simulation, and a complete blow out (lasting 0.50 seconds) of air spring 1 occurs after 2.00 seconds of simulation.

The curve entry shown in Figure 9 is the same for both normal and emergency operation. A straight section of guideway (from zero to 500 m) is connected to a constant radius of curvature guideway section (from 1000 to 2400 m; radius:

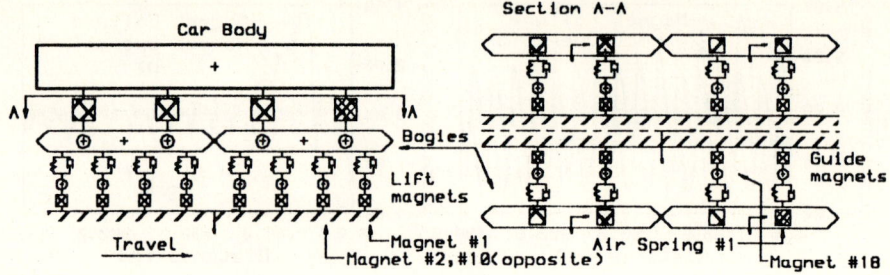

Figure 8. Multibody Model of MAGLEV Vehicle for Curving Behavior Analysis

2000m) by a transition section. The constant radius of curvature guideway is superelevated 0.2 radians about the center line of the guideway.

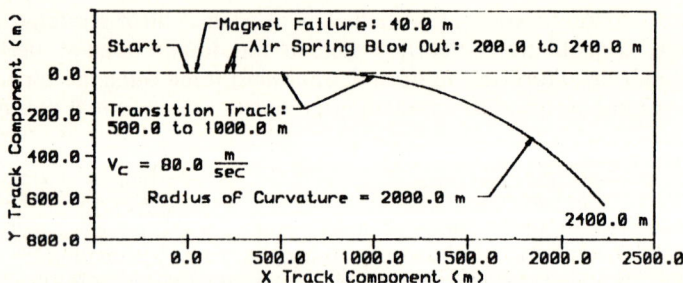

Figure 9. Guideway Shape for Curving Behavior Investigations

The simulation results from MEDYNA contain a large number of system variables. For the sake of brevity Figure 10 shows only plots of two different motions of the car body. Although the effect of magnet failure is hardly noticeable in these plots, the effect of the air spring blow out is very apparent.

In normal operation the normal component of the centrifugal forces causes the car body to "sink" in towards the guideway, see Figure 10a, as the vehicle enters the curve. As expected, the greatest vertical deflection occurs when the vehicle is fully in the stationary curve. The loss of the air spring in the emergency case not only increases the vertical deflection in the stationary curve, but causes transient underdamped vertical oscillations of the car body. These oscillations are, however, isolated from the lift magnets by the primary suspension and thus do not compromise forward travel of the vehicle.

In terms of the car body roll angle, the performance of the system during stationary curving is actually improved by the blow out of the air spring. However, this is not always the case. If the failed air spring is located on the outer side, the performance will be correspondingly worse. Again both the normal and emergency operation are shown in Figure 10b. It is important to note that these plots are made with respect to the superelevated guideway. For the normal operation, the centrifugal acceleration during curving has not been completely compensated by the superelevation of the guideway. Thus there is a 0.06 radian outward roll of the car body in the stationary curve. In the emergency operation it is the inner forward

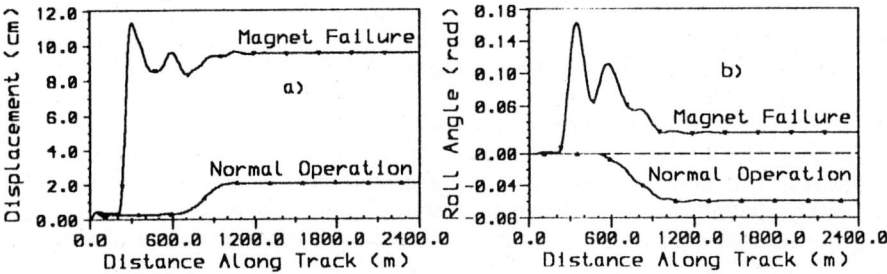

Figure 10. Motion of the Car Body During Curving: a) Vertical Displacement, b) Roll Angle

air spring that has been blown out. For this reason the car body rolls inward and reduces the overall roll angle (with respect to the guideway) in the stationary curve. The clear drawback of this failure is seen during the blow out itself. The air spring failure causes a roll oscillation that peaks at a value (0.16 rad) nearly as great as the superelevation of the guideway. This is certainly detrimental to the safety of the vehicle and might be avoided by a slower blow out of the air spring.

Response curves are also shown in Figure 11 for two different magnets during normal and emergency operation. The first magnet, Figure 11a, is a lift magnet (#10), and the second magnet, Figure 11b, is a guide magnet (#18) for the bogie containing the failed magnet. From these curves it is clear that the failure of magnet 2 has very little effect on the displacement of these magnets. It is the entry into the curve, the transition track, that causes the most significant displacements of the magnets. Even so, these displacements are within the limits of acceptable tolerances. The performance of the guide magnet is especially good, with a maximum displacement of less than 1.0 mm.

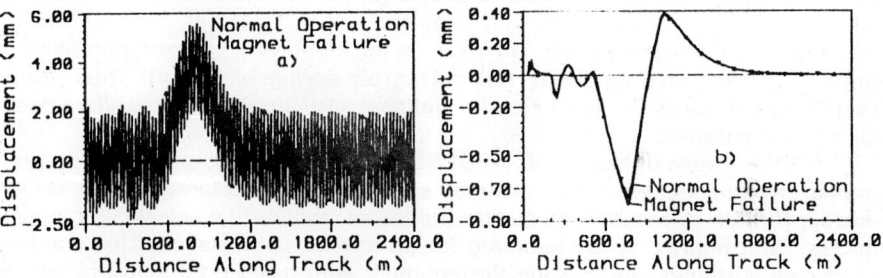

Figure 11. Magnet Displacements During Curving: a) Vertical Displacement of Lift Magnet 10, b) Lateral Displacement of Guide Magnet 18

4. Conclusions

In these investigations several different HSGT vehicles modeled as multi-body systems have been studied and analyzed in the interest of controller design and performance evaluation. This work was not aimed at directly improving the performance of such vehicles, but rather was intended to demonstrate an application of the general purpose multibody program MEDYNA, and its use in the further development of HSGT systems.

This paper in particular has focused on low-order controller design, with a subsequent stability analysis on higher order vehicles interacting with elastic guideways. The final analysis was concentrated on the curving behavior of a completely spatial vehicle model in normal and critical operation.

From controller design studies it is seen that the application of low-order controllers to higher order vehicles is possible. Two different design approaches were studied: pole placement and quadratic synthesis (Riccati design). Design trade-offs were apparent with the analysis of each controller. The Riccati designed controller required less power for normal operation, but proved to be insufficient for magnet failure. This critical case was less problematic for the pole placement method, but normal operation energy demands were higher,

The TR06-C controller demonstrated clearly the advantages of added dynamics and filtering in the controller: lower power demand for normal operation, and rapid response to accomodate magnet failure. Thus, controllers with dynamic feedback (e.g. model-based) may be desireable as future improvements to HSGT systems. With the aid of MEDYNA, the design procedure as proposed here helps the engineer to construct and implement such controllers in a meaningful way.

The stability problem of a hovering vehicle was evaluated in terms of eigenvalues. It was shown that, inspite of the various controller structures evaluated, an actively controlled vehicle has a destabilizing effect when combined with an elastic guideway possessing little or no inherent damping. For this reason it may be advisable to consider either the introduction of alternative means of damping to lightly damped structures, or further analysis aimed at developing more robust control strategies.

The analysis of the complete three-dimensional vehicle during curving shows that the modeling of a spatial vehicle is both possible and meaningful. Normal operation can be analyzed to provide a performance reference basis and to provide data for comparison with experimental tests. It has been shown that analysis of the emergency case can be used for identifying critical operating conditions (e.g. excessive car body roll during air spring blow out). Thus, further straightforward analysis and clearly defined structural or controller design changes are possible.

Further developments of MEDYNA, to enhance its performance as a dynamic system analysis tool, include: simulating of accelerating and braking vehicles, further completion of elastic guideway options (i.e. elastic for moving vehicles), and improving the handling for parameter optimization. This combination of new attributes will provide the engineer with a powerful software tool for further computer-aided design analysis of high speed ground transportation systems. Future work with such systems will certainly improve overall vehicle performance, as well as passenger ride comfort and vehicle safety.

Bibliography

[1] Kortüm, W., Schiehlen, W.; *General Purpose Vehicle System Dynamics Software Based on Multibody Formalisms*, Vehicle System Dynamics, 14, 1983, pp. 229 - 263.

[2] Wallrapp, O.; *MEDYNA - An Interactive Analysis and Design Program for Flexible Multibody Systems with Small Relative Motions*, Third ICTS Course and Seminar on Advanced Vehicle System Dynamics, Amalfi, Italy, May 1986.

[3] Duffek, W., Führer, C., Schwartz, W., Wallrapp, O.; *Analysis and Simulation of Rail and Road Vehicles with the Program MEDYNA*, Proc. 9th IAVSD-Symposium on the Dynamics of Vehicles of Roads and on Tracks, Swets & Zeitlinger B. V. Lisse, 1986.

[4] Jaschinski, A., Kortüm, W., Wallrapp, O.; *Simulation of Ground Vehicles with the Multibody Program MEDYNA*, Symposium on Simulation and Control of Ground Vehicles and Transport Systems AMD - vol. 80, pp. 315 - 341, DSC - Vol. 2., 1986

[5] Kortüm, W.; *Simulation of the Dynamics of High Speed Ground Transportation Vehicles with MEDYNA- Potentials and Case Studies*, Proceedings of International Conference on MAGLEV and Linear Drives, Las Vegas, 19-21 May 1987

[6] Führer, C., Wallrapp, O.; *A Computer Oriented Method for Reducing Linearized Multibody System Equations by Incorporating Constraints*, Computer Methods in Applied Mechanics and Engineering 46, 1984, pp. 169 - 175, North Holland.

[7] Kortüm, W., Wormley, D. N.; *Dynamic Interactions Between Traveling Vehicles and Guideway Systems*, Vehicle System Dynamics 10, 1981, pp. 285 - 317.

[8] Goodall, R., Kortüm, W.; *Active Controls in Ground Transportation Vehicles - A Review of the State-of-the-Art and Future Potential*. Vehicle System Dynamics, 12, 1983, pp. 225 - 258.

[9] MBB-Bericht Nr. TN-NT 242-6/86: *Programmbeschreibung zur Simulation der Längsbewegung ausgedehnten Fahrzeugs TR06/I auf elastischer Fahrbahn*.

[10] Cellier, F., Rimvall, M.; *Computer Aided Control Systems Design*, European Simulation Congress, Aachen 12.-16.9.83

Appendix A: Eigenmodes

I. Eigenmode

$$\varphi_I(x) = 8.5524 \cdot 10^{-3} \sin \frac{\pi}{L_T} \cdot x, \quad \psi_I(x) = \frac{d\varphi}{dx} = -1.081 \cdot 10^{-3} \omega \frac{\pi}{L_T} \cdot x$$

Natural Frequency $\omega_I = 42.447 \text{ s}^{-1}$, Modal Masses $m_{qI} = \int \varphi \varphi dm = 1.3265 kg$

II. Eigenmode

$$\varphi_{II}(x) = 1.06905 \cdot 10^{-3} \sin \frac{2\pi}{L_T} \cdot x, \quad \psi_{II}(x) = -2.7026 \cdot 10^{-4} \cos \frac{2\pi}{L_T} \cdot x$$

Natural Frequency $\omega_{II} = 169.79 \text{ s}^{-1}$, Modal Masses $m_{qII} = 2.0726 \cdot 10^{-2} kg$

III. Eigenmode

$$\varphi_{III}(x) = 3.1676 \cdot 10^{-4} \sin \frac{3\pi}{L_T} \cdot x, \quad \psi_{III}(x) = 1.20116 \cdot 10^{-4} \cos \frac{3\pi}{L_T} \cdot x$$

Natural Frequency $\omega_{III} = 322.03 \text{ s}^{-1}$, Modal Masses $m_{qIII} = 1.8196 \cdot 10^{-3} kg$

For the rigid guideway, the deformation was modeled as a half-sine wave for displacement excitation: $w = 0.00417 \cdot \sin \frac{\pi}{24.854} \cdot x$

Trajectory Planning and Motion Control of Mobile Robots

J-P. LAUMOND, T. SIMEON, R. CHATILA, G. GIRALT

Groupe Robotique et Intelligence Artificielle
LAAS-CNRS
7 avenue du Colonel Roche
31077 Toulouse Cedex, France

Abstract

This paper addresses two aspects of the navigation problem for a two d.o.f mobile robot (non holomic system): trajectory planning and motion control. Trajectory planning concerns the existence and the generation of a feasible collision-free trajectory, and motion control the actual execution of this trajectory.

The problem has to be solved in constrained and non-constrained environment. We summarize some results previously obtained in non constrained space and develop a general approach for finding feasible trajectory in constrained space. This method is based on a result which characterizes the existence of a feasible trajectory by means of the existence of a connected open component in the admissible configuration space. Its current implementation, based on a configuration space structured into hyper-parallelepipeds, is described.

The trajectory is then analyzed in order to smooth it when possible, using clothoid curves. Its execution is controlled by means of comparing sensor readings with the local environment model along it.

1 Introduction

Over the last decade, robotics researchers have had to address the problems of planning and control of robot motions, including issues that range from geometric reasoning to the study of control. To accomplish this, they have developed their own tools [3].

Over the same period, an important research effort in the field of geometry has primarily focused on the design and analysis of efficient algorithms relying on various approaches ranging from real algebraic geometry to computational geometry [26].

Recent developments tend to establish a fruitful synergy between the techniques involved in these two fields. Notice that the desire to build actual physical systems gives rise to novel and challenging issues, as in the case of the problem dealt with in this paper.

The work described has been conducted within the HILARE mobile robot project developed at LAAS. It deals with the navigation of a mobile robot subject to major environmental and kinematic constraints. The problem is the following :

"How to plan and control collision-free trajectories for mobile robots for which the dimension of the configuration space (three) is larger than the number of d.o.f (two) ?"

After a brief review of existent methods for trajectory planning and motion control for mobile robots, we especially investigate the geometric aspects of the question. We mention in the last section the techniques which are in development in order to control robot motion from the sketch of trajectory provided by geometric reasoning.

2 Mobile robot motion planning

Trajectory planning is only one aspect of the global navigation problem that includes also environment perception and modeling, accounting for inaccuracies, real-time decision-making, spatial structure learning... An overall synthesis of such issues is given in [7].

Even if we restrict ourselves to the geometric and control aspects, collision-free motion planning for a mobile robot still remains an open problem, in spite of important partial results. There are four classes of methods to deal with this problem; according as the geometric constraints of the environment are more or less strong, the methods integrate more or less motion control aspects.

The first kind of approaches is applied in highly structured environments. The better known systems concern the road-following problem. [29] [22] study the global architecture needed by a trajectory planning and control system that uses vision for guidance. The most relevant issues in such systems are the real time processing of road feature extraction, and visual feature tracking in order to control vehicle motion.

The second class concerns the local methods. Their principle consists in using only local and poor but quickly acquired information on the environment, in order to plan a trajectory in real time. The potential fields based methods are the most commonly used [15]: the robot is supposed to be moving in a fictive potential field wherein obstacles are associated with a repulsive field and the goal with an attractive one. This method is efficient in numerous situations (convex obstacle avoidance for instance) but not in very constrained space, where the goal and the obstacles are very near. [10] palliates this last drawback by using an approach wherein collision-free constraints appear as linear constraints in a quadratic criterium minimization problem associated to the goal. Because of a local view of the environment these methods are not complete (i.e. they do not guaranty to find a solution if it exists).

Several methods can be gathered in a third class. They deal with unstructured environments, and are based on a structuring of the euclidian free-space using particular approximation shemas. If the robot is assumed to be circular [23] [17] propose as a structure, respectively the Voronoi diagram of the polygonal environment, and a generalized visibility graph in a more general environment; trajectories thus produced are smooth (they do not have angular points nor cusps). Other methods decompose the free-space in elementary places (convex polygons [6], generalized cones [4] ...), which are structured into a graph whose adjacency relation indicates the possibility (and the associated way) of moving from a place to another. [28] associates a local method developed for corners with a decomposition of the free-space into lanes based on its Voronoi diagram. All these methods are only applicable when free-spaces is large with respect to robot geometric and kinematic constraints.

Motion planning in a very constrained environment needs to consider the formalism of the configuration space (CS) [21]. This space is the space of independant parameters that characterize the position and the orientation of a mobile body ($\mathcal{R}^2 * S^1$ in our case, where S^1 is the unit circle). It is divided into the admissible space (ACS) in which the mobile body does not intersect the obstacles, the free space (FCS), defined as the closure of the ACS interior, the occupied space (OCS), defined as complementary to ACS (for an analysis of the connectivity and topology of such spaces, see [18]).

There are two types of methods for configuration space exploration, viz:

- The methods [25] [2] [1] that lead to an exact partitioning of either FCS or its boundary which is constituted by quadratic surface patches [24]. Notice that the most efficient (in $O(n^3 \log n)$) is [1] and has been implemented.

- The numerous methods of *"paving"* CS into *"space quanta"*: cells [5], hyperparallelepipeds [12] [27], one-dimensional slices structured into regions [20], cubes structured into octrees [9]. All these methods have been implemented.

These last class of methods solves (completely or partially) the classical piano-mover problem that assumes the piano to be holonomic. The aim of the next section is to take into account kinematic constraints in such formalisms.

3 Trajectory planning for non-holonomic robots using a configuration space approach

3.1 Position of the problem

The last approaches characterize the existence of a trajectory by means of the existence of a connected component of ACS including the initial and final configurations. Such characterization is a priori valid only for holonomic systems. Let us recall briefly some fundamental concepts of analytical mechanics [30].

The joints expressing the relations between the velocities of the configuration parameters, which cannot be reduced to relations between these parameters (and which therefore cannot be integrated) are called non- holonomic joints. The number d, of degrees of freedom of a system is defined by $n - r$, n representing the number of configuration parameters and r the number of independent non-holonomic joints. A holonomic system is a system without non-holonomic joint, i.e., $d = n$. For such systems, any infinitesimal motion (*i.e.*, any infinitesimal variation of the configuration parameters) can be achieved. This property does not hold for non-holonomic systems.

Let us consider a mobile robot whose locomotion system consists of two independent driving wheels located on a common axis (see Fig. 1). Let (x, y, θ) be the three configuration parameters.

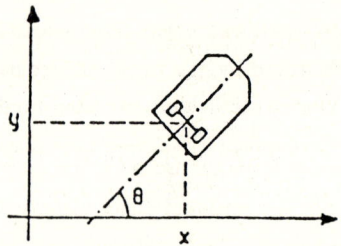

Figure 1: Configuration parameters

The state equations characterizing the system motion are defined by:

$$\begin{aligned} dx &= \frac{1}{2}(v_1 + v_2)\cos\theta \\ dy &= \frac{1}{2}(v_1 + v_2)\sin\theta \\ d\theta &= \frac{1}{2}(v_1 - v_2) \end{aligned} \quad (1)$$

where v_1 and v_2 stand for the velocities of both driving wheels. From these equations we deduce that there exists one (and only one) non-holonomic joint :

$$dy - dx \tan \theta = 0.$$

For such a system all trajectories in ACS are not necessarily feasible (see Fig 2). A *feasible trajectory* is a function of time, piecewise continuous and differentiable (the robot's linear speed vector determines its orientation), the points where the linear speed is zero corresponding to "pure" rotations. In order to distinguish forward and backward motions [13] uses the notion of *tracing* that retains only the topological and geometrical characteristics of the trajectory. With respect to this terminology our problem consists in defining an algorithm for planning polygonal tracings.

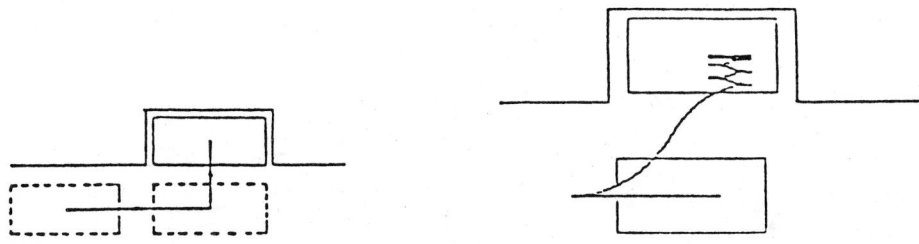

Figure 2:   Non feasible trajectory   Feasible trajectory

3.2   An algorithm schema

[18] establishes that:

Property : If c and c' are two configurations contained in a single connected domain of the interior of ACS, then there exists a feasible collision-free and contact-free trajectory between c and c'.

Remark : this result is established in the more general case where the gyration radius is lower bounded (as for a car).

The proof of this property is based on the existence of a feasible trajectory between any two configurations of an elementary open set of $\mathcal{R}^2 * S^1$. This existence proves that any configuration resulting from a motion consequent to an infinitesimal variation of the configuration parameters can be reached in an open set. Several procedures for searching feasible trajectories between two configurations of an open set can be defined according to the type of open set considered. A detailed proof is given in [18]. It is constructive and leads to the following algorithm:

Input data:

- A contact-free trajectory T (i.e., in an open connected domain D of ACS) between two configurations c and c'.

- A procedure $P(c_1, c_2)$ which produces a feasible trajectory between any two configurations c_1 and c_2 in an open set of given type O.

Output data: A feasible contact-free trajectory T' between c and c'.

Algorithm:

Cover T by a finite sequence of open sets $O_1, \ldots, O_p$ of type O such that:
$$O_i \subset D, O_i \cap O_{i+1} \neq \emptyset, c \in O_1, c' \in O_p.$$
$i \leftarrow 1$
$c_1 \leftarrow c$
While $(i < p)$
    Let c'' be a configuration of $O_i \cap O_{i+1}$
    $\tau_i \leftarrow P(c_i, c'')$
    $i \leftarrow i + 1$
    $c_i \leftarrow c''$
$T' \leftarrow (\bigcup_{1 \leq i < p} \tau_i) \cup P(c_i, c')$

The implementation of this algorithm requires :

- A procedure for computing ACS or FCS.

- A procedure for searching a contact-free trajectory.

- The definition of a type of open sets of $R^2 * S^1$ and the associated procedure P for searching a feasible trajectory.

Notice that the data structures used by the methods representing the configuration space by means of discretization offer the advantage of directly providing a path in the space quantum adjacency graph.

To adapt this algorithm schema, it suffices to define a procedure for searching feasible trajectories within these quanta. (Remark: these space quanta are closed but it can easily be shown that the algorithm holds).

3.3 Planning of polygonal tracing

In this section we present an implementation based on a general software described in [27] and resumed in section 3.3.1, that structures FCS into hyperparallelepipeds (parallelepideds in our case). The procedure for searching feasible trajectories in these parallelepipeds is described in 3.3.2. The results, the extensions currently under study and the elements used for analyzing its complexity are discussed respectively in 3.3.3, 3.3.4 and 3.3.5.

3.3.1 FCS Computation and Exploration

The algorithm is based on some principles established in [12]. It receives as input:

- A mobile body $A_n(q_1, \ldots, q_n)$ (or an articulated system $A(q_1) \ldots A_n(q_1, \ldots, q_n)$) and a set of obstacles O_i, described by assemblies of elementary surfaces or volumes (polygons in this application).

- A CS to be analyzed (interval product $I_i = [q_{i_{min}}, q_{i_{max}}]$).

- A discretization step on each dimension.

From a hierarchical description of the mobile by means of different volumes

$$B_i(q_1, \ldots, q_i) = \bigcup_{i < j \leq n} \{A_j(q_1, \ldots, q_i, \ldots, q_j)/q_j \in [q_{j_{min}}, q_{j_{max}}]\}$$

one gets as output a tree structuring of CS of the form:

- OCS_i = subspace of CS occupied whatever $q_{i+1}, \ldots, q_n$.

- FCS_i = subspace of CS free whatever $q_{i+1}, \ldots, q_n$.

- $MCS_i = OCS_{i+1} \cup FCS_{i+1} \cup MCS_{i+1}$, subspace of CS for which a subspace of dimension $j > i$ had to be analyzed recursively to determine its belonging to either OCS or FCS.

Each component is represented by a set of hyperparallelepipeds (HP) of dimension i. The principle used to analyze a discretized subspace of dimension i relies on :

- The computation of a function $Distance(Q)$ (minimal translation allowing $A_i(q_1,\ldots,q_i)$ to be either put in contact or removed from an obstacle), for discrete values of Q.

- A function for propagating the results on a ball centered on Q and of radius $\mathcal{F}(Distance(Q), dQ)$.

- The use of diverse heuristic techniques permitting reduction of the number of calls of the distance function.

A tree representation of CS under the HP form permits easy superposition of a graph structure whose vertices are the elements of FCS_i and whose arcs reflect a connectivity relationship between two nodes.

For Q_{start} and Q_{end} given, the search carried out in this graph with an algorithm A^* provides a trajectory hull. In the case of a mobile robot with kinematic constraints, the heuristic used involves a weighting between the distance to the goal and a criterion characterizing the robot's maneuverability to traverse the HP.

3.3.2 Procedure P

The trajectories produced by this procedure are polygonal tracings *i.e.*, consisting of rotations and of line segments going either forward or backward.

The input data of **P** are a parallelepiped $Pa = [X_1, X_2] * [Y_1, Y_2] * [\theta_1, \theta_2]$, an initial configuration $c = (p, \theta) \in \mathcal{R}^2 * \mathcal{S}^1$ and a window Wd on the boundary of Pa allowing passage to the adjacent parallelepiped. Wd can be of three types according to whether the adjacency is for a constant x, y or θ_i.

The procedure **P** furnishes a feasible trajectory between c and a configuration of Wd. We denote by Pa^r and Wd^r the projections of Pa and Wd on $\mathcal{R}^2$ (see Fig. 3). δPa^r stands for the Pa^r boundary rectangle.

Given a point p of the plane and a sub-interval $[\theta, \theta']$ of $[\theta_1, \theta_2]$, we call $\text{Sec}(p,[\theta,\theta'])$ the domain swept by the lines passing through p and of orientation $\theta'' \in [\theta, \theta']$. $Rot(p, [\theta, \theta'])$ refers to a rotation at point p, allowing to reach (p, θ') from (p, θ) by a segment included in $\{p\} * [\theta_1, \theta_2]$. $Line(p, p')$ stands for a translation of vector $\vec{pp'}$.

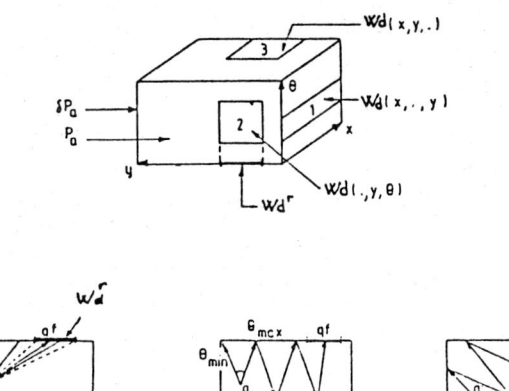

Figure 3: Trajectory planning in a parallelepiped of CS

Procedure $P(Pa, c, Wd)$

$\tau \leftarrow \emptyset$
$c_i \leftarrow c$ /* $c_i = (p_i, \theta_i))$ */
While $(Sec(p_i, [\theta_1, \theta_2]) \cap Wd^r = \emptyset)$
    Compute $Int = \{p_1, p_2, p_3, p_4\} = Sec(p_i, [\theta_1, \theta_2]) \cap \delta Pa^r$
    Choose $p_j \in Int$ such that $Dist(Line(p_i, p_j), Wd^r)$ is minimal
    /* Let θ_j the orientation of the line (p_i, p_j) in $[\theta_1, \theta_2]$ */
    $\tau \leftarrow \tau \cup Rot(p_i, [\theta_i, \theta_j]) \cup Line(p_i, p_j)$
    $c_i \leftarrow (p_j, \theta_j)$
Choose a point $p' \in Sec(p_i, [\theta_1, \theta_2]) \cap \delta Wd^r$
/* Let θ' the orientation of the line (p, p') in $[\theta_1, \theta_2]$ */
$\tau \leftarrow \tau \cup Rot(p_i, [\theta_i, \theta']) \cup Line(p_i, p')$

Figure 3 illustrates this algorithm on an example.

3.3.3 An example

Figure 4 shows an environment with a corridor and a door, and the associated FCS. Figure 5 shows two results furnished by the algorithm starting with the same data, but with two distinct heuristics (in the path search step) : the first one minimizes the angular gap between two adjacent parallelepipeds in $R^2 * S^1$, the second chooses the parallelepipeds whose dimension on S^1 is maximal.

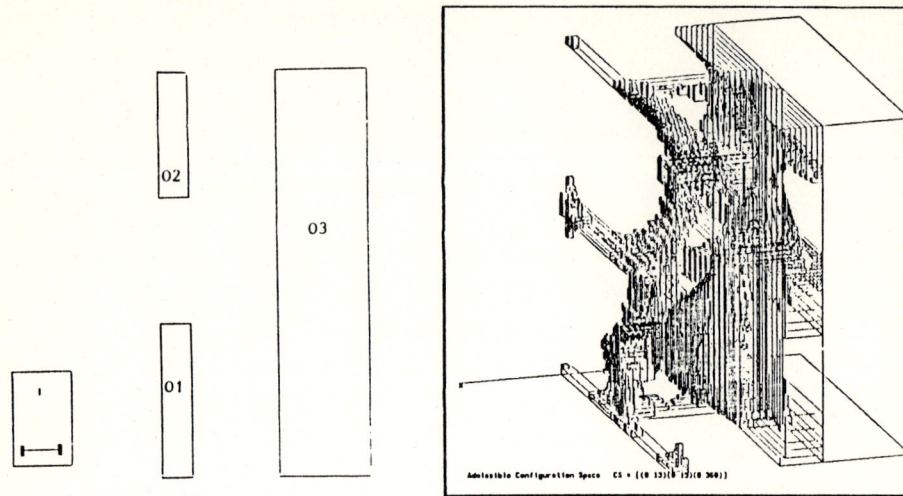

Figure 4:   Environment   Free Configuration Space (x, y, θ)

3.3.4 Complexity

The complexity of the global algorithm is governed by the representation and exploration of CS. It is in $O(n/\varepsilon^3)$ where n represents the total number of obstacle vertices and ε the size of the elementary parallelepiped.

The complexity of the procedure **P** is difficult to assess since it depends on the number of maneuvers (defined as the configurations in which the robot's speed is zero). Evidently, there exists some cases where **P** is optimal (i.e., where no "better" trajectory exists in terms of number of maneuvers). However, in general, one would have to compare this number to the optimum number. Evaluating such optimum is a difficult task which, as yet, has not been performed.

Initial results have been obtained : [19] proposes an algorithm sketch for searching maneuver-free trajectories for a non-holonomic circular robot with a lower bounded gyration radius. [11] shows that the problem for a point in a polygonal environment is decidable in $2^{O(poly(n))}$ where *poly* designates a polynomial function.

3.3.5 Extensions

The extensions under study concern with:

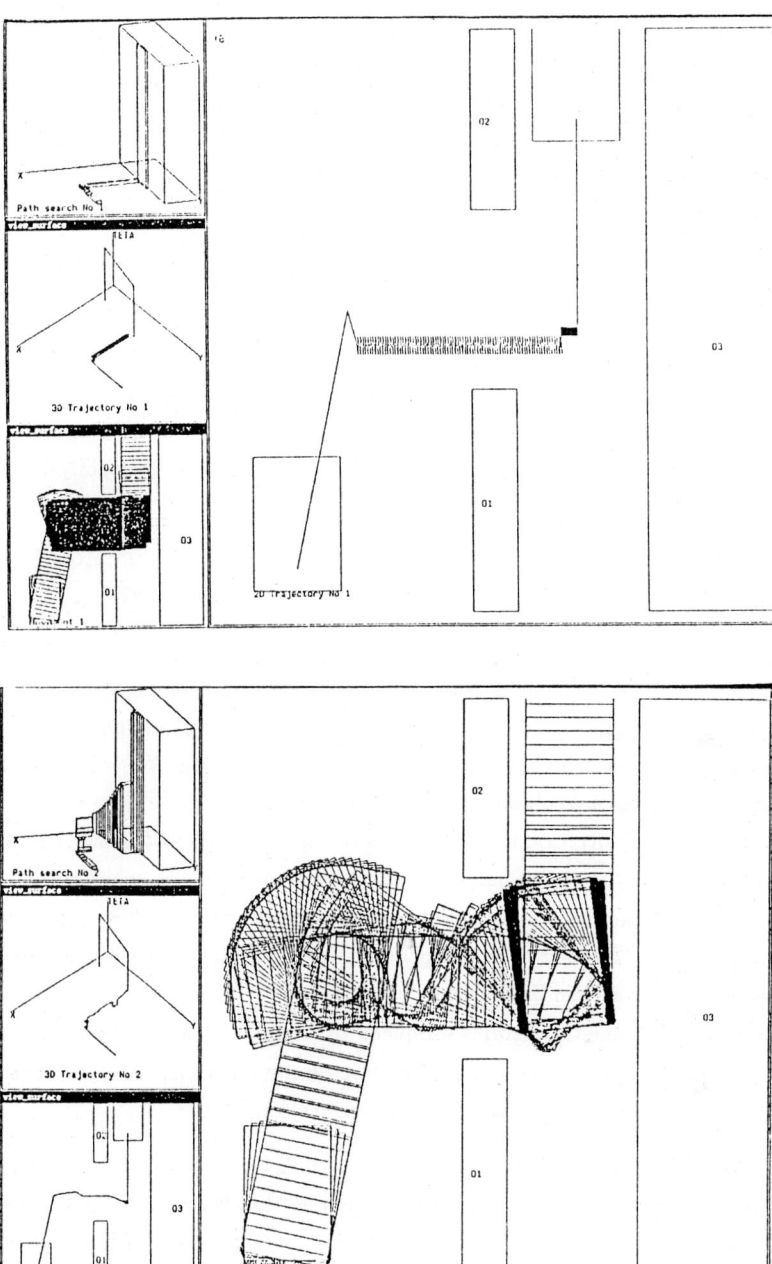

Figure 5: Trajectories produced by the algorithm

- Extending the scope of **P** to parallelepipeds adjacent to the hull provided by algorithm 3.3.1, in order to decrease the number of maneuvers when it is possible.

- Replacing procedure **P** by a procedure allowing helix planning in the parallelepipeds. This new procedure will provide more general trajectories than polygonal tracings and will allow to deal with the lower bounded gyration radius constraint which appears in most mobile robots.

The approach can be applied whenever a $R^2 * S^1$ configuration space representation and exploration system is available. These systems are often complex and highly sophisticated. The choice made in our implementation is certainly not optimal since the tool described in section 3.3.1 is a general purpose tool (valid for spaces of any dimension k, efficient for k= 2,3,4). It has been used for obtaining rapidly the initial feasible trajectories for the mobile robot.

4 Trajectory Execution

As we already mentioned before, while the road-following type of methods rely on *physical* features of the local structured environment (*e.g.*, road boundaries) to guide robot motion, in our case such features are not always available. We will then replace this information by the precomputed trajectory.

After producing the trajectory, the problem is to control robot motion so that it stays on this trajectory. Due to inaccuracies in the measurement of robot position (by odometrical dead-reckoning for example), the movement will not follow the computed trajectory in general. Furthermore, the precomputed trajectory is based on a model of the environment that is also inaccurate.

On the other hand, some trajectories produced by a search in the configuration space, while guaranteeing collision-free motion, are not "easily" feasible, or require very slow movements because of their shape (*e.g.*, saw-like trajectories).

In order to take into account the mentioned two kinds of errors on the one hand, and to smoothen the considered constrained trajectory so that the movement is more continuous on the other hand, we propose to consider the local model of the environment together with the precomputed trajectory as inputs to the control system.

Let us recall that the trajectory planning method produces a polygonal tracing, *i.e.*, a trajectory constituted by straight line segments on which the robot can move either forward (if the motion agrees with its orientation) or backward (if both are opposite), and turns.

If there are turns along the trajectory between two consecutive segments, they correspond to corners or cusps according as the movements on the segments are identical (both forward or backward), or opposite. In each case the velocity vector (derivative vector of the position) must be zero. Cusps impose a mandatory stop.

We show in this section that corners can be smoothed. More precisely we show how to link the two segments with a doubly differentiated curve (*i.e.*, without a zero velocity vector) which is as close as necessary to the corner.

In order to link two straight line segments with a curve C such that the union is a doubly differentiated curve, C (assumed to be parametrized by time) has to pass in finite time from an infinite curvature radius to a finite one.

From equations (1) the curvature radius ρ of robot trajectory is given by : $\rho = \frac{l(v_1+v_2)}{2(v_1-v_2)}$ where l designates the distance between the two wheels and v_1 and v_2 their respective linear velocities. A particular solution is given by *clothoids*.

Several papers investigate the use of clothoids in mobile robot motion planning [8] [14]. The major property of a clothoid is that its curvature is in inverse ratio to the curviline abscisse : $\rho(k,t) = k*t/V$ where k is the proportionnality ratio, V the constant norm of the velocity vector along the clothoid, and t the time parameter.

The advantage of this choice is an easy command of the two driving wheels [16]. Indeed the vehicle describes a clothoid when wheel accelerations are constant and opposite, which furthermore leads to an optimal command *w.r.t* energy consumption.

A clothoid thus permits to pass from an infinite curvature radius to a finite one in finite time (and *vice-versa*). In order to link two segments we must use two tangent clothoids arcs. A property established in [16] shows that it is possible to compute a doubly differentiable curve consisting of a pair of clothoid arcs that connect two intersecting segments such that this curve remains inside *any* given region bounded by the two segments and an arc of circle tangent with them. From this property we deduce easily that the clothoid arcs can be as close to the corner as the environment constraints may impose it.

Because the polygonal tracing we have planed is in an open component of the free-space, we have the guaranty that the corners can be smoothable whitout vehicule stops. The only points where the velocity vector has to be zero are the cusps.

The resulting final trajectory, smoothed when possible, will be executed using sensor data (*e.g.*, ultrasonic sensors). If the trajectory is already smooth (*e.g.*, straight lines, clothoids, *etc.*), then sensor data, matched to the local environment model, will help to localize the robot along its trajectory, thus correcting the dead-reckoning system's error. In the case of *small* variations of the environment with respect to the model, the use of sensor data enables to control the motion in order to avoid collisions, thus departing from the computed trajectory. We rely on a basic assumption: the computed trajectory is not unique but belongs to a family of very close trajectories such that we can actually replace it by the family's envelope. Indeed, we have produced a non-contact feasible trajectory by means of paving the free-space with open cylinders. Within an open cylinder, there exists at least one trajectory between any two configurations, and in general more than one. Therefore, while staying inside the same open cylinder, the robot can actually move within this family of trajectories, using for this purpose sensor readings. Notice that characterizing the amplitude of the small authorized variations is a difficult open problem linked with the precise study of the topological structure of equivalent trajectories.

References

[1] F. Avnaim and J-D. Boissonnat. A practical exact motion planning algorithm for polygonal objects admist polygonal obstacles. In *IEEE, International Conference on Robotics and Automation, Philadelphia (USA)*, 1988.

[2] Zhang Bo, Zhang Ling, and Zhang Jianwei. *An efficient algorithm for findpath with rotation.* Report Department of Computer Science, Beijing Univ., 1986.

[3] M. Brady, J.M. Hollerbach, T.L. Johnson, T. Lozano-Perez, and M.T. Mason. *Robot motion : planning and control.* MIT Press, 1982.

[4] R. A. Brooks. Solving the find-path problem by good representation of free space. *IEEE journal on Systems, Man and Cybernetics*, 2(13), 1983.

[5] R. A. Brooks and T. Lozano-Perez. A subdivision algorithm in configuration space for findpath with rotation. *IEEE journal on Systems, Man and Cybernetics*, 2(15), 1985.

[6] R. Chatila. Path planning and environment learning in a mobile robot system. In *ECAI, Orsay (France)*, Juillet 1982.

[7] R. Chatila and G. Giralt. Task and path planning for mobile robots. In *NATO ARW on Machine Intelligence and Knowledge Engineering, Maratea (Italy)*, Mai 1986.

[8] H. Chochon and Leconte B. *Etude d'un module de locomotion pour un robot mobile*. Rapport de fin d'étude ENSAE, Laboratoire d'Automatique et d'Analyse des Systèmes (C.N.R.S.), Toulouse (France), Juin 1983.

[9] B. Faverjon. Object level programming of industrial robots. In *IEEE, International Conference on Robotics and Automation, San Francisco (USA)*, 1986.

[10] B. Faverjon and P. Tournassoud. A local based approach for path planning of manipulators with a high number of degrees of freedom. In *IEEE, International Conference on Robotics and Automation, Raleigh (USA)*, 1987.

[11] S. Fortune and G.T. Wilfong. *Planning constrained motion*. Technical Report, ATT Bell Laboratories, Murray Hill, Mai 1988.

[12] L. Gouzènes. Strategies for solving collision-free trajectories problems for mobile or manipulator robot. *International Journal of Robotics Research*, 3(4), Winter 1984.

[13] L. Guibas, L. Ramshaw, and J. Stolfi. A kinetic framework for computational geometry. 1983.

[14] Y. Kanayama and N. Miyake. Trajectory generation for mobile robots. In G. Giralt O. Faugeras, editor, *Robotics Research 3*, MIT Press, 1986.

[15] O. Khatib. Real time obstacle avoidance for manipulators and mobile robots. *International Journal of Robotics Research*, 1(5), 1986.

[16] A. Khoumsi. *Pilotage, asservissement sensoriel et localisation d'un robot mobile autonome*. Thèse de l'Université Paul Sabatier, Toulouse (France), Laboratoire d'Automatique et d'Analyse des Systèmes (C.N.R.S.), Toulouse (France), Juin 1988.

[17] J. P. Laumond. Obstacle growing in a nonpolygonal world. *Information Processing Letters*, 25(1), Avril 1987.

[18] J.P. Laumond. Feasible trajectories for mobile robots with kinematic and environment constraints. In *International conference on autonomous systems*, Amsterdam, Netherland, 1987.

[19] J.P. Laumond. Finding collision-free smooth trajectories for a non-holonomic mobile robot. In *10th IJCAI, Milan (Italy)*, 1987.

[20] T. Lozano-Perez. *A simple motion planning algorithm for general robot manipulators*. Robotics Research: The Third International Symposium, O. Faugeras and G. Giralt (Eds), MIT Press, Cambridge, Massachusetts, 1986.

[21] T. Lozano-Perez. Spatial planning : a configuration space approach. *IEEE Transaction Computer*, 32(2), 1983.

[22] B. Mysliwetz and E.D. Dickmanns. A vision system with active gaze control for real-time interpretation of well structured dynamic scenes. In F. C. A. Groen L. O. Hertzberger, editor, *Intelligent Autonomous Systems*, North Holland, 1987.

[23] O'Dunlaing and C. Yap. A retraction method for planning a motion of a disk. *J. of Algorithms*, 6:104–111, 1985.

[24] J. Reif. Complexity of mover's problem and generalizations. pages 421–427, 1979.

[25] J. T. Schwartz and M. Sharir. On the piano mover : the case of a two dimensional rigid polynomial body moving amidst polygonal barriers. *Communication on Pure and Applied Math*, (36), 83.

[26] J. T. Schwartz, M. Sharir, and J. Hopcroft. *Planning, Geometry and Complexity of Robot Motion*. Artificial Intelligence, Ablex, 1987.

[27] T. Siméon. Planification de Trajectoires sans collision. Une approche par Espace des Configurations. In J.P. Laumond J.D. Boissonnat, editor, *Journées géométrie et robotique*, LAAS/CNRS, INRIA, Mai 1988.

[28] P. Tournassoud. Motion planning for a mobile robot with a kinematic constraint. In *IEEE Int. Conf. on Robotics and Automation*, 1988.

[29] R. Wallace, K. Matsuzaki, Y. Goto, J. Crisman, J. Webb, and T. Kanade. Progress in robot road-following. In *IEEE Int. Conf. on Robotics and Automation*, 1986.

[30] E.T. Whittaker. *A treatise on the analytical dynamics of particles and rigid bodies*. Cambridge University Press. 4ème Ed., 1965.

Researches of the Biped Robot in Japan

H. MIURA

Department of Mechanical Engineering
The University of Tokyo
Bunkyo-ku,Tokyo 113,Japan

Summary
In Japan many researches of the biped locomotive robot have been conducted. In this paper the fundamental characteristics of the robots which have been constructed in these researches are listed up first and some interesting robots of these are discussed.

Survey of the biped developed in Japan[1]
In Table 1. almost all biped robots developed in Japan(reported in the scientific paper) are listed up. In Table 2. the developer of each robot is shown.
Until 1973,the mini-computers were used as the processors of the controller. In these years,only Waseda University(Prof.Kato) was very active in this field.
After the micro-computer was widely spreaded in the control application,many researchers challenged this subject. The purposes of the research are various. The followings are some example of the purpose of the research of the biped.
(i)   the developement of an artificial leg
(ii)   showing the efficiency of the newly constructed control
        theory using the biped as one example of applied system
(iii)   the analysis of walking motion of the human from the
        standpoint of biomechanism
(iv)   the educational material for training of mechatronics

It is interesting that the researchers are not so serious for an actual application of the biped. It seems to the author that many researchers consider that the quadruped is better for the actual application system of the legged machine than the biped. These two or tree years, some researchers listed up in Table 2.

Table 1-1. The Bipeds Developed In Japan

year	Name of robot	① DOF	② P	③ kg	④ cm	⑤ cm	⑥ speed	⑦ D	purpose, control, etc.
70	WAP-2	10	p	5	84	18		3	static walk
71	WAP-3	10	p	5	83	18		3	static walk(plane,step, slope),adaptive control
72	WL-5	10	h	130	125	15	45	3	static walk,payload 30kg
73	Asshy-3	12	h						stable standing,a little swing(left and right)
79	Asshy-10	17	h	200	200				stands on one leg,bends both knees,pump is mounted
	MEG-1	1	e					3	synthesis of link mechanism,control of balancing weight
	Biper-1 Biper-2	2	e					2	dynamic walk,walks only sideward(pitch motion is constrained)
80	Bipman-2	4	e	37	150			2	piston-cylinder leg,dynamic walk
	Biper-3	3	e	1.8	33		0.5	3	stilts-type,dynamic walk ,time-sharing control of pitch and roll motion
81	WL-9DR	10	h	43	100	45	10	3	semi-dynamic walk

to be continued

Table 1-2. The Bipeds Developed In Japan

year	Name of robot	Characteristics							
		DOF	P	kg	cm	cm	speed	D	purpose,control,etc.
	N-1	5	e					2	control of two leggs supporting phase
	Idaten-1	5	e				1	2	dynamic steady walk
	Kenkyaku-1	4	e	30	110	30	0.45	2	dynamic steady walk(2 dimensional walk without kick)
	Biper-4	9	e	2.5	33		1	3	dynamic steady walk
	CW-1	6	e	15	75	12 30	1	2	dynamic steady walk, Optimum regurator
82	WL-9DR mkII	10	h	43	100	45	6	3	semi dynamic walk,torque control of ankle
	N-2	7	e						aiming at autonomous walk
	Idaten-2	7	e	29	140	25	0.8	3	dynamic steady walk,head motion control by translation of balancing mass
	Biper-5	7	e	2.3	37				control(including three processors) was mounted
83	WL-10R	12	h	70	120	45	6	3	computer is mounted,aiming at turning
	Asshy-13	17	h						computer is mounted,aiming at static walk

to be continued

Table 1-3. The Bipeds Developed In Japan

year	Name of robot	Characteristics							
		DOF	P	kg	cm	cm	speed	D	purpose,control,etc.
	MEG-2	2	e	22	95	26	0.8-0.5	3	dynamic steady walk with swinging body following the curve of "8"
	Kenkyaku-2	6	e	40	120	35-45	0.7-1	2	kick motion
	CW-2		e	20	70	35	3	2	aiming at kick
	SMA-LEGS	6	e	2	37			2	shape memory alloy is used for actuator
	Strider-2	7	e	15	66			2	control by robust-servo system
84	WL-10RD	12	h	85	144	40	1.3 plane 1.5 0.5 5(10cm step)	3	dynamic walk in not-flat environment, controller(without power supply) is mounted
	CW-3D	11	e	40	100			3	aiming at 3-dimensional dynamic walk
	AYUMI	5	e	20	70			2	dynamic steady walk
	KRL-1	8	e	13	48			3	non-interference control

to be continued

Table 1-4. The Bipeds Developed In Japan

year	Name of robot	Characteristics							
		DOF	P	kg	cm	cm	speed	D	purpose,control.etc.
	Asshy-15	24	h	250	250			3	static walk, installing the arm for anti-falling down, multi-mode attitude control
	Kenkyaku -3	8	e	47	147			3	aiming at 3-dimensional walk with kick motion

① :degree of freedom---the number of the joints
② :power source---"p" is pneumatic
　　　　　　　　"h" is hydraulic
　　　　　　　　"e" is electric
③ :weight of the robot
④ :height of the robot
⑤ :length of one step of walking
⑥ :walking speed(sec/one step)
⑦ :dimension of the space of motion of the robot

Table 2. Institutes which developed the robots in table 1.

Institute	Robot
Waseda University (Kato)	WAP-2·3,WL-5,WL-9DR,WL-mkII, WL-10R,WL-10RD
Shibaura Institute of Technology(Sato)	Asshy-3·10·13·15
Tokyo Institute of Technology(Mori)) Toha University(Kato)	Bipman-2
Tokyo Institute of Technology(Funahashi)	MEG-1·2
Nagoya University (Ito)	N-1·2,AYUMI
Osaka University (Arimoto)	Idaten-1·2
Gifu University (Furushou)	Kenkyaku-1·2·3
The University of Tokyo(Miura,Shimoyama)	Biper-1·2·3·4·5
Chiba University(Mita)	CW-1·2·3D
The University of Tokyo(Morishita)	M-1
Kobe University (Kitamura)	KRL-1
Electric and Information University(Sato)	SMA-LEGS
Kumamoto University (Kawaji)	Strider-2

are conducting researches of the quadruped.The author succeeded
in the costruction of the quadruped which can walk dynamically.
In this research the experience obtained during the research of
the biped was very helpful.

Dynamic Walk of The Biped (Biper-3·4) [2]
Basical idea of control algorithm for Biper-3(Fig.1.) and Biper-
4(Fig.2.) is the same. Walking motion is planned(the left part
of Fig.4. and Fig.5.) considering the actual walking motion of
the human. Joint angles are designated in Fig.3.
The neccesary torque at each actuator to realize this motion can
be caliculated following the technology of the inverse dynamics.
These torques are used for the feedforward control. The feedback
control is also applied using the diffrence between the planned
motion and the actual angles detedted by the potentiometers at
the joints.
At the reverse side of the foot,the touch sensor is installed to
detect which leg is contacting the floor.

The Biped with Hydraulic Actuators(WL-10RD) [3]
WL(Waseda Leg)-10RD(Refined Dynamic) is shown in Fig.6.
One step of this biped is divided into two phases---the single
leg support phase and the change over phase. In the single leg
support phase,the programmed control is used following the pre-
desinged walking pattern. In the change over phase,the sequence
control is used with the variable torque and the variable mechanical impedance of the ankle joint.

Synthesis of Link Mechanism for The Biped(MEG-2) [4]
MEG-2 (shown in Fig.7.) is the very unique biped. Fig.7.(a) is
the leg mechanism. It seems the open mechanism but actually all
elements(links) are constructed by the closed four-bar linkages
as shown in Fig.8.(a). The desired motion of the ankle (J_A in
Fig.7.(a)) and the desired angle (ϕ_A in Fig.7.(a)) are decided
from the walking pattern of the human. To realize this motion
the link mechanism Fig.8.(a) is synthesized. In this mechanism
there are two prime links(shown with arrow in Fig.8.(a)). These
two links are drived synchronously by one actuator.

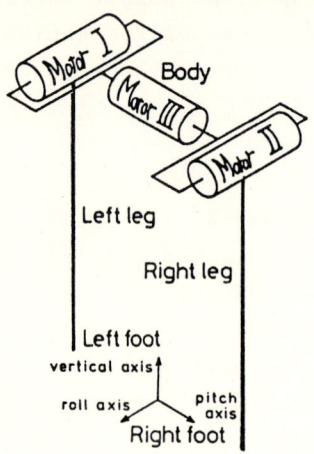
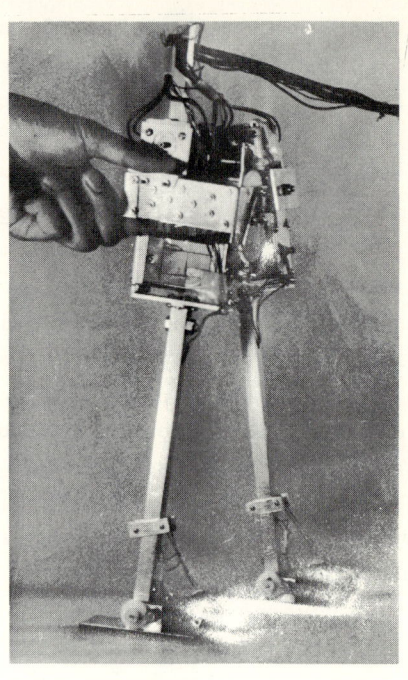

Fig.1. Biper-3
The board at the bottom of the leg is for getting the angle between the floor and the leg and not for stability.

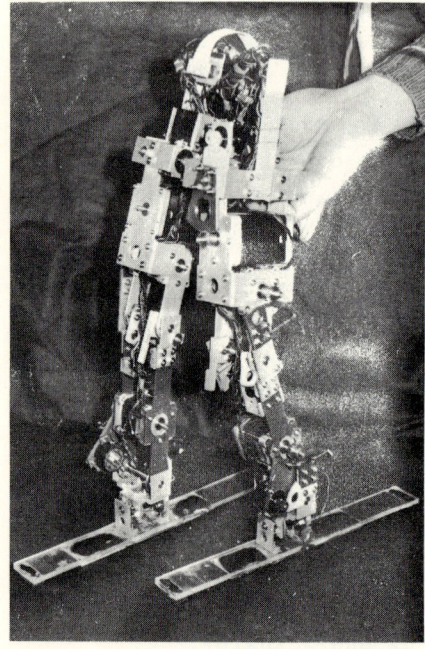
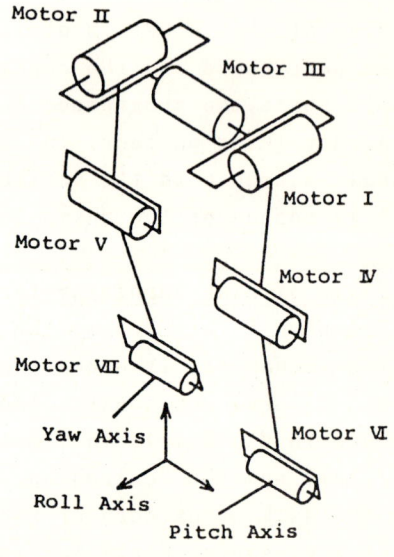

Fig.2. Biper-4

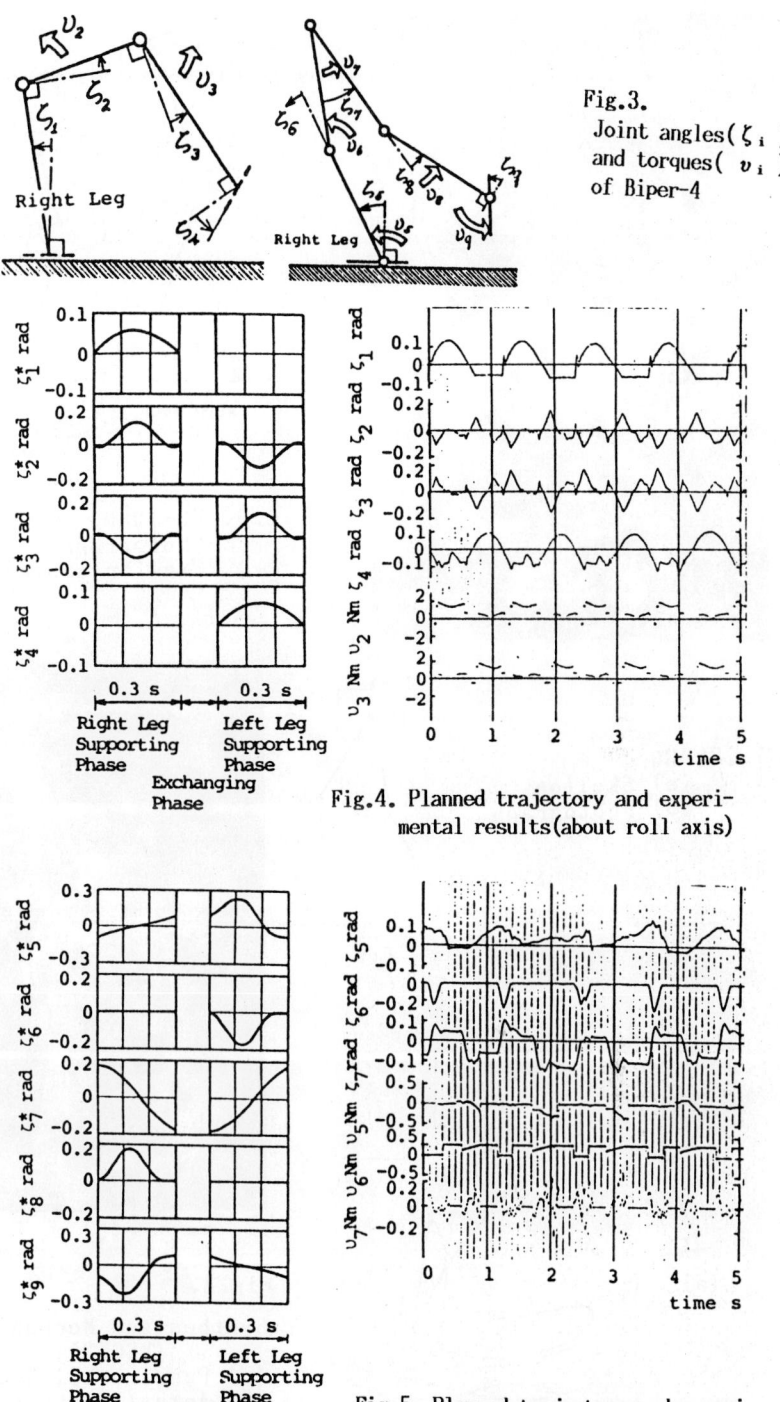

Fig.3. Joint angles (ζ_i) and torques (v_i) of Biper-4

Fig.4. Planned trajectory and experimental results (about roll axis)

Fig.5. Planned trajectory and experimental results (about pitch axis)

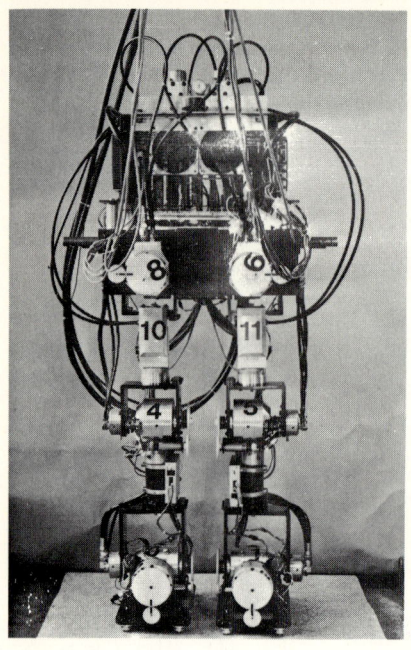

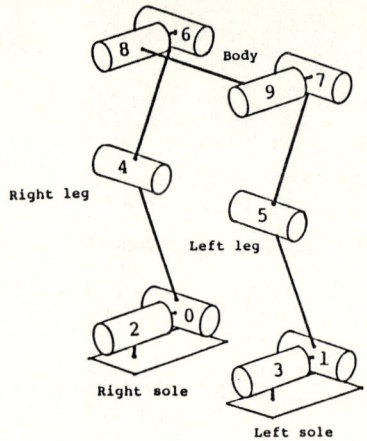

Fig.6. WL-10RD
Photograph(left)
Skelton(above)

Fig.7. MEG-2
(a) Skelton
(b) Photograph

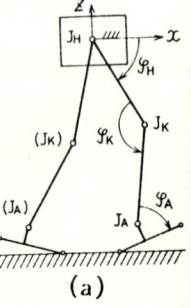

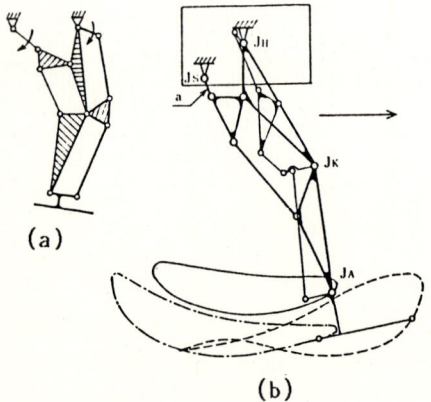

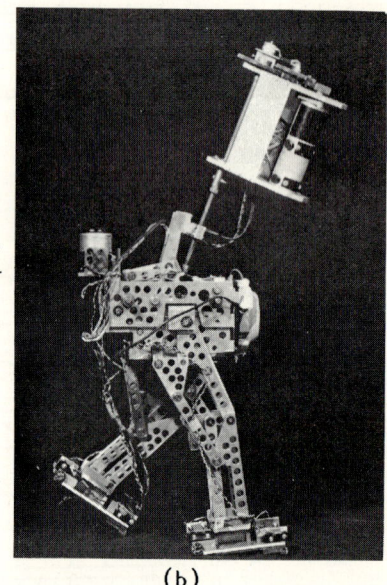

Fig.8.
Synthesized Mechanism
(MEG-2)
(a) Fundamental Mechanism
(b) Actually Synthesized Mechanism

377

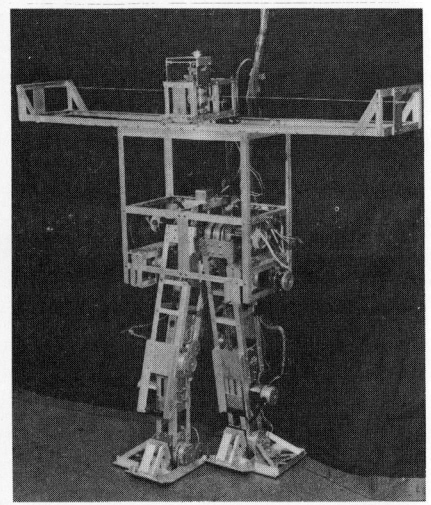

Fig.9. Idaten-2

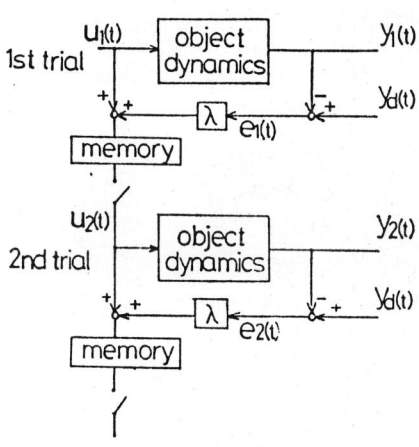

Fig.10. Learning Control

Learning Control of the Biped (Idaten-2) [5] [6]

Idaten-2(Fig.9.) walks by learning control. $y_d(t)$ in Fig.10 is the desired walking pattern. $u_i(t)$ is the input to the servo-system of the biped. $u_i(t)$ is corrected at every trial of the control. After several trial the suitable input $u_i(t)$ can be obtained. In Fig.10. the control mechamism is simply shown.

References
1. The report of the research group of "The Study of Mechanism and Control of The Biped" under Grant in Aid for Co-operative Research(No.60302045) supported by Ministry of Education,Science and Culture.1987
2. Miura,H.;Shimoyama,I.: Dynamic Walk of a Biped. The International Journal of Robotics Research vol.3,no.2 (1984) 60-74
3. Takanishi,A.; Ishida,M.; Yamazaki,Y.; Kato,I.:The Realization of Dynamic Walking by the Biped Walking Robot WL-10RD.Journal of Robot Society of Japan vol.3,no.4 (1985) 325-336 (in Japanese)
4. Funabashi,H. et al.: Synthesis of Leg Mechanism of the Biped. Trans. of Japan Society of Mechanical Engineers 50-455(1984) 1285 (in Japanese)
5. Kawamura,S. et al.: Realization of Biped Locomotion by Motion Pattern Learning.Journal of Robot Society of Japan vol.3,no.3 (1985) 177-18 (in Japanese)
6 Arimoto,S;Kawamura,S.;Miyazaki,F:Can Mechanical Robots Learn By Themselves?. Robotics Research(2nd. International Symposium) 127-134: MIT Press 1985

TJ 212.2 .I87 1989
IUTAM/IFAC Symposium (1988 :

Dynamics of controlled
 mechanical systems

MAR 3 1 1989